Buckling of Ship Structures

Mohamed Shama

Buckling of Ship Structures

Springer

Author
Mohamed Shama
Faculty of Engineering
Naval Architecture & Marine Engineering
Alexandria University
Alexandria
Egypt

ISBN 978-3-642-17960-0 e-ISBN 978-3-642-17961-7
DOI 10.1007/978-3-642-17961-7
Springer Heidelberg New York Dordrecht London

Library of Congress Control Number: 2012932486

Cover design: WMX, Heidelberg

Printed on acid-free paper

Springer is part of Springer Science+Business Media (www.springer.com)

To my wife

For her love, patience, encouragement and support

To my late parents

For their continuous care and encouragement

To my students

Whose enthusiasm and hard work have encouraged me to prepare the course material of this book

Mohamed Shama

Preface

Strength members of ship hull girder are subjected to several types of static and dynamic stresses. The main stress components are the primary, secondary and tertiary stresses. Bottom plating of ship structure are subjected to additional local bending stresses induced by the external water pressure. This complex system of stresses when compounded over the thickness or cross section of a strength member could produce unacceptable high values of equivalent stresses. Because of the hostile and corrosive environment of ship operation, the strength of ship hull girder and its structural members deteriorate with time coupled with the possibility of occurrence of overloading, these strength members may fail to perform satisfactorily. Buckling of ship strength members represent one of the main modes of ship structure failure.

Because the use of the finite element method is more costly and time consuming, the introduction of simplified methods to improve design of ship strength members is always welcome. In order to achieve this goal, a comprehensive analysis is given for the determination of the compound stresses induced by the various types of loadings imposed on the different strength members of ship structure. The compounding of stresses takes account of the primary, secondary and tertiary stresses. The compounding of stresses of bottom plating takes account of the local stresses induced by the local loading of the external water pressure. The compounding of stresses is carried out at the locations expected to reveal the highest values of compound stresses. Because of the inevitable presence of geometric imperfections in ship strength members due to fabrication processes, the actual buckling strength of these members may not attain their design values.

The assessment of buckling of web plates and face plates of deck and bottom girders is presented. The assessment of buckling of side shell plating for the various induced in-plane loading conditions is presented. Assessment of buckling strength of plating for different end support conditions and for a variety of loading patterns is given. The prime aim of the book is to cover an area of ship structure analysis and design that has not been exhaustively covered by most published text books on the strength of ship structures. The book presents a practical approach for the analysis and assessment of ship strength members with particular emphasis on the buckling strength of ship structural members. The book, therefore, contains the main equations required to determine the critical buckling stress for both ship plating and the primary and secondary stiffening members. The critical buckling stress is given for ship plating having the most common boundary conditions encountered in ship structures and subjected to the compound in-plane stresses. The methods commonly used to control buckling failure are introduced.

The book should bridge the gap existing in most books covering the subject of buckling of ship structures by putting the emphasis on the practical methods required to ensure safety against buckling of ship strength members. The book should be very useful to ship designers, shipyard engineers, naval architects, international classification societies and also to students studying naval architecture, marine engineering and offshore structures. The book is enhanced with a set of some solved and unsolved problems.

Outline of the Book

The book is composed of **14 Chapters**. The first 13 Chapters are divided into **Three Parts**. The last Chapter presents a set of problems on the subject material given in the book.

Part I

Part I is Composed of Four chapters

Chapter 1 presents the basic configurations and structural features of some ship types. The main design features of single and double side bulk carriers are presented. The main types and categories of bulk carriers are classified. The structural components of single and double skin bulk carriers as well as the construction of double bottom are specified. The commonly used abbreviations to describe the different types and sizes of bulk carriers are enumerated. The main types and structural characteristics of general cargo and container ships are highlighted. The basic arrangements and design features of Ro-Ro ships are described. The structural systems and design features of single and double hull tankers are clarified. The advantages and drawbacks of the double-hull design are clarified.

Chapter 2 presents the main configurations and characteristics of ship structural assemblies. Transversely and longitudinally stiffened bottom, deck and side shell structure assemblies are considered. The main structural features of transverse bulkheads are described. A brief description of the scantlings of ship strength members is introduced. The basic role of classification societies is clarified. Some ship structural connections and details such as frame brackets, beam and tripping brackets, etc. are illustrated.

Chapter 3 presents the main configurations and geometrical properties of ship structure members. The structural features of ship stiffened panels and frameworks are clarified. Standard and fabricated sections commonly used as stiffening members in ship construction are addressed. The geometrical properties of the various stiffening members with attached plating are presented. The effect of variation of thickness of attached plating on the magnitude of section modulus and second moment of area is quantified. Geometrical and flexural properties of curved plates are given. Equivalent rolled and fabricated sections of commonly used sections of ship strength members are explained. Rational shapes of cross-sections of beams and columns are given. Procedures for the design of fabricated T and I-sections are presented.

Chapter 4 introduces the application of the simple beam theory to ship structural members. The limitations of application of the simple beam theory to thin-walled asymmetrical sections are highlighted. A full explanation of the idealization of beam elements is illustrated. The effective breadth concept is explained for uniform and curved structural members. The concept of modeling ship structure assemblies by beam elements is introduced. Several examples of 2D and 3D modeling of deck, bottom and side structures using beam and plate elements are illustrated. The various types of boundary end conditions commonly used in ship structure analysis are given. The concept of span points and effective span of a beam are clarified. A method is given to determine the optimum span length of a beam and the size of the attached brackets. The influence of the type of end support on the magnitude and distribution of the bending moment are presented. Bending stresses in beams constructed with high tensile steel is clarified. Flexural stresses in fabricated symmetrical and asymmetrical sections are presented. The importance of calculating the equivalent stress is highlighted. A simple procedure for calculating flexural warping stresses is given. The main parameters affecting the magnitude and distribution of flexural warping stresses for asymmetrical sections are explained. The basic concept of effective breadth of uniform symmetrical and asymmetrical face plates is introduced.

Part II

Part II is Composed of Five Chapters

Chapter 5 presents the main components of hull girder bending moments, shear forces and torsion moments. Design values given by Classification societies for still water and wave induced vertical, horizontal and torsional moments are given. An approximate estimate of the maximum value of the wave induced bending moment is given. The distribution of the largest expected vertical wave-induced shearing force is presented. A method is given to determine an approximate value of the maximum vertical shear force. Hull girder dynamic shear force and bending moment components are clarified. The dynamic loading due to shipping green seasis described. The phenomenon of springing resulting from the hydrodynamic loadings induced by the periodic loads generated by the wave actions is explained. The terms "slamming" and "pounding" describing forward bottom impact are explained. The effect of ship hull girder longitudinal vertical deflection on the distribution of shear force and bending moment along ship length is studied. The short and long term predictions of hull girder loadings is given together with an approach to predict their extreme values.

Chapter 6 presents the methods commonly used for the calculation of the primary stresses induced by the vertical and horizontal hull girder bending moments. The strength members of ship hull girder sustaining the primary hull girder stresses are specified. The common procedures used for the calculation of ship section geometrical and flexural properties are given. The method of calculating hull girder bending stresses when the ship is in the inclined condition is presented.

Chapter 7 gives a full analysis of the secondary loadings and stresses induced in ship structural assemblies of cargo ships and conventional oil tankers. Strength members sustaining secondary loadings and stresses of transversely and longitudinally stiffened bottom and deck ship structure assemblies are specified. The secondary stresses induced by the bending of double and single bottom structures are presented for hogging and sagging conditions. The loadings and stresses induced in deck and bottom girders, longitudinals and plating are highlighted. Secondary loadings and stresses induced in tank top longitudinals and plating are given. Secondary loading and stresses in bottom structure assemblies of oil tankers are identified.

Chapter 8 gives a comprehensive analysis of the tertiary loadings and stresses induced in the various strength members of longitudinally and transversely stiffened ship structure. The tertiary strength members of longitudinally and transversely stiffened deck and bottom structures are specified. The tertiary loadings and stresses induced in deck, bottom and tank top longitudinals and plating are presented. The local loadings and stresses induced in transversely and longitudinally stiffened bottom plating is explained. A method is given to determine the minimum required thickness of bottom plating of ship structure.

Chapter 9 gives a full analysis of the compounding of stresses induced in the various ship strength members of transversely and longitudinally stiffened double bottom and deck structures.The compounding of stresses is carried out for the main strength members of a ship which includes girders, longitudinals and plating. The compounding of stresses induced in any strength member takes account of the primary, secondary and tertiary stresses. The primary stresses included in the compounding process are calculated when the ship is in sagging condition when compounding is carried out for deck strength members and for the strength members of the bottom structure when the ship is in hogging condition. The compounding of stresses in tank top longitudinals and plating are also considered. The compounding of stresses of bottom plating takes account of the local stresses induced by the local loading of water pressure. The locations of compounding of stresses expected to reveal the highest values of stresses are identified.

Part III

Part III is Composed of Four Chapters

Chapter 10 presents the stability phenomenon of ship structure. The basic equations of buckling of columns and beam columns are given. The most common classes of perturbations experienced by beam columns are identified. The physical problem of stability is explained and is defined by its state of equilibrium. The concept of critical force and critical stress are explained. The effect of eccentric loading on the critical buckling critical stress is given. The load-deflection relationship of beam columns is introduced. The behavior of beam columns under various loading conditions and different types of end supports are investigated.

Chapter 11 presents comprehensive analysis of buckling of stiffened panels. Global and local buckling modes of deformation of stiffened panels are given. The commonly used idealizations of boundary support conditions of ship plating are clarified. The general equations for plate buckling under single and combined loading patterns are considered. The basic and Interaction equations of buckling of plating subjected to a variety of combined system of loadings are given. The concept of effective width of plating is clarified. The non-uniform in-plane compressive loadings are idealized by the combined loadings of uniform compression and pure bending. The various boundary conditions commonly assumed for girders and plating is given. The general modes of buckling of face plates and web plates of girders are illustrated. Post–buckling strength of plating is introduced. Ultimate stress of simply supported plate panels is given.

Chapter 12 presents simplified procedures for the assessment of buckling of ship strength members. The main strength members of longitudinally and transversely stiffened bottom and deck structures sustaining compressive forces are identified. The compressive loadings are the compound in-plane stresses induced by the hull girder, secondary, and tertiary stresses. The assessment of buckling of web plates and face plates of deck and bottom girders is presented. Assessment of buckling strength of plating for different end support conditions and for a variety of loading patterns is given. The assessment of buckling strength of strength members of bottom structure is carried out when the ship is in hogging condition and for strength members of deck structure when the ship is in sagging condition. For both cases, the compounding of stresses is carried out when the secondary and tertiary loadings are inducing compressive stresses. The assessment of buckling of side shell plating for various induced in-plane loading conditions is presented. The importance of ensuring that ship strength members sustaining compressive forces have adequate buckling strength against buckling failure is stressed.

Chapter 13 gives a detailed analysis of the various measures commonly used to control buckling of ship structural members. Reliability basis of ship structural safety is introduced. The role of Classification societies in controlling failure of ship structural members is explained. The deleterious effects of deterioration of strength of ship structural members with time are highlighted. The effect of corrosion of ship strength members on the flexural rigidity and buckling strength are described. Linear and exponential equations are used for modeling the variation of the rate of corrosion with time. Improved designs to control buckling failure of ship structural details and connections are presented. Commonly used owners approach for improving ship safety is given. The most common design and construction measures adopted to control welding distortions are specified. Measures to control fabrication deformations and warping of steel sections are clarified. The importance of improving control on tolerances of ship structure members and quality of ship fabrication processes is stressed.

List of Symbols

A,a	Sectional area
A_f	Flange area
Ap	Area of attached plating
At	Total area
Aw	Web area
b	Flange width, frame spacing, longitudinal spacing
b_e	Effective breadth, effective width
b_f	Width of flange
C	Torsion rigidity
C_1	Warping constant
C_b	Block coefficient at summer load waterline
C_w	Warping constant
D	Ship depth
d	Web depth
E	Modulus of elasticity
e	Distance of shear center, eccentricity of loading
EI	Flexural rigidity
e_y	Vertical distance of the shear center
f	A factor representing the degree of constrained
F	Shear force
F_s	Stillwater shear force
F_w	Wave induced shear force
F_d	Dynamic shear force
F_V	Vertical shear force
F_{hog}	Shear force in hogging condition
F_{sag}	Shear force in sagging conditions
$f_X(x)$	Probability density function of X
G	Shearing modulus of elasticity
h	depth of girder

h_w	Wave height
I, i	Moment of inertia
i_g	Moment of inertia of girder section
I_x	Moment of inertia about x-axis
i_l	Moment of inertia of longitudinal section
I_y	Moment of inertia about the y-axis
i_p	Polar moment of inertia
i_l	Moment of inertia of longitudinal
I_p	Polar moment of inertia
J	Torsion constant
L	Ship length
Le	Effective length
M	Bending moment
M_h	Horizontal bending moment
m	Bending moment, Intensity of torque load
M_{sw}	Still water bending moment
M_{Dyn}	Dynamic bending moment
M_V	Vertical bending moment
M_S	Stillwater bending moment
M_W	Wave bending moment
m_o	Variance, Area under the spectrum curve
N	Normal force
P_{cr}	Critical force
P_E	Euler buckling load
P_F	Probability of failure
P_u	Ultimate load
p	Pressure load
q	Uniform load, shear flow
Q	Reaction force, shear force
R, r	Radius
R_m	Mean radius
S.S.	Simple support
S	Transverse web spacing
s	Longitudinal spacing
T	Torque
t	Thickness
t_f	Flange thickness
t_o	Original thickness
T_p	Thickness of attached plating
t_t	Total plate thickness

t_w	Web plate thickness
y_D	Distance of N.A. from base line
y_p	Distance of attached plating from N.A
y_f	Distance of flange from N.A.
Z_f	Section modulus at flange
Z_B	Section modulus at base line
Z_D	Section modulus at deck
Z_p	Section modulus at attached plating
δ_o	Initial deflection
δ_t	total deflection
λ_{cr}	Critical slenderness ratio
σ	stress
σ_E	Euler buckling stress
σ_{cr}	Critical stress
σ_{ec}	Euler buckling stress for compression
σ_{eb}	Euler buckling stress for bending
σ_b	Bending stress
σ_p	Stress in plating
σ_{loc}	Local stress
σ_{pl}	Local stress in plating
τ_y	Shear yield stress
τ_{cr}	Critical stress for shear
τ_{es}	Euler buckling stress for shear

SI Units

International System of Units

This system can be divided into basic units and derived units as given in tables (1 and 2).

Table (1), Basic Units

Quantity	Unit	Symbol
Length	Meter	m
Mass	Kilogram	kg
Time	Second	s
Eclectic Current	Ampere	A
Thermodynamic Temperature	Degree Kelvin	°K
Luminous Intensity	Candela	cd

Table (2), Derived Units

Quantity	Unit	Symbol
Force	Newton	$N = kg.m/s^2$
Work, Energy	Joule	J = N.m
Power	Watt	W = J/s
Stress, Pressure	Pascal	$Pa = N/m^2$
Frequency	Hertz	$H_z = s^{-1}$
Acceleration	Meter per second squared	$g = m/s^2$
Area	Square meter	m^2
Volume	Cubic Meter	m^3
Density	Kilogram per cubic meter	$\rho = kg/m^3$
Velocity	Meter per second	$\upsilon = m/s$
Angular velocity	Radian per second	rad/s
Dynamic Viscosity	Newton second per meter squared	$N.sec/m^2$
Kinematic Viscosity	Meter squared per.second	m^2/s
Thermal Conductivity	Watt per (meter degree Kelvin)	W/(m.deg.k)

Table (3), General units

Gravity acceleration: g	9.807 m/sec^2
Water density (salt water): ρ_{sw}	1.025 tonne/m^3
Modulus of elasticity: E	20.9 MN/cm^2
Atmospheric pressure: p_{at}	10.14 kN/m^2
1.0 ton displacement	9964 N

Table (4), Conversion table

Quantity	Units		
	United States Customary System	Metric	International System (SI)
Weight or force	1 lb (force)	0.4536 kg	4.448 N
Beam loads	1000 lb/in.	178.6 kg/cm 1488 kg/m	175.13 kN/m 14.59 kN/m
Moment	1 in.-lb 1 ft-lb	11.52 g·m 0.1383 kg·m	0.1130 N·m 1.356 N·m
Weight/volume (water)	1 lb/in^3 (62.4 lb/ft^3)	27.68 g/cm^3 16 kg/m^3 (1000 kg/m^3)	271.4 kN/m^3 157.1 N/m^3 (9.81 kN/m^3)
Moment of inertia	1 in^4	41.62 cm^4	41.62 cm^4
Section modulus	1 in^3	16.39 cm^3	16.39 cm^3
Mass	lb (mass)	0.4536 kg	0.4536 kg
Stress or pressure	1 psi 1000 psi (1 ksi)	0.070 31 kg/cm^2 70.31 kg/cm^2	6.895 kN/m^2 (kPa) 6.895 MN/m^2 (MPa)
steel stress *E* of steel *E* of aluminum	20,000 psi 29(10^6) psi 10.1(10^6) psi	1406 kg/cm^2 20 390 kg/mm^2 7100 kg/mm^2	137.9 MN/m^2 199.9 GN/m^2 69.6 GN/m^2

Contents

Part I: Chapter 1 – Chapter 4

Chapter 1: Ship Structure Configurations and Main Characteristics............ 3
1 Introduction 3
2 Ship Types and Main Characteristics 3
2.1 Bulk Carriers 3
2.2 Double Sides Bulk Carriers 6
2.3 Bottom Structure of Bulk Carriers 7
2.4 Types and Categories of Bulk Carriers 7
2.5 Main Structural Components of Single Skin Bulk Carriers 8
3 General Cargo Ships 10
4 Container Ships 14
5 RoRo Ships 17
5.1 Ramps Types and Arrangements in Ro-Ro Ships 18
6 Tankers 19
6.1 Single Hull Tankers (Conventional Construction) 19
6.2 Design Features of Double Hull Tankers 21
6.3 Structural System of Double Hull Structure 22
6.4 Double Bottom and Double Side Construction of Oil Tankers 23

Chapter 2: Configurations and Characteristics of Ship Structural Assemblies 25
1 Introduction 25
2 Ship Structural Assemblies 25
3 Bottom Structure 25
3.1 Single Bottom Structure 25
3.2 Double-Bottom Structure 26
3.3 Transversely Framed Double Bottom 27
3.4 Longitudinally Framed Double Bottom 28
4 Side Shell Structure 29
4.1 Transversely Framed Side Shell Structure Assemblies 30
4.2 Longitudinally Framed Side Shell Structure 31
5 Deck Structure 31
5.1 Deck Plating 32
5.2 Transversely Stiffened Deck Plating 33
5.3 Longitudinally Stiffened Deck Structure 33

6 Transverse Bulkheads 34
7 Scantlings of Ship Structural Members 38
8 Ship Structural Connections and Details 38
8.1 Frame Brackets 38
8.2 Beam Brackets 39
8.3 Tripping Brackets 41
8.4 Connection between Bottom Longitudinals and Bottom Transverses 42

Chapter 3: Configurations and Geometrical Properties of Ship Structure Members 45
1 Introduction 45
2 Structural Units of a Ship 45
2.1 Stiffened Panels 45
2.2 Frameworks 47
2.3 Hull Fittings 48
3 Configurations and Geometrical Properties of Ship Structure Members 48
3.1 Standard Rolled Sections with Attached Plating 48
3.2 Fabricated Sections 49
3.3 Geometrical Properties of Fabricated Symmetrical Sections with Attached Plating 50
3.3.1 Flat-Bar 51
3.3.2 Standard Angle Sections 52
3.3.3 Offset Bulb 54
4 Flexural Properties of Fabricated Sections with Attached Plating 55
5 Equivalent Section Modulus 56
6 Effect of Variation in Thickness of Attached Plating on the Section Modulus and Second Moment of Area 59
7 Geometrical and Flexural Properties of Curved Plates 59
8 Rational Selection of Equivalent Rolled and Fabricated Sections of Ship Strength Members 60
9 Scantlings of Ship Structural Members 61
10 Rational Shapes of Cross-Sections of Beams 62
11 Rational Shapes of Column Sections in Compression 63
12 Design of Girders Having Fabricated T-Sections 64
13 Determination of Optimum Depth of I-Section Girders 65

Chapter 4: Bending of Beams and Girders 71
1 Introduction 71
2 Subdivision of Ship Structure into Members and Assemblies 71
3 Representation of Structure by Elements 72
4 Modeling of Structure 72
4.1 Forces and Moments on a Beam Element 72
4.2 Modeling of Ship Structural Members 73
4.3 Boundary Conditions of Idealized Beam Elements 74
4.4 Modeling 2D Frame Structures Using Beam Elements 74

4.5 Modeling 2D Grillage Structure 75
4.6 Modeling 2D Deck Structure 75
4.7 Modeling 2D Bottom Structure 75
4.8 Modeling 2D Side Structure 76
4.9 Modeling 2D Transverse Bulkhead 76
4.10 Modeling 3D Space Frame Structures Using Beam Elements 77
5 Modeling by Using Plate Elements 77
5.1 FEM Idealization Using Plate Elements 78
6 Boundary Conditions of Beams and Columns 80
7 Effective Span of a Beam 80
8 Determination of the Optimum Span Length and Size of Bracket 82
9 Simple Beam Theory 83
9.1 Beam Loading and Response 83
9.2 Beam Deflections 87
10 The Influence of the Type of End Support on the Magnitude and Distribution of the Bending Moment 89
10.1 Effect of Degree of Constraint at the End Support on the Magnitude and Distribution of the Bending Moment 90
10.2 General Case of Uniform Loading and Constrained End Supports 91
11 Beam Stresses 93
11.1 Beam under Normal (Axial) Loading 93
11.2 Beams Subjected to Bending Stresses 93
11.2.1 Bending of Symmetrical Sections 93
11.2.2 Bending of Sections with One Axis of Symmetry 94
11.3 Bending of Asymmetrical Sections 95
12 Bending Stresses in Beams Constructed with High Tensile Steel 96
13 Equivalent Stress 97
14 Flexural Stresses in Fabricated Asymmetrical Sections 98
14.1 A Simple Procedure for Calculating Flexural Warping Stresses 99
14.2 Main Parameters Affecting the Magnitude and Distribution of Flexural Warping Stresses 103
15 Effective Breadth Concept 105
15.1 Effective Breadth of Uniform Symmetrical Sections 106
15.2 Effective Flexural Properties of Sections 107
15.3 Effective Breadth of Asymmetrical Face Plates 110
15.4 Effective Breadth of Curved Face Plates 112
15.4.1 Symmetrical Face Plates 112
15.4.2 Asymmetrical Face Plate 113

Part II: Chapter 5 – Chapter 9

Chapter 5: Hull Girder Loading 117
1 Introduction 117
2 The Nature of Hull Girder Loads 117
3 Classification of Hull Girder Loads 118

4 Hull Girder Longitudinal Vertical Bending Moments 118
4.1 Stillwater Shear Force and Bending Moment.................................. 118
4.2 Wave-Induced Component.. 120
5 Effect of Hull Girder Vertical Deflection on the Distribution of Shear Force and Bending Moment along Ship Length........................ 120
5.1 Shear Force and Bending Moment Correction due to Ship Deflection 123
6 Hydrodynamic Loads.. 124
6.1 Dynamic Loadings due to Shipping Green Seas 126
7 Hull Girder Dynamic Shear Force and Bending Moment 127
8 Hull Girder Design Vertical Bending Moment... 128
8.1 Standard Still Water Bending Moments.. 128
8.2 Vertical Wave Bending Moment.. 129
8.3 An Approximate Estimate of the Maximum Value of the Wave Induced Bending Moment M_W 130
9 Horizontal Bending Moment 130
10 Hull Girder Shearing Forces 131
10.1 Total Vertical Shearing Force F_V 131
10.2 Stillwater Shear Force Component F_S 132
10.3 The Distribution of the Vertical Wave-Induced Shearing Force ... 133
10.4 Approximate Value of the Maximum Vertical Shear Force.......... 133
11 Wave Induced Torsion Loading 134
12 Probabilistic Prediction of Hull Girder Loading.................................. 137
12.1 Short Term Prediction of Loading.. 137
12.2 Long Term Predictions.. 139
12.3 Extreme Value Distributions 139

Chapter 6: Hull Girder Bending Stresses 141
1 Introduction 141
2 Hull Girder Bending Stress Components... 141
2.1 Hull Girder Primary Stresses Induced by Longitudinal Vertical Bending Moments.. 142
3 Geometrical and Flexural Properties of Ship Sections 145
3.1 Flexural Properties of Longitudinally Framed Deck and Bottom Structures 147
4 Hull Girder Stresses When the Ship Is In the Inclined Condition 149
5 Hull Girder Stresses due to Horizontal Bending Moment 150
6 Hull Girder Shear Stresses.. 151

Chapter 7: Secondary Loading and Stresses.. 153
1 Introduction 153
2 Strength Members of Ship Bottom Assemblies Sustaining Secondary Loadings.. 153
2.1 Secondary Loading and Stresses in Bottom Assemblies.................. 154
2.2 Secondary Loading and Stresses in Transversely Stiffened Bottom Assemblies 154

2.2.1 Secondary Loading and Stresses in Bottom Girders 155
2.2.2 Secondary Stresses in Bottom Plating 156
2.2.3 Secondary Stresses in Tank Top Plating 157
2.3 Secondary Stresses in Longitudinally Stiffened Double Bottom Structure 157
2.3.1 Secondary Stresses in Bottom Girders 158
2.3.2 Secondary Stresses in Bottom Longitudinals 159
2.3.3 Secondary Stresses in Bottom Plating 160
2.3.4 Secondary Stresses in Tank Top Longitudinals 161
2.3.5 Secondary Stresses in Tank Top Plating 161
3 Secondary Loading and Stresses in Deck Structure Assemblies 162
3.1 Secondary Stresses in Transversely Stiffened Deck Structure 162
3.1.1 Secondary Stresses in Deck Girders 162
3.1.2 Secondary Stresses in Deck Plating 163
3.2 Secondary Stresses in Longitudinally Stiffened Deck Structure 163
3.2.1 Secondary Stresses in Deck Girders 163
3.2.2 Secondary Stresses in Deck Longitudinals 164
4 Secondary Stresses in Bottom Structure Assemblies of Oil Tankers 165
4.1 Secondary Stresses in Bottom Girders 165
4.2 Secondary Stresses in Bottom Longitudinals 166
4.3 Secondary Stresses in Bottom Plating 168
5 Grillage Structure 169

Chapter 8: Tertiary Loading and Stresses in Strength Members of Ships 171
1 Introduction 171
2 Local Loading in Transversely Stiffened Bottom Plating 171
3 Local Stresses in Transversely Stiffened Bottom Plating 172
4 Tertiary Loading and Stresses in the Strength Members of Longitudinally Stiffened Bottom Structure 174
4.1 Tertiary Loading on Bottom Longitudinals 174
4.2 Tertiary Stress in Bottom Longitudinals 176
4.3 Tertiary Loading on Tank Top Longitudinals 178
4.4 Tertiary Loading on Bottom Plating 179
4.5 Tertiary Stress in Bottom Plating 181
4.6 Local Stresses in Bottom Plating 182
4.7 Minimum Required Thickness of Bottom Plating 184
4.8 Tertiary Loading and Stresses in Tank Top Longitudinals 185
5 Tertiary Loading and Stresses in Longitudinally Stiffened Deck Structure 186
5.1 Tertiary Loading on Deck Longitudinals 186
5.2 Tertiary Stresses in Deck Longitudinals 187
6 Local Loading and Stresses in Side Longitudinals 189

Chapter 9: Compounding of Stresses in Ship Strength Members 191
1 Introduction 191
2 Various Stresses in Strength Members of Ship Structure 191
2.1 Total Stress Induced in Ship Structural Members 192
3 Compounding of Stresses in Ship Strength Members 193
3.1 Compounding of Stresses in Strength Members of Transversely Stiffened Double Bottom Assembly 194
3.1.1 Locations of Compounding of Stresses 194
3.1.2 Compounding of Stresses in Bottom Girders of Transversely Stiffened Double Bottom Structure 195
3.1.3 Compounding of Stresses in Bottom Plating 199
4 Compounding of Stresses in Tank Top Plating 202
5 Compounding of Stresses in the Strength Members of Longitudinally Stiffened Double Bottom Structure 204
5.1 Locations of Compounding of Stresses for Longitudinally Stiffened Bottom Structure 205
5.2 Compounding of Stresses in a Bottom Girder 206
5.3 Compounding Stresses in Bottom Longitudinals 208
5.4 Compounding of Stresses in the Bottom Plating 212
5.5 Compounding of Stresses in the Tank Top Longitudinals 215
5.6 Compounding of Stresses in the Tank Top Plating 215
6 Compounding of Stresses in Longitudinally Stiffened Deck Structure 217
6.1 Compounding of Stresses in Deck Girders 217
6.2 Compounding of Stresses in Deck Longitudinals 219
6.3 Compounding of Stresses in Deck Plating 222
7 Compounding of Stresses for Oil Tankers 225
7.1 Compounding of Stresses in the Bottom Girder 226
7.2 Compounding of Stresses in Bottom Longitudinals 228
7.3 Compound Stress in Bottom Plating of an Oil Tanker Subjected to Sagging Moment 230

Part III: Chapter 10 – Chapter 14

Chapter 10: Columns and Beam Columns 237
1 Introduction 237
2 Structural Members Subjected to Compressive Loadings 237
3 Classes of Perturbations 238
4 The Problem of Stability 240
4.1 Critical Force and Critical Stress 241
4.2 Effect of Eccentric Loading 248
5 Beam Columns 251
5.1 Load-Deflection Relationship of Beam Columns 252
6 Stresses in Beam Columns 254

Chapter 11: Buckling of Stiffened Panels 267
1 Introduction 267
2 Basic Configurations of Stiffened Panels 267
3 Modes of Deformation of Transversely Stiffened Plate Panels 268
3.1 Modes of Buckling Deformation of Stiffeners 269
3.2 Global Mode of Deformation of Stiffened Panels 272
4 Assessment of Buckling Strength of Plating 273
4.1 Commonly Used Idealized Plate Boundary Support Conditions 274
4.2 General In-Plane Loading Conditions 275
4.2.1 Single Loading Conditions 275
4.2.2 Combined Loading Conditions 276
4.3 Modes of Buckling Deformation 278
4.3.1 Mode of Buckling Deformation of a Long Plate for the following Conditions 278
4.3.2 Mode of Buckling Deformation of a Plate Fixed at the Long Edges is as Shown in Fig. (11,28) 279
4.3.3 Mode of Buckling Deformation of a Long Plate for the following Edge Supports and Loading Conditions is shown in Fig. (11.29): 279
4.3.4 Mode of Buckling Deformation of a Long Plate for the following Edge Supports and Loading Conditions is shown in Fig. (11.30): 279
4.3.5 Mode of Buckling Deformation of a Long Plate for the following Edge Supports and Loading Conditions is Shown in Fig. (11.32): 280
4.3.6 Mode of Buckling Deformation of a Long Plate for the following Edge Supports and Loading Conditions is Shown in Fig. (11.33) 280
4.3.7 Mode of Buckling Deformation of a Long Plate for the following Edge Supports and Loading conditions is Shown in Fig. (11.34) 281
4.3.8 Mode of Buckling Deformation of a Long Plate for the following Edge Supports and Loading Conditions is Shown in Fig. (11.35) 281
4.3.9 Mode of Buckling Deformation of a Long Plate for the following Edge Supports and Loading conditions is Shown in Fig. (11.37) 282
5 Basic Equations of Plate Buckling for Various Boundary Conditions and Different Loading Combinations 282
5.1 Plate Fixed at all Edges and Subjected to In-Plane Compressive Stresses over the Short Edges 282
5.2 Plate Simply Supported at All Edges and Subjected to In-Plane Compressive Stresses over the Long Edges 283
5.3 Plate Has One Long Edge Free and All Other Edges Simply Supported 284

5.4 Plate has One Long Edge Free and the Other Long Edge Clamped ... 285
6 Buckling of Simply Supported Plate under Various in Plane Loading Conditions ... 285
6.1 Plate Subjected to Pure Bending Stresses ... 285
6.2 Plate Subjected to Pure Shear Stresses ... 286
7 In-Elastic Bucking ... 287
7.1 Inelastic Buckling due Shear Loading ... 288
8 Assessment of Buckling Strength of Plating Subjected to Combined Loading ... 288
8.1 Combined Shear and Compressive Stresses on the Short Edges ... 289
8.2 Combined Shear and In-Plane Bending on the Short Edge ... 290
8.3 Combined In-Plane Bending and Compression ... 291
8.4 In Plane Compression in Two Orthogonal Directions ... 292
8.5 Combined Shear, In-Plane Bending and Compression ... 293
9 Critical Buckling Stress of Plating of Stiffened Panels ... 294
9.1 Longitudinally Stiffened Panels ... 294
9.2 Transversely Stiffened Panels ... 295
10 Post–buckling Strength of Plating ... 298
10.1 Ultimate Stress of Simply Supported Plate Panels ... 300
10.1.1 Long Edges Loaded ... 300
10.1.2 Short Edges Loaded ... 301
11 Buckling Limit State of Plate Panels ... 302
11.1 Uncertainty Modeling of Buckling Safety Margin ... 303

Chapter 12: Assessment of Buckling of Ship Structure ... 305
1 Introduction ... 305
2 Ship Strength Members Sustaining Compressive Forces ... 305
3 Basic Equations of Buckling of Plate Panels Subjected to Non-uniform In-Plane Compression ... 307
3.1 Idealization of the In-Plane Compressive Loadings ... 307
3.2 Critical Buckling Stress ... 309
3.3 Boundary Conditions ... 309
3.3.1 Boundary Conditions of Girders ... 309
3.3.2 Boundary Conditions of Plating ... 310
4 Modes of Buckling ... 312
4.1 General Modes of Buckling of Girders ... 312
4.2 Modes of Buckling of Face Plates ... 312
4.3 Modes of Buckling of the Web Plate ... 314
5 General Mode of Buckling of Longitudinals ... 314
6 General Mode of Buckling of Plating ... 315
7 Assessment of Buckling of Girders and Longitudinals ... 316
7.1 Torsion Buckling ... 316
7.2 Lateral Buckling of Flanges ... 318
8 Assessment of Buckling of Deck Girders ... 319
8.1 Assessment of Buckling of the Web Plate ... 319
8.2 Assessment of Buckling of the Face Plate ... 321

9 Assessment of Buckling of Longitudinals ... 322
10 Assessment of Buckling of Plating ... 326
10.1 Buckling of Bottom Plating ... 326
10.2 Buckling of Deck Plating ... 328
10.3 Buckling of Side Shell Plating ... 332
10.3.1 Configurations of Side Shell Plate Panels ... 332
10.3.2 Induced Stresses in the Side Shell Plating ... 332
10.3.3 Compounding of Stresses in Side Shell Plate Panels ... 334
10.3.4 Assessment of Buckling of Side Shell Plating ... 335

Chapter 13: Control of Buckling Failure of Ship Structure ... 339
1 Introduction ... 339
2 Reliability Basis of Ship Structural Safety ... 339
3 Deterioration of Structural Capability with Time ... 342
4 Responsible Authorities for Ensuring Structural Safety ... 343
5 Main Causes of Buckling Failure ... 344
6 Control of Ship Structure Failure by Improving Design ... 345
6.1 Improving Design of Plate and Tripping Brackets ... 345
6.2 Using Symmetrical Face Plates of Girders ... 347
6.3 Improving Design of Curved Part of Web Frame Brackets ... 347
6.4 Improving Design of Plate Panels Loaded by Compressive Forces ... 348
6.5 Improving Design of Plate Panels Loaded by Shear Forces ... 348
6.6 Improving Design of Local Structural Connections ... 349
6.7 Improving Design of the Connection between the Web Plate Stiffeners and Longitudinals ... 349
6.8 Improving Design of Web Plating of Top Wing Tanks ... 350
6.9 Improving Design of the Ends of Side Girders in Oil Tankers ... 350
6.10 Improving Design of Web Plating of Deep Girders ... 351
7 Owners Approach for Improving Ship Structure Operational Life and Safety ... 351
7.1 Impact of Corrosion on Strength of Ship Structure Members ... 352
8 Control of ship Structure Failure by Improving Quality of Ship Fabrication Processes ... 356
8.1 Control of the Out-of Straightness of Stiffeners/Girders and Plate Panels ... 356
8.2 Control of Fabrication Deformations of Ship Structure Members ... 356
8.2.1 Out-of-Straightness ... 357
8.2.2 Warping of the Whole Section of the Strength Member ... 358
8.2.3 Warping of Face Plate ... 358
8.2.4 Lateral Deviations between Centerline of the Web and Centerline of the Flange ... 358
8.2.5 Inclination of the Web Plate of Section with Respect to the Attached Plating ... 359
8.2.6 Deformations and Deviations of Face Plates or Flanges ... 359
8.2.7 Gap Between Beam and Frame, see fig. (13.40) ... 360

8.3 Control of Welding Distortions 360
8.3.1 Avoiding Over-Welding 360
8.3.2 Placing Welds near the Neutral Axis or the Center of the Part 360
8.3.3 Balancing Welds Around the Neutral Axis 360
8.3.4 Control of Alignment of Butt and Fillet Welds 361
9 Improving Control of Corrosion 361

Chapter 14: Problems 363

References 377

Index 381

Introduction

A major requirement for any marine structure is to have low initial and operational costs, to be reasonably safe, not to have catastrophic failure nor to have much trouble in service due to frequent minor failures. Safety is concerned not only with the structure itself, but also with external damage that may result as a consequence of failure. Ship structural analysis and design involves the determination of the design loads, defining acceptable criteria and conducting strength assessment. The establishment of acceptable criteria should be based on the accumulated knowledge and expertise from several sources such as owners, builders, classification societies, and researchers. Once the loads and acceptance criteria are defined, the assessment of strength can be carried out.

The development of a satisfactory ship structure involves the determination of the scantlings (sizes) of all its strength members. These scantlings should provide adequate strength to resist the various types of hull girder and local loads imposed on the ship during her operation among sea waves. These loads include longitudinal and transverse bending, torsion, and shear in still-water and among waves and also the static and dynamic loads resulting from the weight of cargo, hydrostatic and hydrodynamic pressures, impact forces and other local loads such as heavy machinery and equipment. Classification societies have developed techniques for calculating the loads on a ship and evaluating the structural integrity of ship hulls.

Ship strength members sustaining compressive forces should have adequate buckling strength to ensure safety against buckling failure. The assessment of buckling strength of bottom structural members is to be carried out when the ship is in hogging condition and for deck structure members when the ship is in sagging condition. For both cases, the compounding of stresses is to be carried out when the secondary and tertiary loadings are inducing compressive stresses. The computational complexity has been greatly simplified and solved with the introduction of rational analysis procedures. These rational procedures include modeling of the complex ship structure, load modeling and application and modeling of boundary conditions. The analysis of structural members aims at checking the strength, stability and rigidity of the preliminarily selected scantlings of the structure. The strength members of ship structure are analyzed on the basis of the theory of the strength of materials and structural mechanics to determine the induced internal stresses under the action of the applied loads. The assessment of buckling of ship strength members should include the web plates and face plates of deck and bottom girders. The assessment of buckling strength of shell plating should cover the various types of induced in-plane loading conditions and the different end support conditions.

Part I
Chapter 1 – Chapter 4

Chapter 1
Ship Structure Configurations and Main Characteristics

1 Introduction

This chapter presents the basic configurations and structural features of some ship types. The main design features of single and double side bulk carriers are clarified. The main types and categories of bulk carriers are classified. The structural components of single and double skin bulk carriers as well as the construction of double bottom are specified. The main types and structural characteristics of general cargo and container ships are highlighted. The basic arrangements and design features of Ro-Ro ships are described. The structural systems and design features of single and double hull tankers are presented. The commonly used abbreviations to describe the different types and sizes of bulk carriers are enumerated. The advantages and drawbacks of the double-hull design are clarified.

2 Ship Types and Main Characteristics

2.1 Bulk Carriers

A bulk carrier is a single deck cargo ship designed to transport dry bulk cargos, such as grains, coal, ore, and cement, see Fig. (1.1). Today, about 40% of the world's cargo fleet is bulk carriers. Bulk carriers are designed either with single skin or with double skin. Most existing bulk carriers are of the single skin type.

The International Convention for the Safety of Life at Sea defines a bulk carrier as "a ship constructed with a single deck, top side tanks and hopper side tanks in cargo spaces and intended to carry dry cargo in bulk", see Fig. (1.2).

The top wing tanks, hopper tanks and the double bottom tanks are used as ballast tanks. The angle of the sloping plate of the top wing tanks should be less than the angle of repose of the anticipated cargoes to be transported. This greatly reduces shifting of cargo which can endanger the ship, see Fig. (1.4). Grain carriers introduce special stability problems due to the free surface effect of the grain, see Fig. (1.3). Shifting boards are used to reduce the free surface effects. The cargo-carrying length of the ship is divided into a number of holds depending upon the types of cargoes to be carried. Bulk carriers carry ore in alternate holds, see Fig.(1.5), to improve ship stability, ship motions and also to reduce dynamic loadings.

Bulk carriers are usually constructed of mild steel. However, high-tensile steel is used to reduce the vessels light weight by using thinner thicknesses for the plates and stiffening members. The use of high tensile steel can reduce the rigidity of the ship hulls and may develop earlier fatigue cracks. Transverse bulkheads are constructed of corrugated steel plates, reinforced at the bottom and top connections with bottom and top stools, see Fig. (1.6).

Fig. 1.1 A picture of a conventional bulk carrier

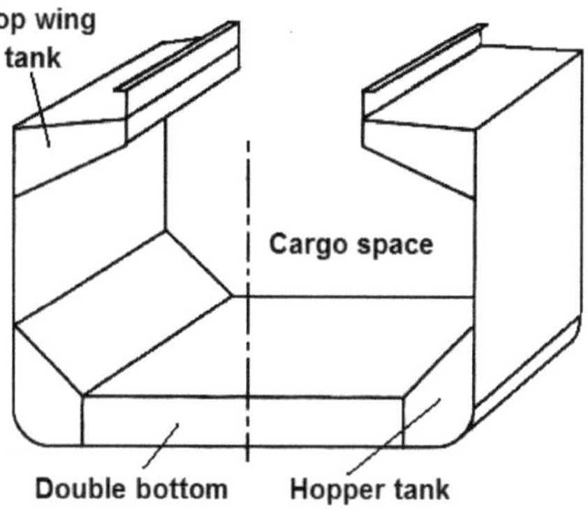

Fig. 1.2 Configuration of a ship section of a bulk carrier

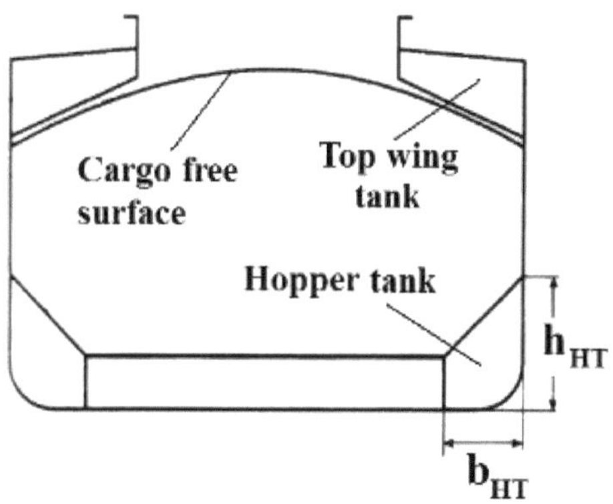

Fig. 1.3 Role of top wing tanks in reducing free surface effect

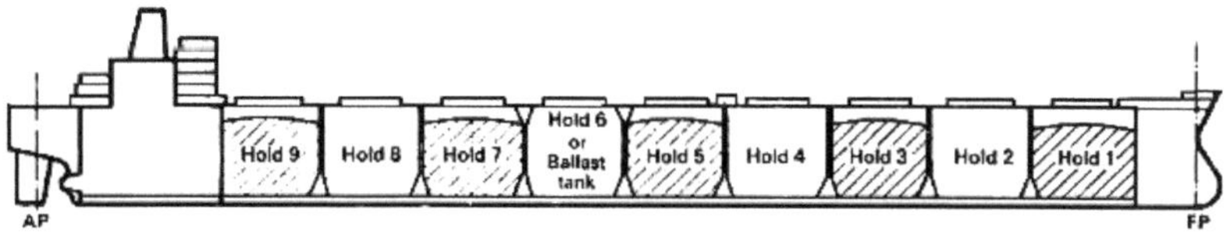

Fig. 1.4 Alternate cargo hold loading

A large proportion of bulk carriers do not carry cargo-handling equipment, because they trade between special terminals and ports which are equipped with effective and adequate facilities for loading and unloading bulk commodities, see Fig. (1.5).

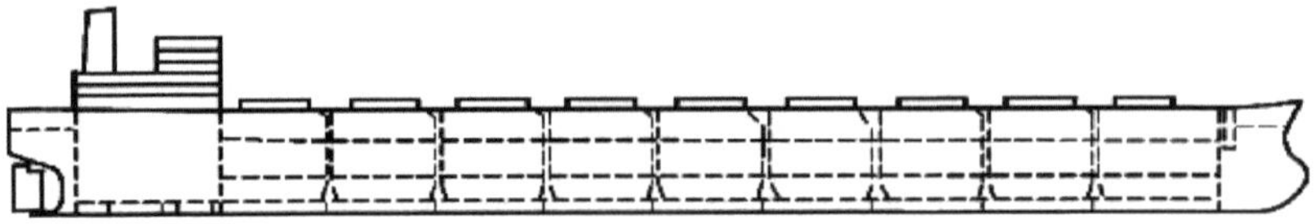

Fig. 1.5 A bulk carrier without cargo handling facilities

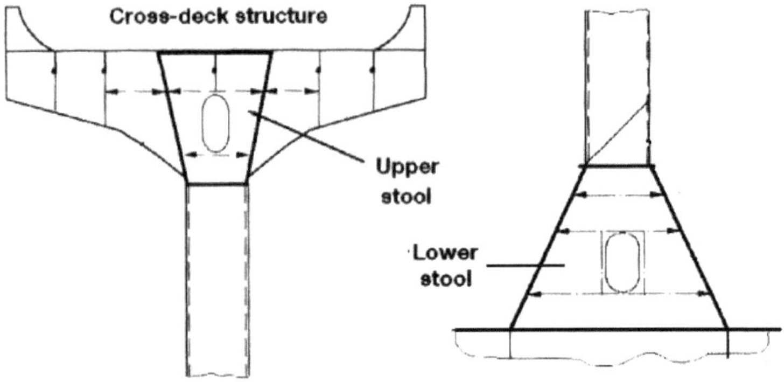

Fig. 1.6 Upper and lower stools of transverse bulkheads

2.2 Double Sides Bulk Carriers

In the past ten years, the new international regulations of bulk carriers have greatly improved the ship design for safety and inspection. Bulk carriers are designed with double sides, see Fig. (1.7).

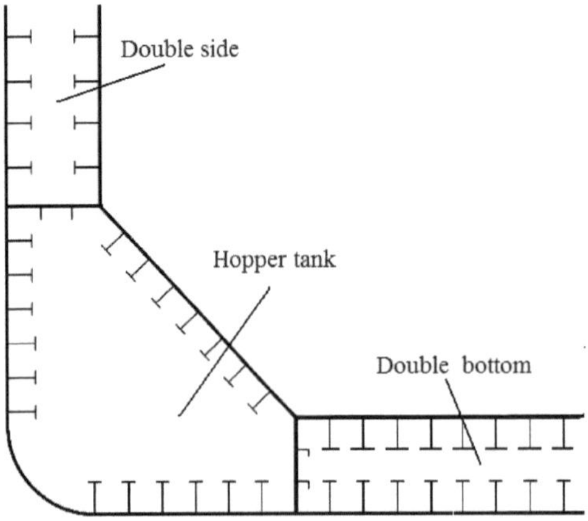

Fig. 1.7 A section of a double side structure

The double side structure of a bulk carrier can prevent the water coming into the hold when the outer side shell is fractured or punctured by collision or any other cause.

All the strength members of the double sides are accommodated in the double side structure. This simplifies the loading, unloading, and cleaning of the cargo holds. Double sides also improve ship's capacity for ballasting, which is useful when carrying light cargoes. The increased capacity of ballast water is used to increase the vessels draft for providing adequate stability and improving sea-keeping.

2.3 *Bottom Structure of Bulk Carriers*

The double bottom is normally deeper than in conventional cargo ships so as to provide the required strength and ballast capacity. Longitudinal framing is adopted for the design of double bottom of bulk carriers, see Fig. (1.8). The maximum spacing of solid plate floors is 2.5 m. and additional intercostal side girders are also provided. Bulk carriers classed for the carriage of heavy cargoes, or ore, will have substantial scantlings for the inner bottom plating, floors, and girders so as to support the heavy cargo loads.

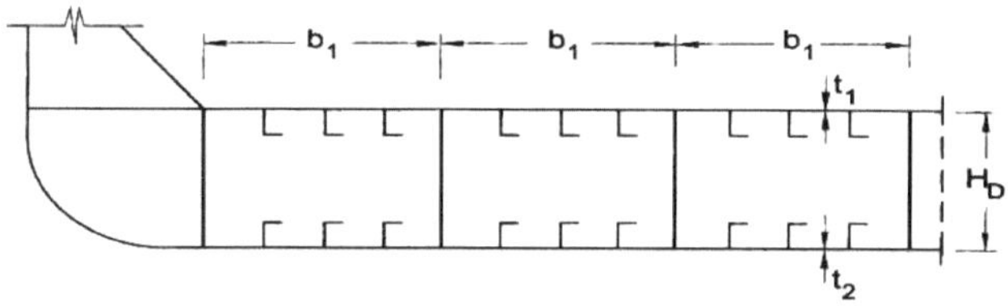

Fig. 1.8 Double bottom structure of a bulk carrier

2.4 *Types and Categories of Bulk Carriers*

A number of abbreviations are used to describe the different types of bulk carriers. "OBO" describes a bulker which carries a combination of ore, bulk, and oil. "O/O" is used for bulk carriers transporting a combination of oil and ore. For very large and ultra large ore and bulk carriers, the terms "VLOC," "VLBC," "ULOC," and "ULBC" are applied.

Bulk carriers are categorized into six major sizes: small, Handysize, Handymax, Panamax, Capesize, and very large. Very large bulk and ore carriers fall into the Capesize category. Handysize and Handymax are general purpose bulk carriers. These two types represent about 71% of all bulk carriers over 10,000 tons DWT and also have the highest rate of growth. This is partly due to the new mandatory regulations which put greater constraints on the design and building of larger bulk carriers. Handymax bulk carriers have five cargo holds and cargo capacity of 52,000 – 58,000 tons DWT. These ships are also general purpose in nature and having lengths 150–200 m. The size of a Panamax bulk carrier is limited by the Panama Canal Locks. These locks can accommodate ships with a beam of up to 32.31 m, length overall of up to 294.13 m, and a draft of up to 12.04 m.

Capesize ships are too large to travel through the Suez or Panama canals and must travel round the Cape of Good Hope or Cape Horn.

Since April 1, 2006, the International Association of Classification Societies (IACS) has adopted the *Common Structural Rules for the design of bulk carriers.* These rules apply to bulk carriers having more than 90 meters in length and require that scantlings calculations should take into account items such as the effect of corrosion, the hostile sea conditions of the North Atlantic, and the induced dynamic stresses during loading.

A type of bulk carrier of increasing importance is the liquefied natural gas carrier (LNG). The cargo tanks are isolated from the ship's structure by very thick insulation and the ship is fitted with double bottom and side protection.

2.5 Main Structural Components of Single Skin Bulk Carriers

Longitudinal framing is adopted in larger bulk carriers, and it is common within the hopper and topside wing tanks of these vessels. Transverse frames are used to stiffen the side shell between the hopper and topside tanks, see Fig. (1.9).

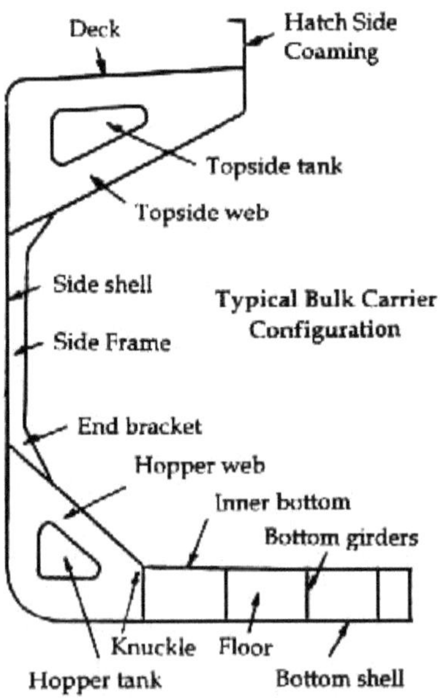

Fig. 1.9 Main structural components of single skin bulk carrier

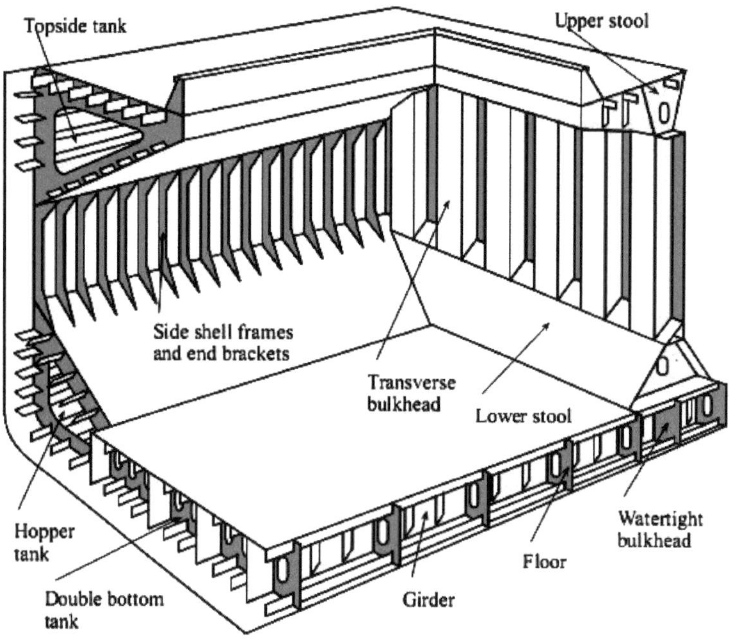

Fig. 1.10 Main structure members of a cargo hold

The main structural components of a typical section of a cargo hold of a single skin bulk carriers are, see Figs. (1.9, 1.10):

- Topside tank
- Transverse bulkhead upper stool
- Transverse bulkhead lower stool
- Double bottom tank
- Hopper tank
- Side shell frames
- Side shell frame end brackets
- Corrugated transverse bulkhead.

The nomenclature of the strength members of a typical cargo hold of a conventional bulk carrier is as follows, see Fig. (1.10):

- Strength deck plating
- Strength deck longitudinals
- Transverse web frame in topside tank
- Side shell longitudinals
- Side shell plating

- Transverse web frame in hopper tank
- Bilge plating
- Bottom shell longitudinals
- Bottom shell plating
- Double bottom floor
- Keel plate
- Duct keel
- Double bottom girders
- Inner bottom plating (tank top)
- Inner bottom longitudinals
- Hopper tank sloping plating longitudinal
- Hopper tank sloping plating
- Side shell frames
- Topside tank sloping plating longitudinal
- Topside tank sloping plating
- Topside tank longitudinal plating (hatch side girder)
- Hatch side coaming.

3 General Cargo Ships

Cargo ships could be single decker or multi-decker. Fig. (1.11) shows the structural arrangement of a cargo hold of a single decker vessel and Fig. (1.12) shows the structural configuration of a tween decker vessel.

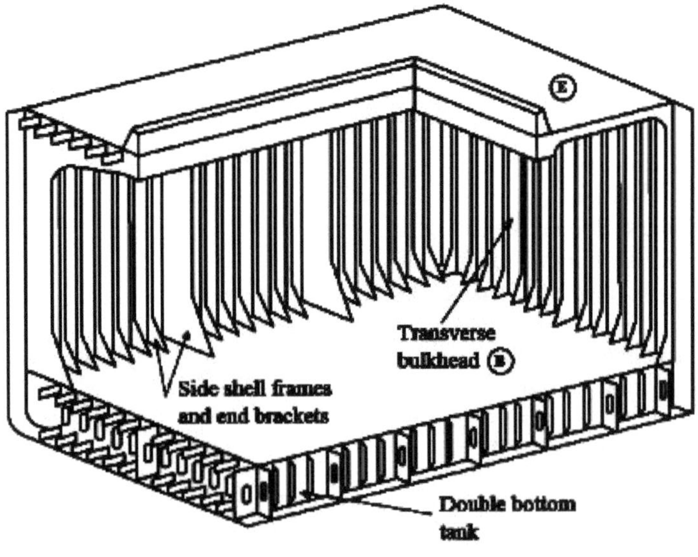

Fig. 1.11 Structural arrangement of a single decker vessel

Cargo ships are usually designed and equipped with cranes and other means of cargo loading and unloading to suit the required trade. Various combinations of derricks, winches and deck cranes are used for the handling of cargo, see Fig. (1.13). Many modern ships are fitted with deck cranes which reduce cargo-handling times and manpower requirements. A special heavy-lift derrick may also be fitted, covering one or two holds. The accommodation and machinery spaces are usually located either aft or with one hold between them and the aft peak bulkhead.

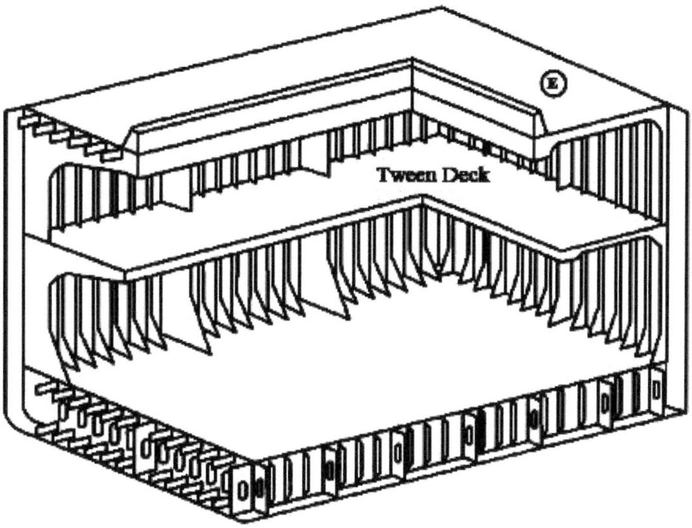

Fig. 1.12 Structural configuration of a tween decker vessel

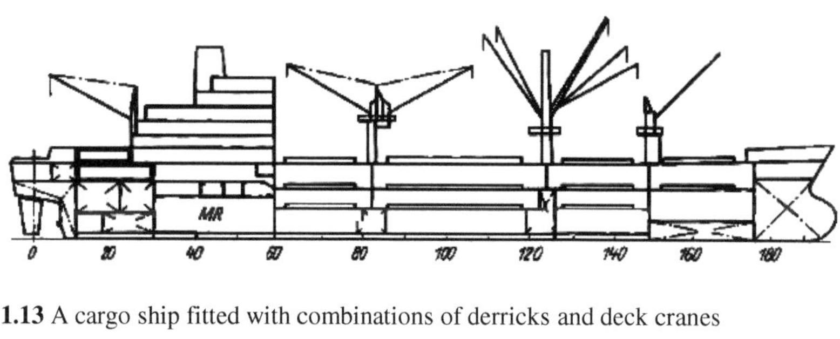

Fig. 1.13 A cargo ship fitted with combinations of derricks and deck cranes

The current range of cargo capacities of general cargo ships is from 2000 to 20000 tons DWT with lengths ranging from 80 m to 160 m and speeds ranging from 12—18 knots. Cargo ships are designed to have a life expectancy of 25 to 30 years before scrapping. Access to the cargo holds is provided by openings in the deck called hatches, see Fig. (1.14). Hatches are made as large as hull girder and local strength considerations allow. Hatch covers are used to close the hatch openings when the ship is at sea. The hatch covers are made watertight and lie upon hatch coamings.

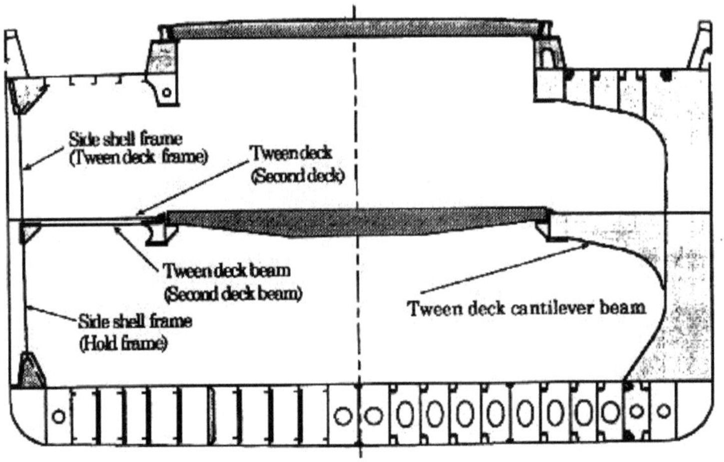

Fig. 1.14 A section of a cargo ship with single deck hatch opening

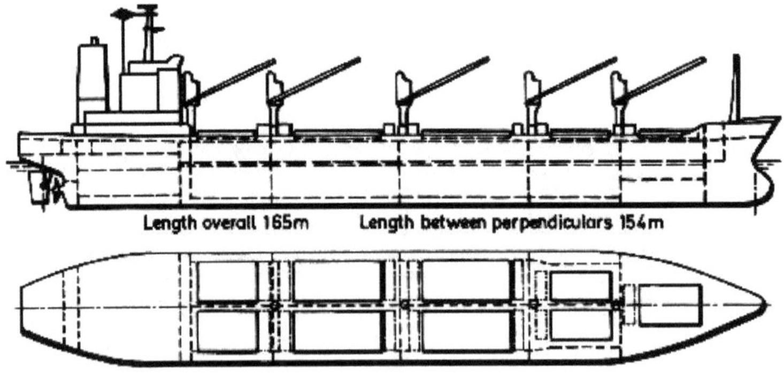

Fig. 1.15 A cargo ship fitted with twin hatches

The hatch coamings are constructed around the hatch openings at some distance from the upper or weather deck so as to reduce the risk of flooding in heavy seas.

The decks could have either single hatch openings or twin hatch openings, see Fig. (1.15).

Since full cargoes cannot be guaranteed in all loading conditions of this type of ship, adequate capacity of ballast tanks must be provided, particularly in the double bottom. The ballast capacity should be adequate to ensure that the vessel has a sufficient draught for stability and total propeller immersion. The fore and aft peak tanks are also used to assist in adjusting the trim of the ship.

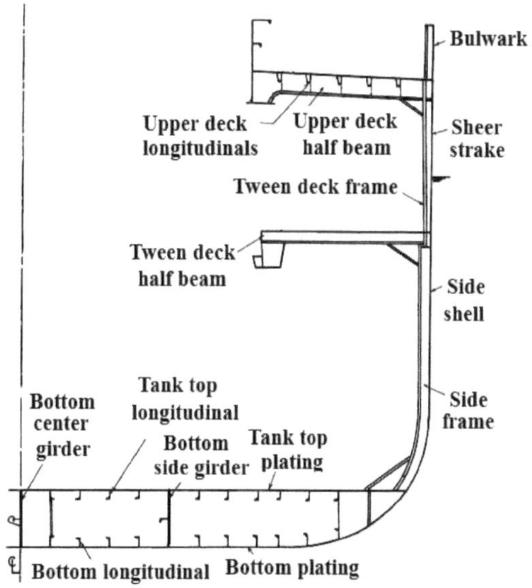

Fig. 1.16 Main structural members of the midship section of a typical tween deck general cargo vessel

The main structural members of the midship section of a typical tween deck general cargo vessel are shown in Fig. (1.16). The hull girder of a general cargo ship could be either transversely framed or could have the deck and bottom structure longitudinally framed and the side structure transversely framed, see Fig. (1.16).

Cargo ships are either liners or tramps reflecting the services they offer to the maritime industry. Cargo liners carry general cargo and operate on fixed schedules and fixed tariff rates. Tramp ships do not have fixed schedules.

The double bottom extends almost the full length of the ship and is divided into separate tanks, some of which carry fuel oil and fresh water. The remaining tanks are used for ballast when the ship is sailing empty or partly loaded. Deep tanks may be fitted which can carry liquid cargoes or water ballast.

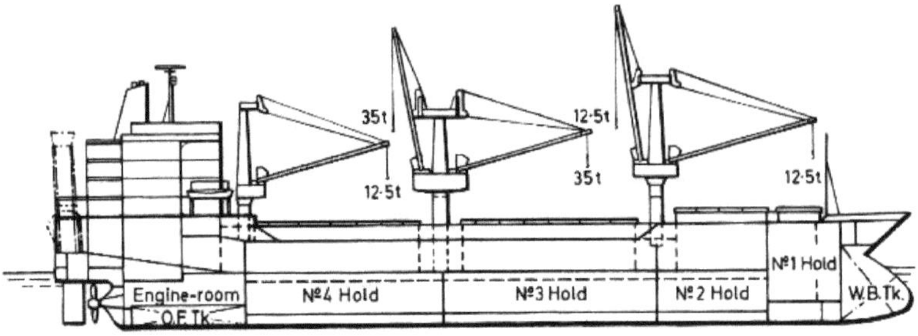

Fig. 1.17 Multipurpose vessel with twin deck hatches

Multipurpose cargo ships can carry containers and bulk cargoes as well as general cargo, see Fig. (1.17). The Hamlet Multipurpose design is a versatile cargo ship having a volumetric bale capacity of 20,000 m^3 and a speed of 15-knot.

Refrigerated cargo ships usually run at higher speeds than general cargo ships, often having speeds up to 22 knots. The fitting of refrigeration plants for the cooling of cargo holds enables the carriage of perishable food cargos by sea. Refrigerated cargo ships may have more than one tween deck, and all hold spaces will be insulated to reduce heat transfer and power consumption. Cargo may be carried frozen or chilled depending upon its nature.

4 Container Ships

The container ship is a cargo ship designed for the carriage of cargo in containers, see Fig. (1.18). Engine room and accommodation are usually located aft to provide the maximum length of the full-bodied ship for container stowage, see Fig. (1.18).

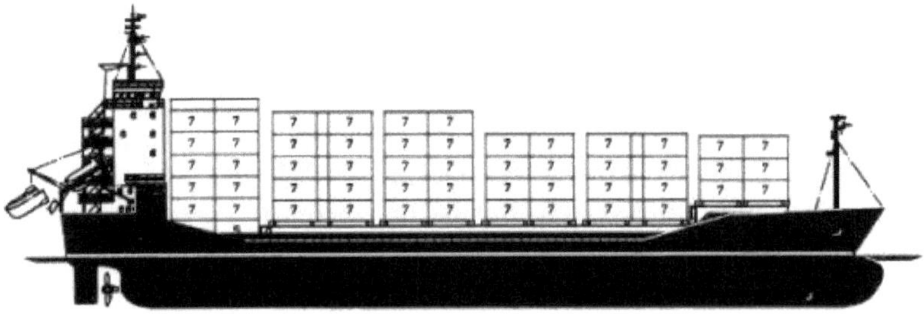

Fig. 1.18 A container ship

A container is a re-usable box having a square section 2435 mm by 2435 mm and lengths of 6055, 9125 and 12190 mm. Most containers used today measure 40 feet in length. Containers are used to transport most general cargoes including refrigerated cargoes.

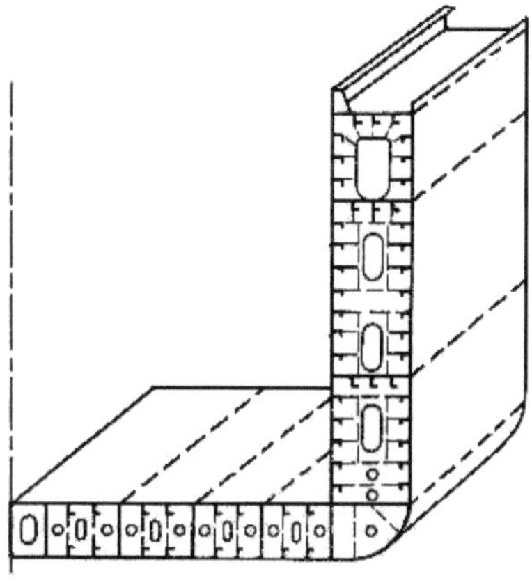

Fig. 1.19 A typical section of a container ship

The cargo-carrying section of the ship is divided into several holds. The ship section has a box-like arrangement of wing tanks which provide the required longitudinal strength of the ship hull girder and space for additional ballast capacity, see Fig. (1.19). Container handling consists only of vertical movement of the cargo in the hold.

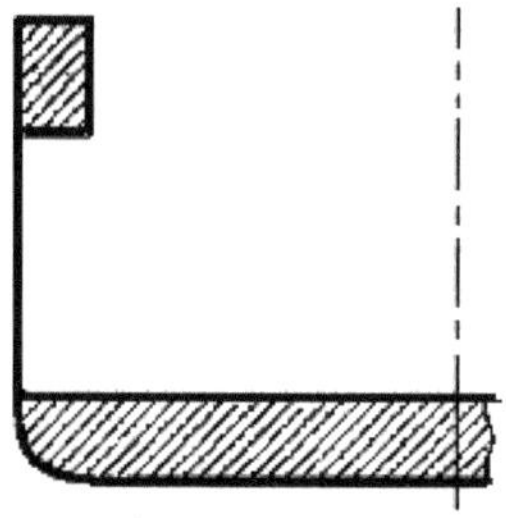

Fig. 1.20 A ship section of a small container ship

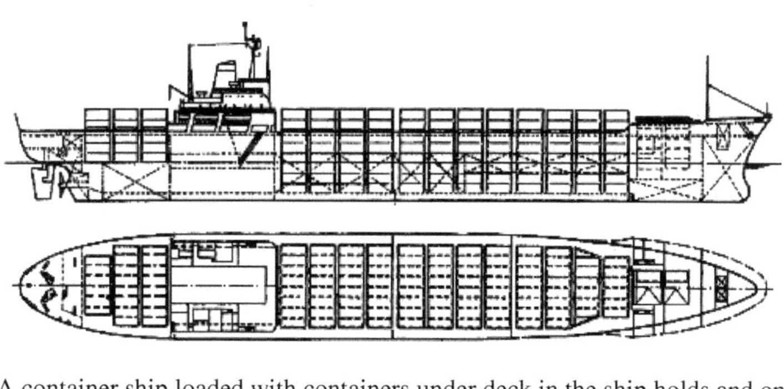

Fig. 1.21 A container ship loaded with containers under deck in the ship holds and on deck

The hatch openings take the full internal width and length of the hold. The double bottom is the main structural component providing the required longitudinal strength and the necessary additional ballast space. The wing tanks may be utilized for water ballast and can be arranged to counter the heeling of the ship when discharging containers. Small container ships are designed with a single hull structure, see Fig. (1.20). The world's largest container ship has a capacity of 15,200 TEU containers. Small container feeder ships are used to transport containers from container terminals to large ships at centralized hub ports.

Ship cargo capacity is measured by Twenty Foot Equivalent unit (TEU), the number of standard 20-foot containers measuring 20 × 8.0 × 8.5 feet (6.1 × 2.4 × 2.6 m).

Containers are carried under deck in the ship holds and can also be stacked on the hatch covers where a low density cargo is carried, see Fig. (1.21). Special lashing arrangements are provided for transporting deck containers.

The double side hull structure of a container ship provides the required hull girder strength and is divided into watertight compartments used as ballast tanks see Fig. (1.22). A container ship has large upper deck openings, see Fig. (1.22). These large deck openings have deleterious effects on the hull girder torsional rigidity. Inadequate torsional rigidity may cause high warping stresses to be developed at the corners of the deck openings. Torsion boxes are added to the double hull side structure to provide the required hull girder torsional rigidity, see Fig. (1.23).

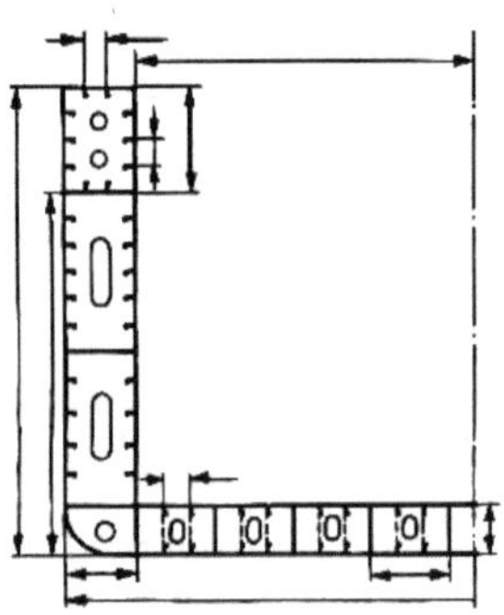

Fig. 1.22 Double side hull structure of a container ship

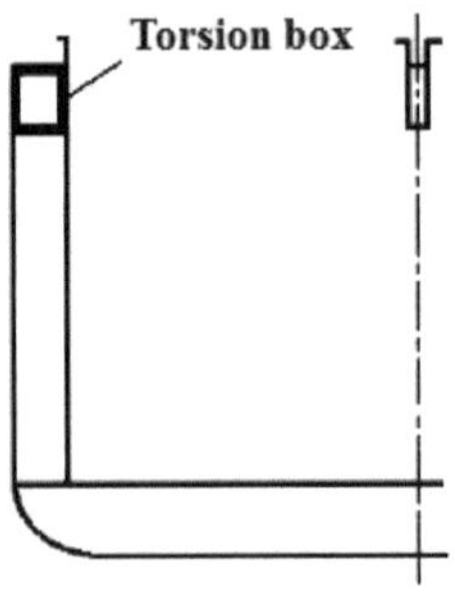

Fig. 1.23 Torsion boxes fitted to the double hull side structure

Fig. 1.24 Large container ship without loading/unloading gear

Large container ships do not carry their own loading gear, see Fig. (1.24). Loading and discharging of large container ships can only be done at ports and terminals having adequate cranes and facilities for rapid loading and discharging of containers.

Smaller ships with capacities up to 3,000 TEU are often equipped with their own cranes, see Fig. (1.25).

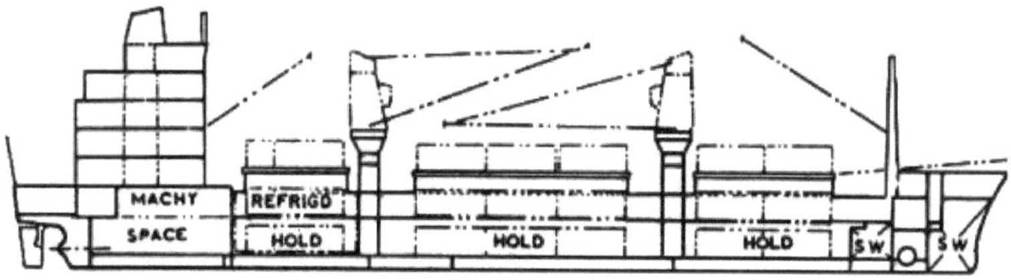

Fig. 1.25 Small container ship equipped with own cranes

5 RoRo Ships

Roll-on/roll-off (Ro-Ro) ships are vessels designed to transport wheeled cargo such as vehicles, cars, trucks, etc. that are driven on and off the ship on their own wheels using external and internal ramps. This is different from lo lo (lift on-lift off) vessels which use cranes to load and discharge cargo. Ro-Ro ships have inherent stability risks. The provision of very high freeboard improves the seaworthiness of these vessels.

The main types of Ro-Ro vessels include ferries, cruse ferries, cargo ships, and barges. The largest Ro-Ro passenger cruise ferry is 75,100 GT and 223.70 m long, 35 m wide and can carry 550 cars. The roll on/roll off passenger (ROPAX) is a Ro-Ro vessel provided with passenger accommodation for transporting passengers and RORO cargo. New automobiles are transported by a large type of Ro-Ro ships called Pure Car Carrier (PCC) or Pure Car Truck Carrier (PCTC). Cargo capacity of Ro-Ro Car Carriers could be measured by car equivalent units (CEU).

5.1 Ramps Types and Arrangements in Ro-Ro Ships

Ro-Ro vessels have built-in ramps which allow the cargo to be efficiently "rolled on" and "rolled off" the vessel when in port. The ramps and doors may be stern-only, see Fig. (1.26), or bow and stern for quick loading, see Fig. (1.27). The bow and stern axial ramps and the quarter ramp are situated on the port side. Internal ramps and elevators provide access to the internal main decks.

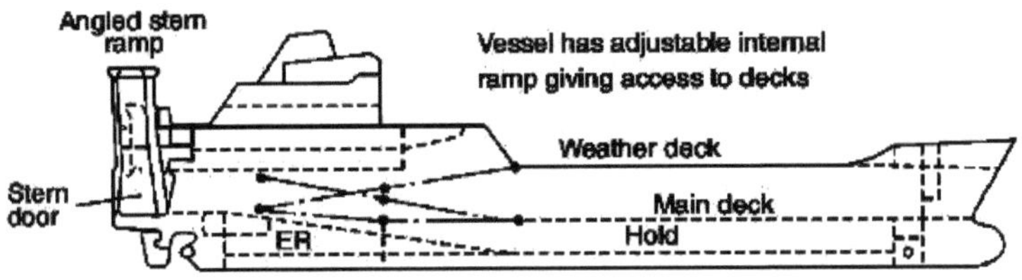

Fig. 1.26 General layout of a Ro-Ro vessel with stern ramp

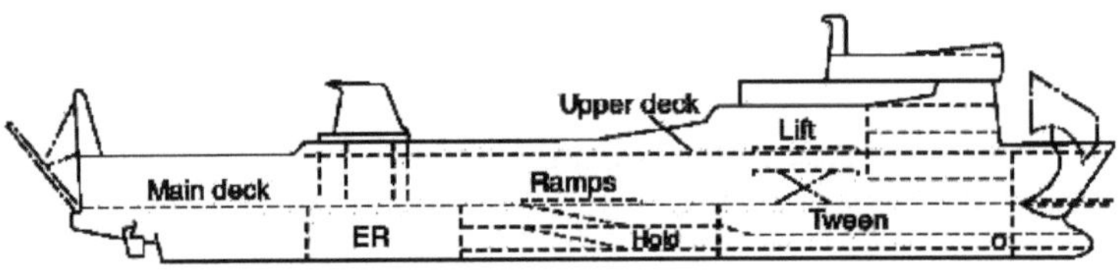

Fig. 1.27 General layout of a Ro-Ro vessel with bow and stern ramp

A ship section of a Ro-Ro vessel is shown in Fig. (1.28).

A hybrid container RORO vessel called ConRo is a Ro-Ro container ship used to transport vehicle Ro-Ro cargo in the internal decks and the container cargo are stacked on the top decks.

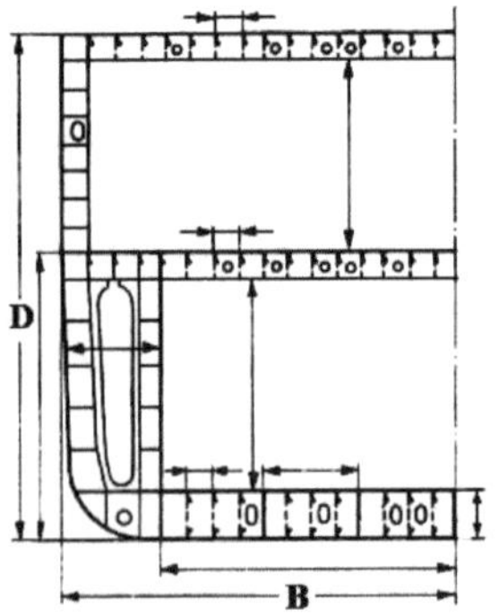

Fig. 1.28 A typical ship section of a RoRo vessel

6 Tankers

Tankers are used for transporting bulk liquid cargoes such as crude oil, chemicals, liquefied petroleum gas, liquefied natural gas, etc. There are two basic types of oil tankers: the crude tanker and the product tanker. Crude tankers transport large quantities of unrefined crude oil from its point of extraction to refineries. Product tankers, generally much smaller in size, are designed to transport petrochemicals from refineries to terminals near consuming markets. The size classes of tankers range from inland or coastal tankers of a few thousand metric tons of deadweight (DWT) to the ultra large crude carriers (ULCCs) of 550,000 tons DWT.

6.1 Single Hull Tankers (Conventional Construction)

The accommodation and machinery spaces are always located aft, see Fig. (1.29). The cargo is discharged by cargo pumps fitted in one or more pump rooms either at the ends of the cargo tank section or sometimes in the middle. The cargo-carrying length of an oil tanker is divided up into individual tanks by transverse bulkheads, see Fig. (1.29), and the ship section is also divided into individual tanks by longitudinal bulkheads, see Fig. (1.30). No double bottom is fitted in the cargo-carrying section of conventional oil tankers.

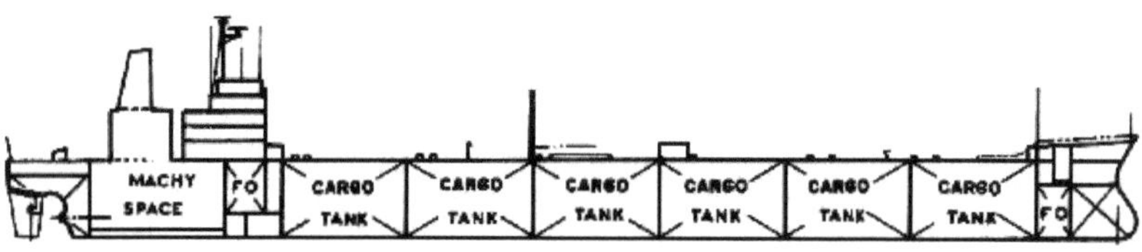

Fig. 1.29 A profile of an oil tanker

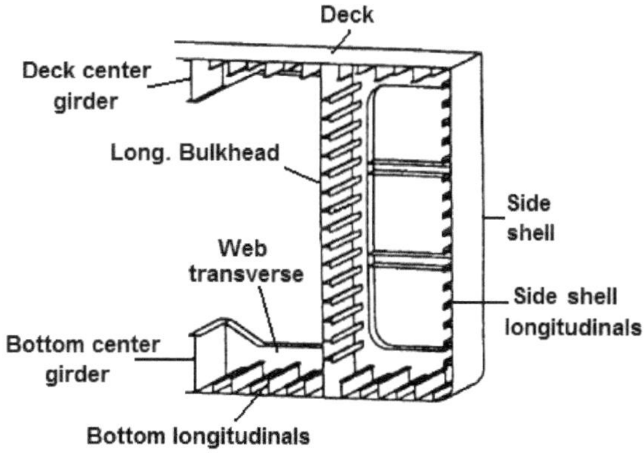

Fig. 1.30 Main structural members of a single hull oil tanker

In accordance with the International Convention for the Prevention of Pollution from Ships, 1973 (MARPOL)' all single-hulled oil tankers will be phased out worldwide. The main structural members of a single hull oil tanker are shown in Figs. (1.30). The main structural members of the bottom structure and transverse bulkhead in a single hull oil tanker are shown in Fig. (1.31).

The primary structural members of the center tank of a single hull oil tanker is shown in Fig.(1.32) and for the wing tank are shown in Fig. (1.33).

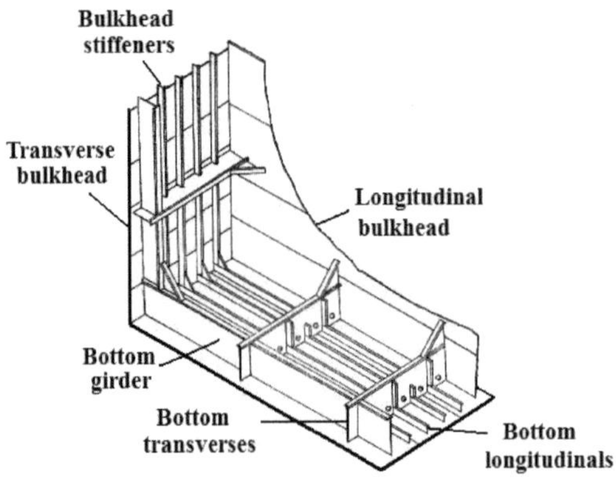

Fig. 1.31 Main structural members of bottom structure and transverse bulkhead in a single hull oil tanker

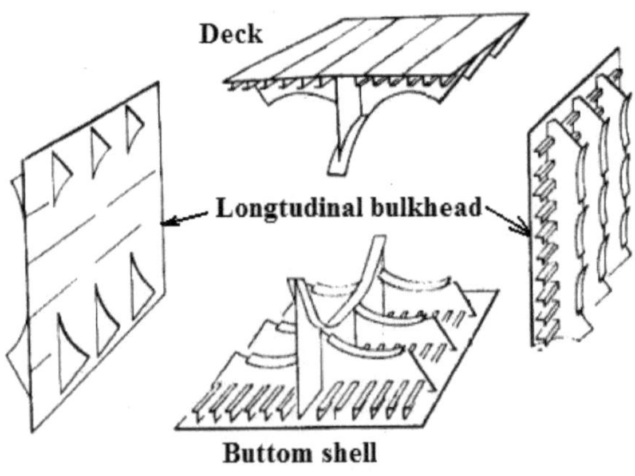

Fig. 1.32 Primary structural members of the center tank of single hull tanker

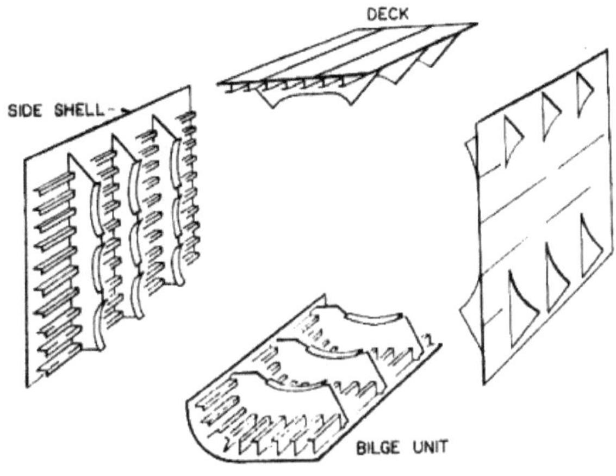

Fig. 1.33 Primary structural members of the wing tank of a single hull tanker

6.2 Design Features of Double Hull Tankers

The double hull tanker is a tanker designed for the carriage of MARPOL Annex I cargoes. The cargo space is protected from the environment by a double hull consisting of double side and double bottom spaces. These double spaces are dedicated to the carriage of ballast water for ships of 5,000 tons DWT and above, see Fig. (1.34). Most new tankers are constructed with double hull.

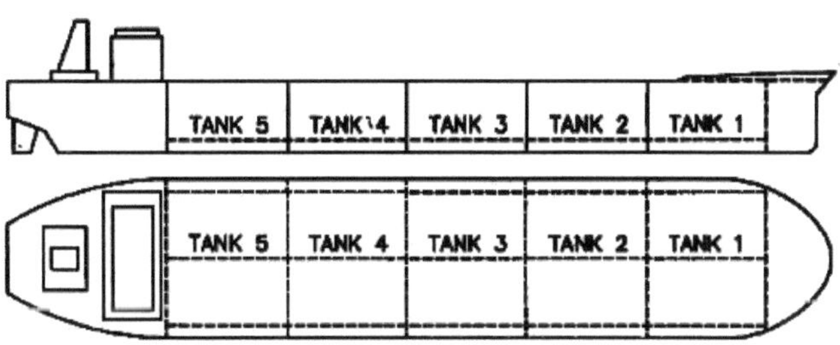

Fig. 1.34 A double hull tanker

Some of the advantages of the double-hull design include ease of ballasting in emergency situations, decreased corrosion problems because of the elimination of saltwater ballasting in cargo tanks, increased environmental protection, cargo discharge is quicker, more complete and easier, tank washing is more efficient, and better protection in low-impact collisions and grounding. Double-hull design is superior in prevention of spillage in low energy small casualties.

The main drawbacks of using the double-hull design are higher build costs, greater operating expenses (e.g. higher canal and port tariffs), ventilation difficulties of ballast tanks, continuous monitoring and maintenance of ballast tanks, increased transverse free surface, much larger surfaces to maintain, greater risk of explosions in double-hull spaces due to failure of vapor detection and difficult of cleaning of ballast tanks. In high energy casualties spills can be significantly high when both hulls are breached.

6.3 Structural System of Double Hull Structure

Tankers less than 5,000 tons DWT will generally have a double bottom in conjunction with either a double side or centerline bulkhead. Above 5,000 tons dwt the double side becomes mandatory, and the minimum required breadth of wing ballast tank will increase from 1 meter to 2 meters as the deadweight rises to 20,000 tons DWT.

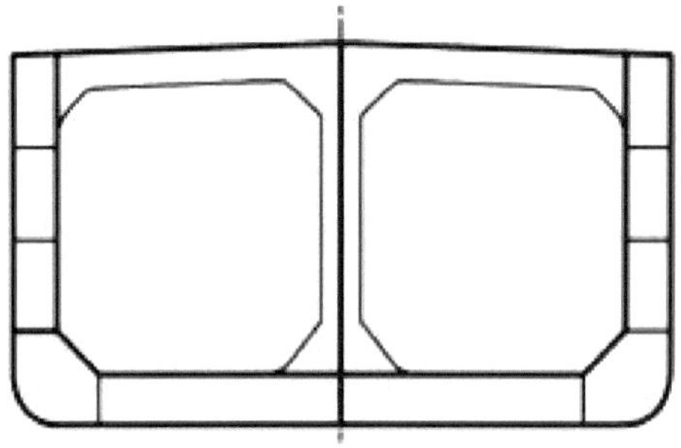

Fig. 1.35 A ship section of a small double hull oil tanker with a centerline longitudinal bulkhead

Fig. (1.35) shows a ship section of a small double hull oil tanker with a centerline longitudinal bulkhead. Fig. (1.36) shows a ship section of a large double hull oil tanker with two longitudinal bulkheads.

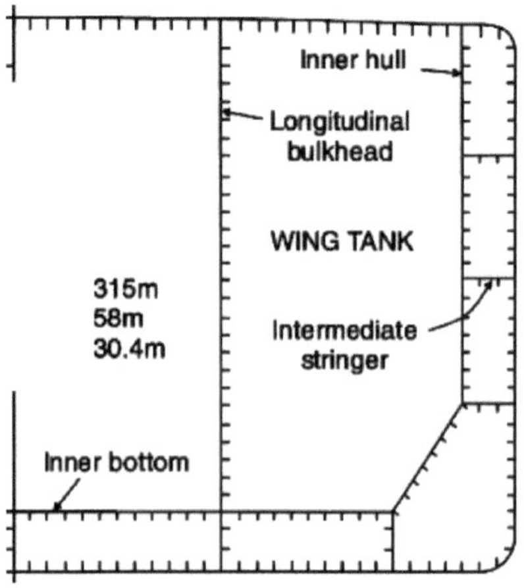

Fig. 1.36 A ship section of a large double hull oil tanker with two longitudinal bulkheads

6.4 Double Bottom and Double Side Construction of Oil Tankers

Other designs of oil tankers are the double-bottom and double-sided construction. These designs combine aspects of single and double-hull construction. Fig. (1.37) shows a ship section of a small oil tanker with a double bottom and a centerline bulkhead. A ship section of an oil tanker with double bottom construction and two longitudinal bulkheads is shown in Fig. (1.38).

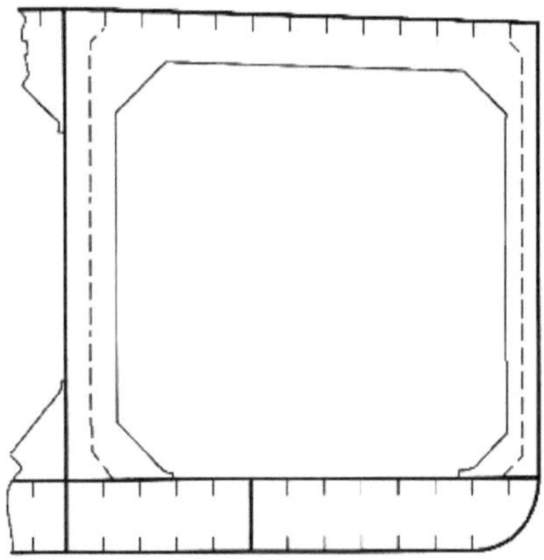

Fig. 1.37 A ship section of a small oil tanker with a double bottom and a centerline bulkhead.

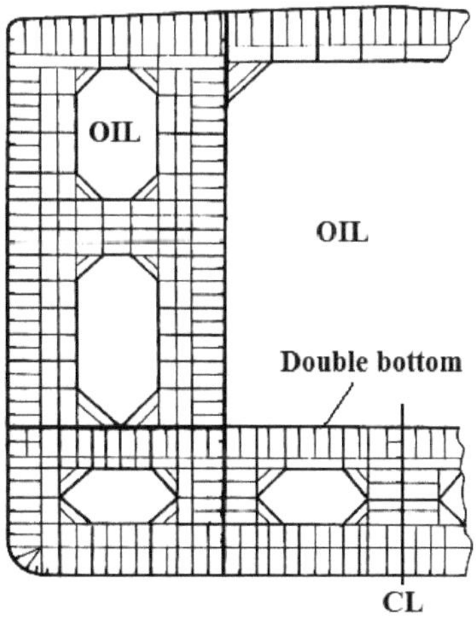

Fig. 1.38 A ship section of an oil tanker with double bottom construction and two longitudinal bulkheads.

Chapter 2
Configurations and Characteristics of Ship Structural Assemblies

1 Introduction

This chapter presents the main configurations and characteristics of ship structural assemblies. Both transversely and longitudinally stiffened bottom, deck and side shell structure assemblies are considered. The main structural features of transverse bulkheads are described. A brief description of the scantlings of ship strength members is introduced. The basic role of classification societies is clarified. Some ship structural connections and details such as frame brackets, beam brackets, tripping brackets and connection between bottom longitudinals and bottom transverses are accentuated.

2 Ship Structural Assemblies

The main ship structural assemblies are:

- Bottom structure
- Side structure
- Deck structure
- Transverse bulkheads

3 Bottom Structure

3.1 Single Bottom Structure

In small vessels having single bottom, open floors are fitted at every frame and are stiffened at their upper edges. When a bar keel is fitted, the floors are continuous from bilge to bilge and when using a flat plate keel a continuous longitudinal center keelson is fitted. In this case the center keelson is continuous and runs as far forward and aft as possible, the floors are made intercostal. Side keelsons are made intercostal and extend as far forward and aft as possible. In way of the machinery space additional side keelsons should be fitted to support the heavier machinery weights.

3.2 Double-Bottom Structure

The double bottom structure is composed of outer and inner watertight bottom plating to provide complete watertight integrity should the outer shell plating be pierced in way of the double bottom. The minimum depth of the double bottom depends on the size of the vessel and is determined by the rule requirements of classification societies. The actual depth is sometimes increased in places to provide adequate capacity of the double-bottom tanks. One or two side girders are fitted port and starboard depending on the width of the ship. The side girders could be either continuous or intercostal, see Fig. (2.1).

Watertight floors are fitted beneath the main bulkheads and are also used to subdivide the double-bottom space into tanks for ballast water, fuel oil or fresh water. Solid plate floors of non-watertight construction are lightened by manholes, see Fig. (2.2). Manholes are provided for access through the tanks and lightening holes are cut to reduce the steel weight of these floors. Also, small air and drain holes may be drilled at the top and bottom edges of the solid plate floors in the tank spaces. The spacing of the solid plate floors varies according to the loads supported and the induced local stresses. Bracket floors are fitted between solid plate floors. Bracket floors consist of plate brackets attached to the center girder and the side shell with bulb plate or angle stiffeners running between. The stiffeners are supported by angle bar struts at intervals and any side girders, see Fig. (2.3). The inner bottom provides a considerable margin of safety, since in the event of bottom shell damage only the double bottom space may be flooded. The double bottom space is utilized to carry oil fuel and fresh water as well as providing the required ballast capacity.

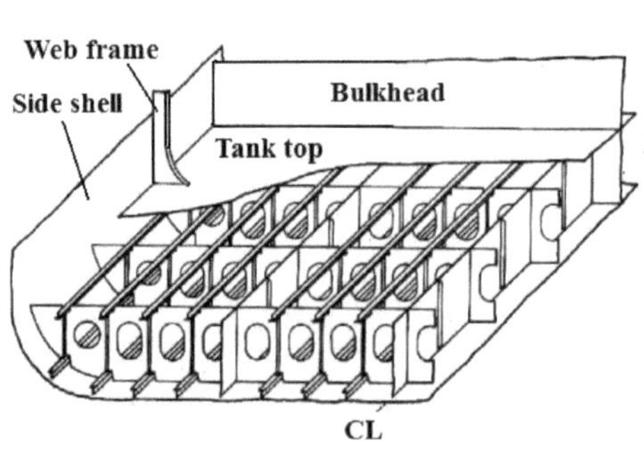

Fig. 2.1 Longitudinally stiffened double bottom structure

Water ballast bottom tanks are commonly provided right forward and aft for trimming purposes and the depth of the double bottom may be increased in these regions. In way of the machinery spaces the double bottom depth is also increased to provide adequate capacities for lubricating oil and fuel oil. The increase in the inner bottom height is always made by a gradual taper in the longitudinal direction to avoid sudden discontinuities in the structure.

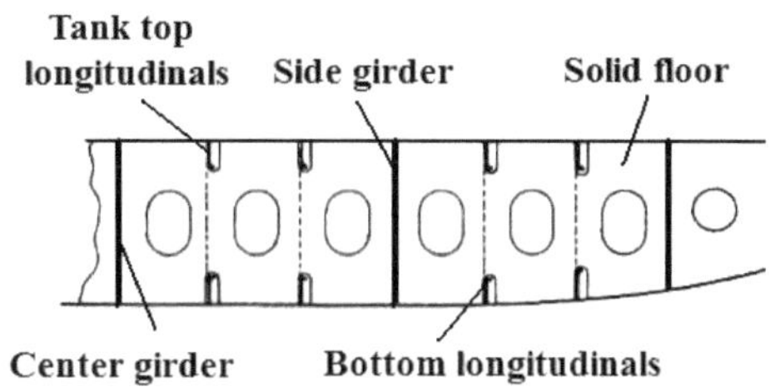

Fig. 2.2 Solid floor in longitudinally stiffened bottom structure

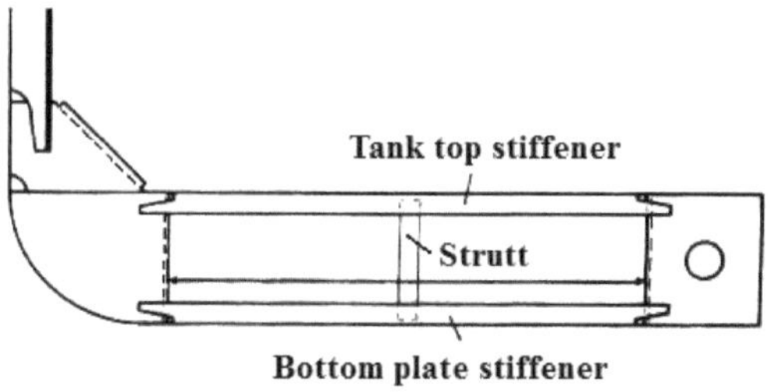

Fig. 2.3 Open floor in a transversely stiffened bottom structure

3.3 Transversely Framed Double Bottom

Transversely stiffened double bottom structure is made up of vertical floors which may be watertight, solid or of bracket construction. The floor structure is continuous from the center girder to the side shell and supports the inner bottom plating. The number of intercostal side girders fitted is determined by classification society rules. Additional side girders are provided in the engine room and also in the pounding region. In way of the forward and after peak tanks, the frame spacing does not exceed 610 mm. The transversely framed double bottom structure is constructed with transverse solid plate floors and bracket floors. This construction provides the principal support for the inner bottom and the outer bottom shell plating. To reduce the span of the frames at the bracket floor, vertical angle bar struts may be fitted, see Fig.(2.3). Solid plate floors are fitted at every frame space in the double bottom structure of the engine room and the forward pounding region. Vertical stiffeners of welded flats are attached to the solid plate floors, see Fig. (2.4).

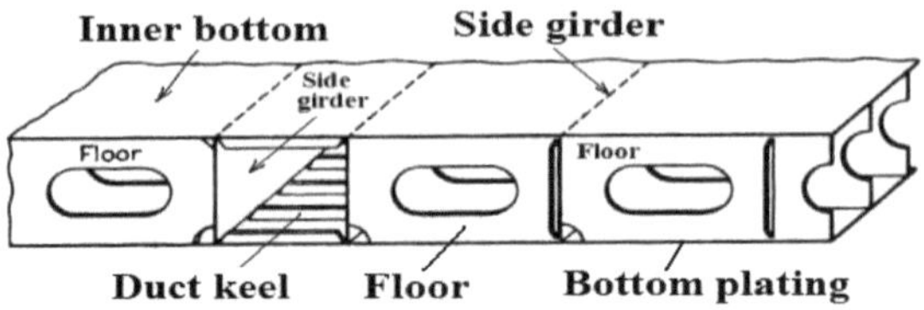

Fig. 2.4 Transversely stiffened double bottom structure

3.4 Longitudinally Framed Double Bottom

For vessels exceeding 120 m in length, longitudinal framing is used in the double bottom structure, see Figs. (2.5, 2.6). Angle sections and offset bulb plates (OB) are used as longitudinal stiffeners on the bottom shell plating and inner bottom plating, see Fig. (2.7).

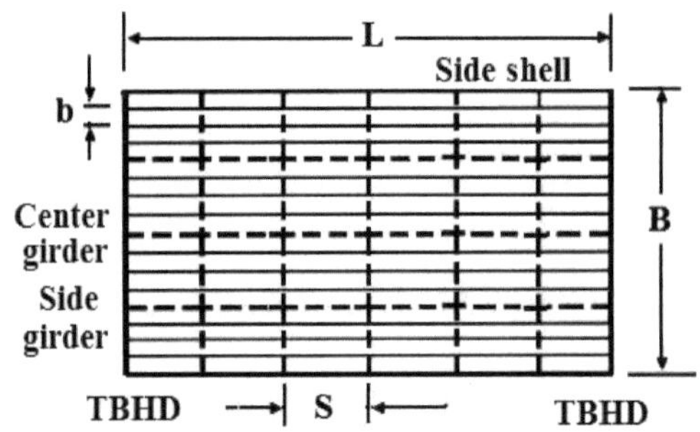

Fig. 2.5 Longitudinally stiffened bottom structure

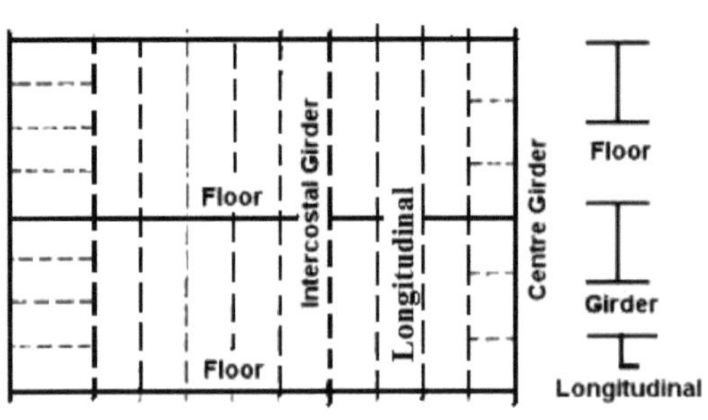

Fig. 2.6 Longitudinally stiffened bottom structure

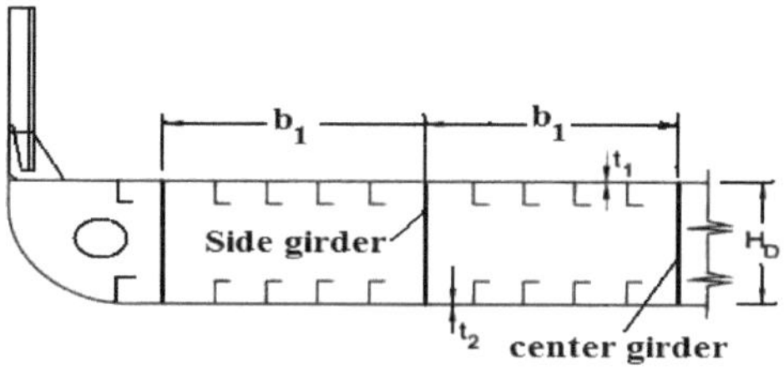

Fig. 2.7 Main structure members in a longitudinally stiffened bottom structure

Brackets are fitted at the center and side girders and side shell at intermediate frame spaces between solid floors. These brackets are flanged at the free edge and extend to the first longitudinal. The number of intercostal side girders depends upon the requirements of classification society rules. In the panting region, solid plate floors are fitted at alternate frame spaces, and intercostal side girders are provided. Where the ballast draft forward is inadequate (less than 4 per cent of the ship's length) the thickness of bottom plating over the flat of bottom in the pounding region is also increased. For heavily loaded bottom structures, extra stiffening is used to provide the required strength to cope with the excessive loading, see Fig. (2.8).

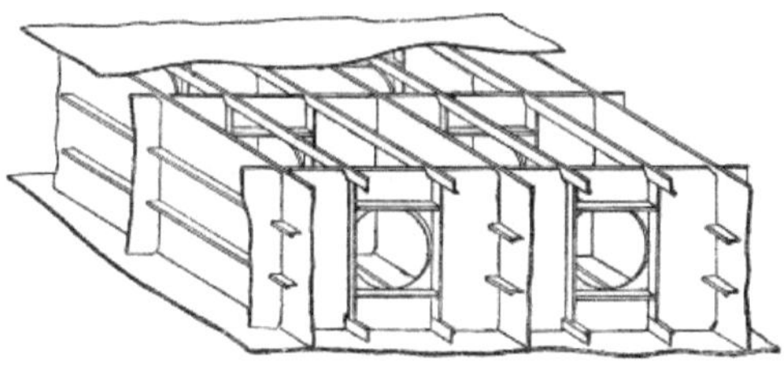

Fig. 2.8 Heavily stiffened longitudinally framed double bottom structure

4 Side Shell Structure

The side shell structure may either be transversely or longitudinally framed. In many conventional cargo ships, transverse framing is adopted to provide the required bale capacity, see Fig. (2.9). Bale capacities are considerably reduced when deep web frames or transverses are fitted. Scantlings of the main frames are primarily dependent on their lengths and also on the rigidity of the end connections.

The sheer strake is the upper strake of plating adjacent to the strength deck, see Fig. (2.9). The thickness of the sheer strake is the largest of the side shell strakes as it is the furthest strake from the neutral axis of the ship section. Rounded sheer strake is used for connecting the side shell plating with the strength deck plating particularly in large tankers and bulk carriers.

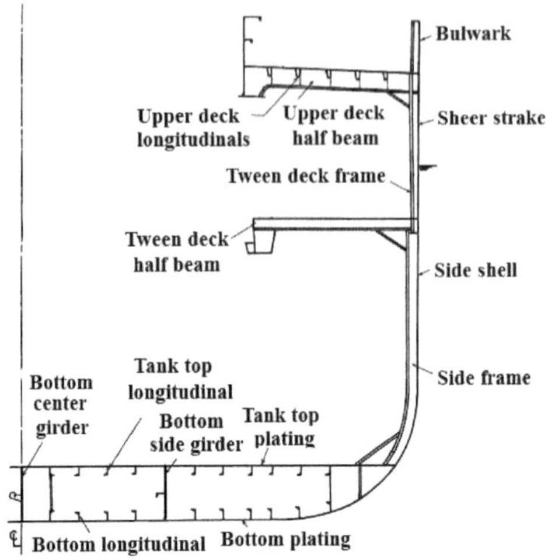

Fig. 2.9 Main strength members of a tween deck cargo ship

4.1 Transversely Framed Side Shell Structure Assemblies

Transverse framing of the side shell plating consists of vertical frame stiffeners, either of bulb plate or angle sections which are attached by brackets to the deck beams and the flooring structure, see Fig. (2.10). A number of web frames of increased scantlings are used to strengthen the side shell assemblies in particular locations along the hold length, see Fig. (2.10). Side stringers of offset bulb plates or inverted angle sections are used to support side frames for large hold depths.

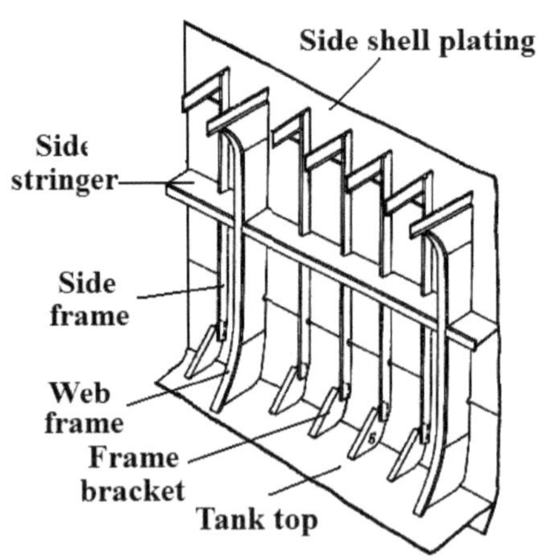

Fig. 2.10 Transversely stiffened side shell structure

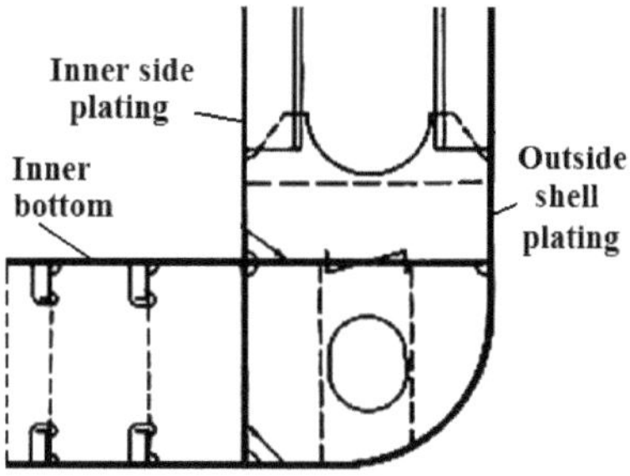

Fig. 2.11 Transversely framed double skin structure

Side shell structure assemblies are either single skin or double sided, see Fig. (2.11). Double sided structure assemblies are used in container ships, double sided oil tankers and new designs of bulk carriers. Side structure assemblies may be also composed of a combination of single and double side assemblies. The double sided part in this case is used as tanks for fuel or ballast.

4.2 Longitudinally Framed Side Shell Structure

Longitudinally framed side structures for single and double side construction is shown in Fig. (2.12). Transverse web frames are used to support the side longitudinals. Their spacing is dependent upon the type of ship, height and length of hold.

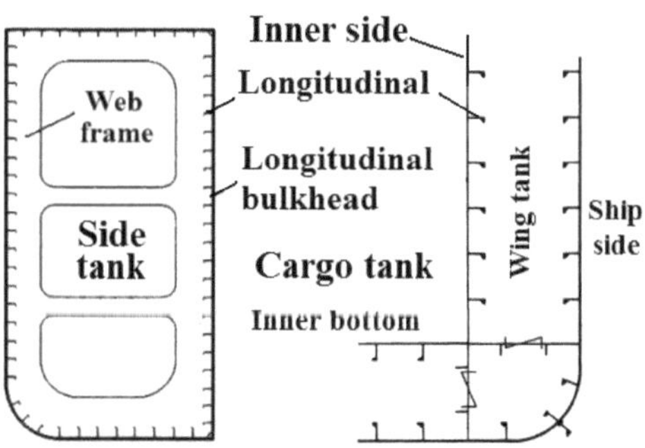

Fig. 2.12 Longitudinally framed single and double side structure

5 Deck Structure

Ships are fitted with one or more decks depending on ship type and trade. The most important decks are the strength deck which forms the upper flange of the main hull girder and the freeboard deck which is the uppermost deck having permanent means of closing all openings in the exposed portions of that deck, see Fig. (2.14).

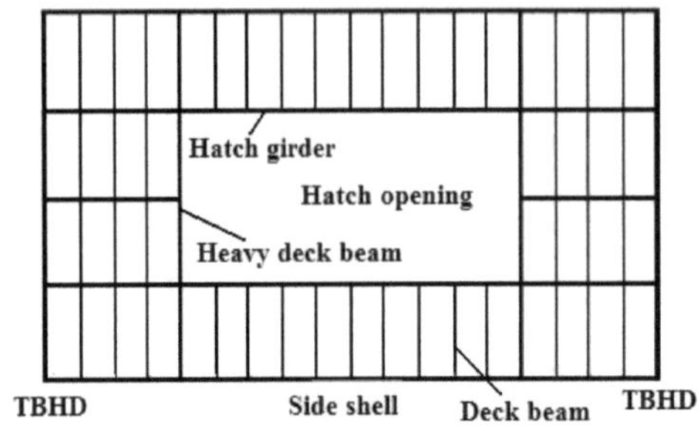

Fig. 2.13 A transversely framed deck structure showing the cargo hatch openings

Decks are main ship structural assemblies maintaining the integrity of the structure strength of the hull girder. The main structural members of the deck are shown in Fig. (2.13). All decks contribute to some extent to the strength of the ship hull girder. Lighter decks may be fitted to provide platforms for passenger accommodation and permit more flexible arrangements for cargo loading.

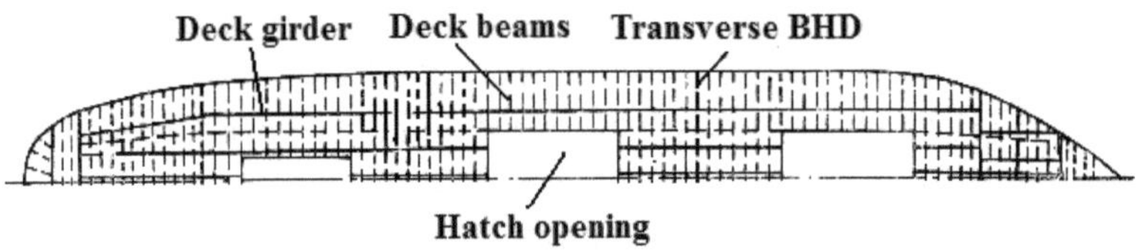

Fig. 2.14 A transversely framed deck structure showing the cargo hatch openings

Large deck openings are used for loading and discharging of cargo. These deck openings may be closed by watertight hatch covers. Other openings are provided in way of the machinery space casing for crew access through the decks. Openings are also provided to allow the removal of machinery items when necessary and to provide adequate light and air to this space. All large openings in the decks have well rounded corners with insert plates fitted.

5.1 Deck Plating

As the largest hull girder bending stresses occurs over the midship region, the largest deck plate thickness is maintained over 40 per cent of the ship length amidships, and it tapers to a minimum thickness permitted at the ends of the ship. When the deck plate thickness within the midship region and at hatch corners exceeds 20.5 mm grade D plating is used. Both Grade D and Grade E plating are used in the decks of refrigerated cargo ships if the temperature falls below certain limits in the region of the deck.

5.2 Transversely Stiffened Deck Plating

Deck plating is stiffened with transverse beams or longitudinals and local stiffening in way of any openings. The transverse framing system is supported by a system of longitudinal deck girders. Fig. (2.15) shows the main structural members and hatch opening of a transversely framed deck structure of a cargo hold of a general cargo ship.

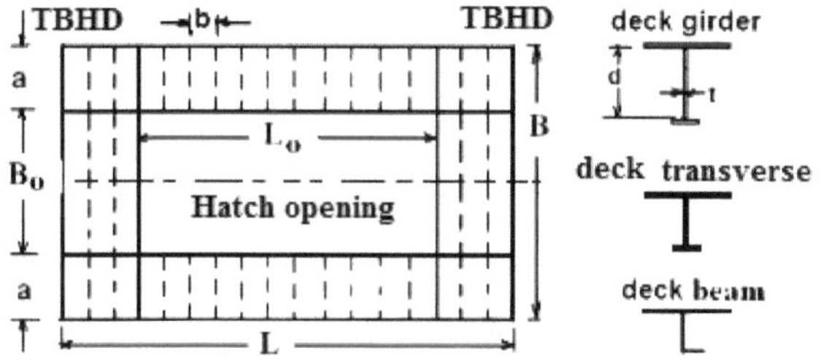

Fig. 2.15 Main dimensions of a transversely framed deck structure of a general cargo ship

The deck structure is composed of deck beams, deck girders and deck transverses. Hold pillars may be used as main structural elements to reduce the scantlings of the deck transverses and deck girders by reducing the effective spans of these heavy structural members. Two arrangements of fitting deck pillars in a transversely framed deck structure is shown in Fig. (2.16).

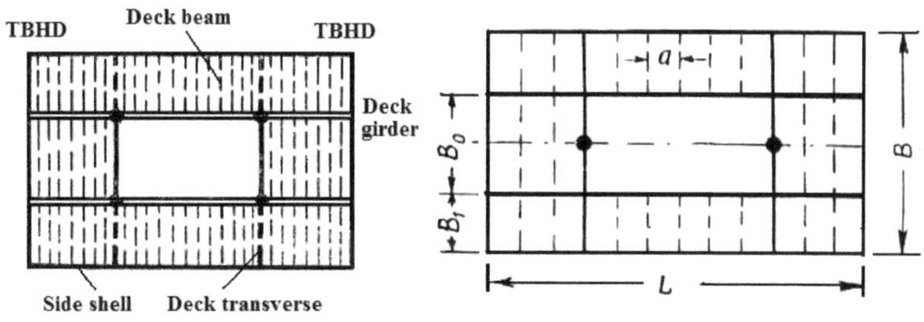

Fig. 2.16 Pillar arrangements for transversely stiffened deck structure

5.3 Longitudinally Stiffened Deck Structure

In longitudinally stiffened strength decks, see Fig. (2.17), the scantlings of the longitudinals depend on several factors such as the magnitude of the longitudinal spacing, the spacing of deck transverses, the length of ship, etc. Deck transverses are used to support the longitudinals and are constructed of built sections composed of a deep web plate and face plate, see Fig. (2.18).

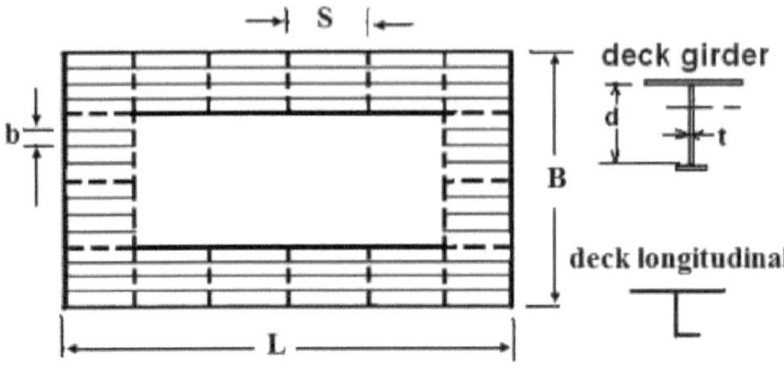

Fig. 2.17 The main dimensions of a longitudinally framed deck structure of a cargo hold in a general cargo ship

The fitting of deck pillars in the four corners of the hatch opening of a longitudinally framed deck structure is shown in Fig. (2.19). This structural arrangement reduces the effective spans of the deck girders. This will lead to a significant reduction of the deck steel weight as a result of the reduction of the scantlings of the deck girders and deck transverses.

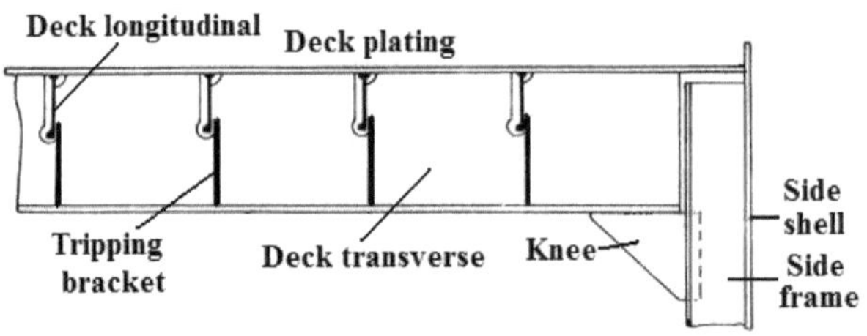

Fig. 2.18 A transverse member in a longitudinally framed deck structure

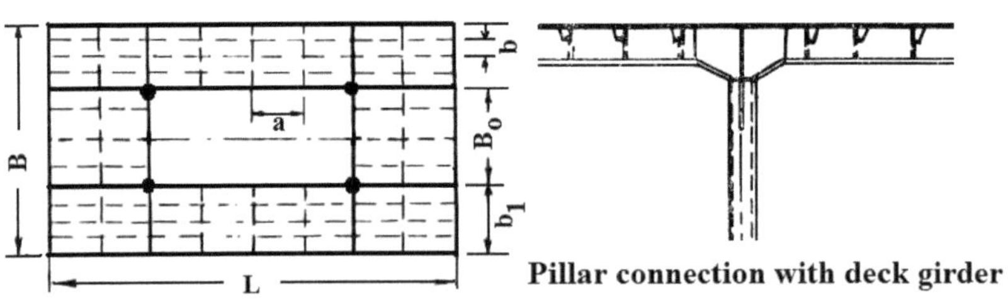

Fig. 2.19 A longitudinally framed deck structure fitted with deck pillars

6 Transverse Bulkheads

Depending on ship type and size, type of cargo carried and length of the ship, the hull girder is subdivided by a number of transverse and longitudinal bulkheads. The minimum number of transverse bulkheads is three when the ship has an aft engine room and four when the engine room is not at the aft end of the ship.

Transverse bulkheads are the main structural members of ship hull girder supporting the external and structural transverse loading. They provide support for deck and bottom girders and longitudinals. Transverse bulkheads are constructed of stiffened panels of plating. Bulkhead plating could be stiffened by welded stiffeners or by using corrugated plating, see Fig. (2.20).

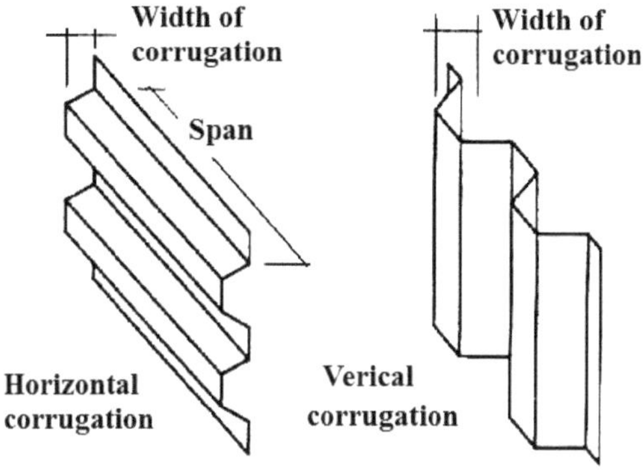

Fig. 2.20 Horizontal and vertical systems of corrugation of the plate panels of transverse bulkheads

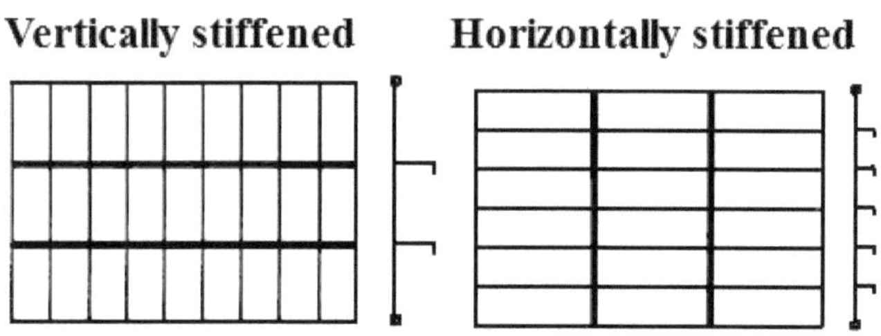

Fig. 2.21 Vertical and horizontal stiffening of bulkhead plate panels.

Bulkhead plate panels could be stiffened with horizontal or vertical unidirectional stiffeners. To reduce the scantlings of the welded vertical stiffeners, one or more horizontal girders with larger scantlings are fitted, see Fig. (2.21).

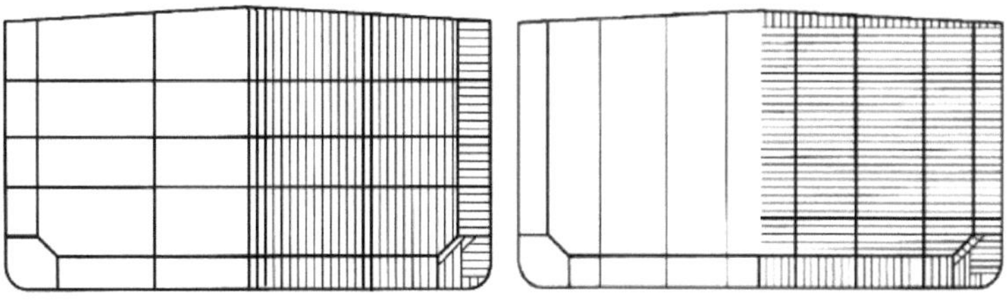

Fig. 2.22 Transverse bulkhead stiffened with welded vertical and horizontal stiffeners

Transverse bulkheads in oil tankers could be stiffened by welded vertical/horizontal stiffeners and three supporting horizontal/vertical girders, see Fig. (2.22).

Bulkheads could be also constructed with horizontal or vertical corrugated plate panels. Vertically and horizontally corrugated transverse bulkheads of oil tanker are shown in Fig. (2.23).

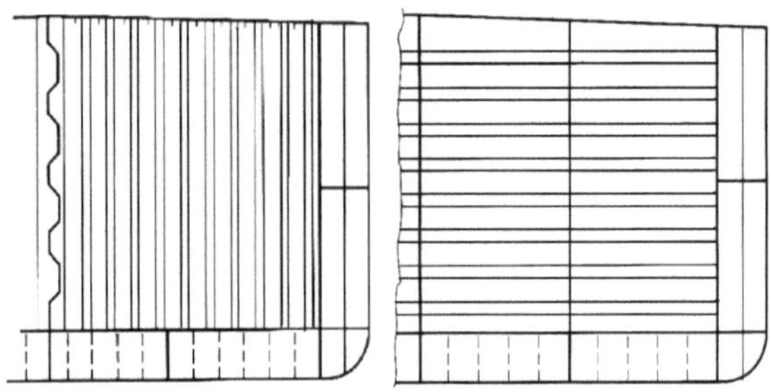

Fig. 2.23 Transverse bulkheads stiffened with vertical/horizontal corrugation

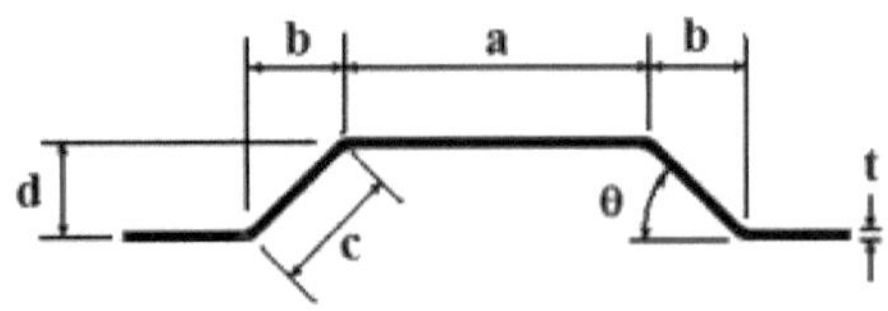

Fig. 2.24 Geometrical configuration of corrugation of bulkhead plate panels

A geometrical configuration and main dimensions of corrugation of bulkhead plate panels is shown in Fig. (2.24).

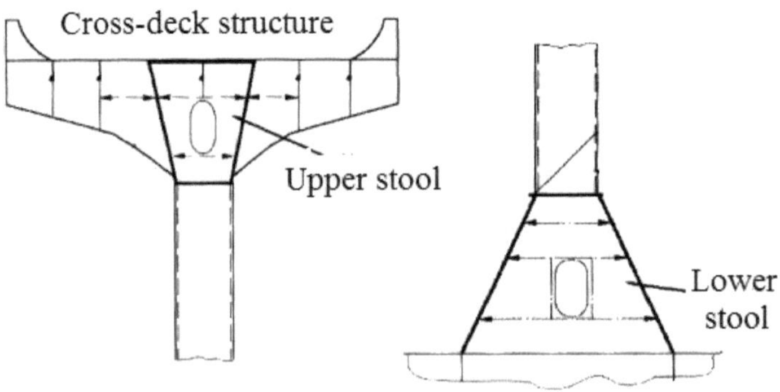

Fig. 2.25 The top and bottom stools of a corrugated transverse bulkhead of a bulk carrier

The corrugated bulkheads used in bulk carriers are normally fitted with top and bottom stools, see Fig. (2.25).

Transverse bulkhead panels of plating could be also stiffened with a combination of welded horizontal and vertical stiffeners see Fig. (2.26). A vertically stiffened transverse bulkhead with three horizontal girders in an oil tanker with double bottom is shown in fig. (2.27).

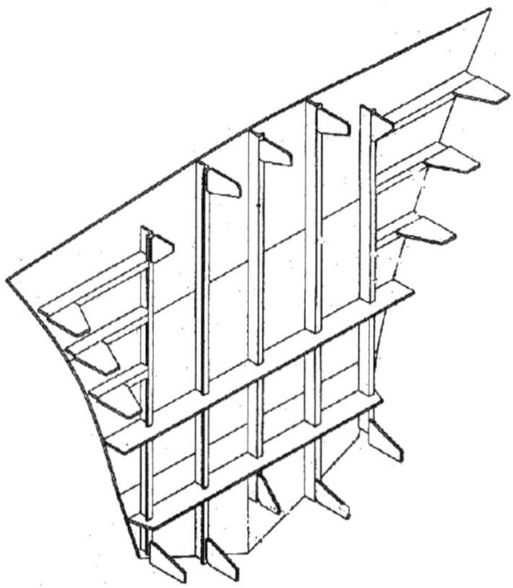

Fig. 2.26 A transverse bulkhead stiffened with a combination of welded horizontal and vertical stiffeners

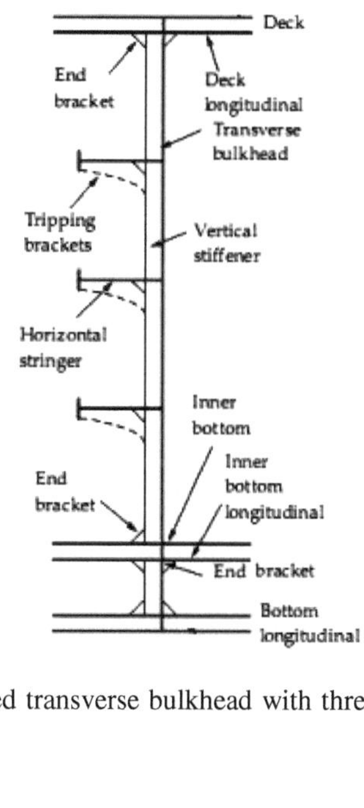

Fig. 2.27 A vertically stiffened transverse bulkhead with three horizontal girders in an oil tanker with double bottom

7 Scantlings of Ship Structural Members

The published rules of Classification societies are primarily simple semi-empirical in format and are calibrated to ensure successful past operational experience. These rules and other conventional codes and guides are used to determine the minimum required scantlings of ship structural members, see Fig. (2.28). However during the last 40 years ship sizes and types have experienced drastic changes. These changes include the complex ship structures of the high speed crafts, very large oil tankers, large LNG carriers and the complex structural assemblies of container ships with large deck openings, etc. The complete reliance on the empirical formulae given in the rules of classification societies cannot give economical, reliable and safe scantlings for these complex ship structures. The design of these new ship structures requires rational methods of stress analysis using the Finite element method, FEM. The computational complexity has been greatly simplified and solved with the introduction of rational analysis procedures using personal computers and the well-known software packages of structural analysis. These rational procedures include modeling of the complex ship structure, load modeling and application, modeling of boundary conditions, and selection of type and mesh size of the finite elements.

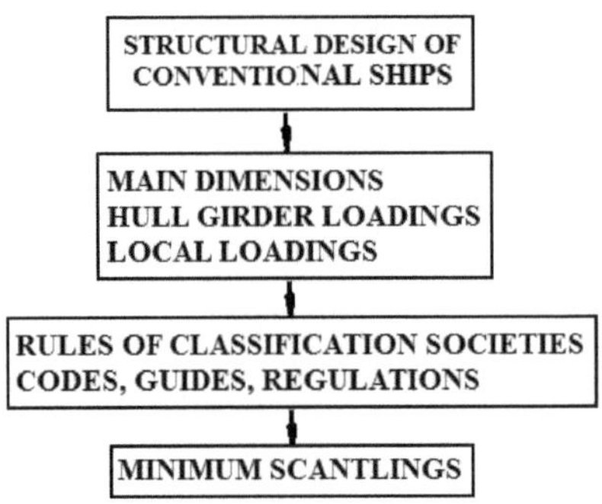

Fig. 2.28 Conventional procedure for determining ship minimum scantlings

8 Ship Structural Connections and Details

8.1 Frame Brackets

Frame brackets are the structural members connecting the lower end of side frames, or web frames with the tank top plating, floors or bottom plating. These structural members transfer the loading between side structure assemblies and bottom structure assemblies. Fig. (2.29) shows a frame bracket connecting the lower end of a side frame with the tank top plating of a transversely and longitudinally framed double bottom structures. Fig, (2.29) shows a frame bracket connecting the lower end of a side frame with bottom plating of a longitudinally framed bottom structure.

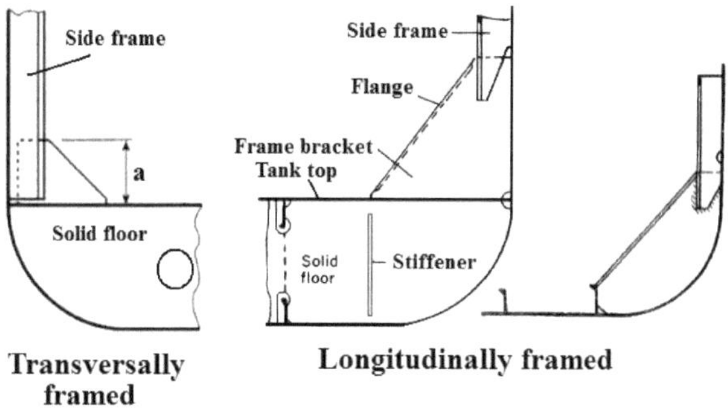

Fig. 2.29 A typical frame bracket connection at the lower end of a side frame for transversely and longitudinally framed bottom structure

8.2 Beam Brackets

These are the structural members connecting the top end of side frames, or web frames, with the deck plating or deck beams. Two designs of a beam bracket connecting the top end of a side frame with a deck longitudinal in longitudinally framed deck structure is shown in Fig. (2.30). Three different designs of beam brackets connecting a deck beam with the top end of a side frame in a transversely framed deck structure is illustrated in Fig. (2.31).

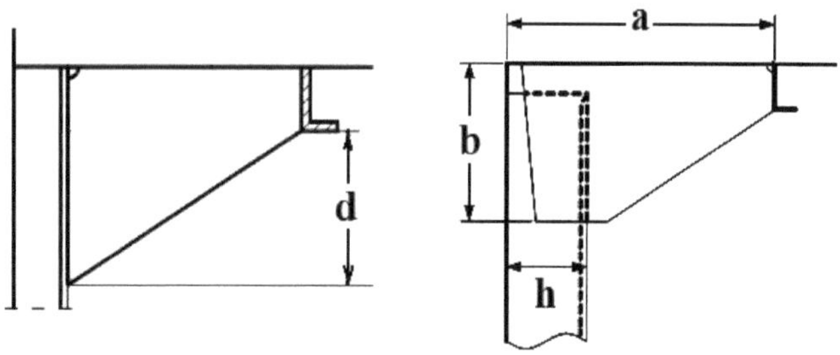

Fig. 2.30 Two designs of a beam bracket connecting the top end of a side frame with a deck longitudinal in a longitudinally framed deck structure

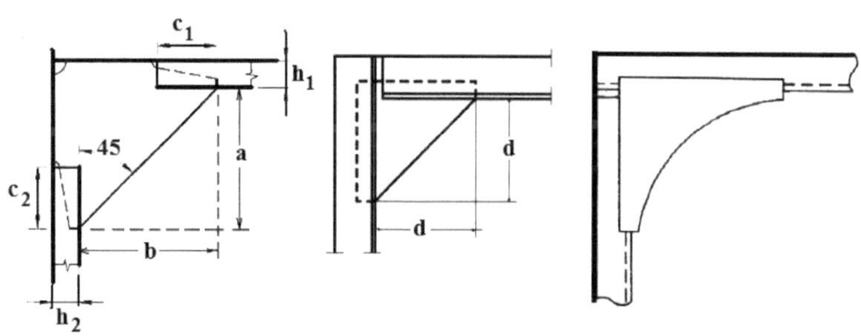

Fig. 2.31 Different design options of beam brackets connecting a deck beam with the top end of a side frame

The connection of tween deck beam with the top end of the main side frame and the lower end of the tween deck frame is shown in Fig. (2.32).

These structural members transfer the loading between deck structure assemblies and side structure assemblies, see Fig. (2.33). An idealized stress distribution over a beam bracket connection due to bending and shear loading is shown in Fig. (2.33).

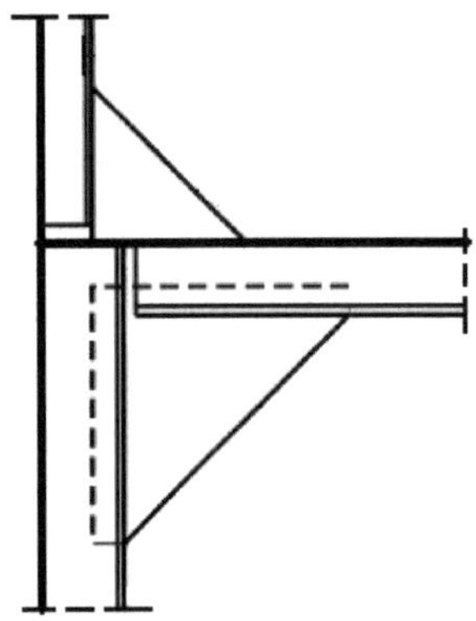

Fig. 2.32 Connection of the top end of the main side frame and tween deck beam with the lower end of the tween deck side frame

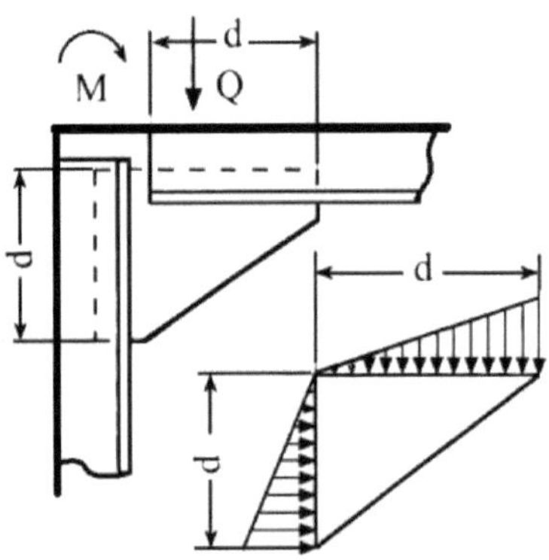

Fig. 2.33 Idealized stress distribution over a beam bracket

Poor design of these brackets could create hard spots which will cause stress concentration and subsequently initiation of cracks, see Fig. (2.34). The initiation of cracks at the toe of bracket could be significantly reduced by improving the design of the bracket ends as shown in Fig. (2.35).

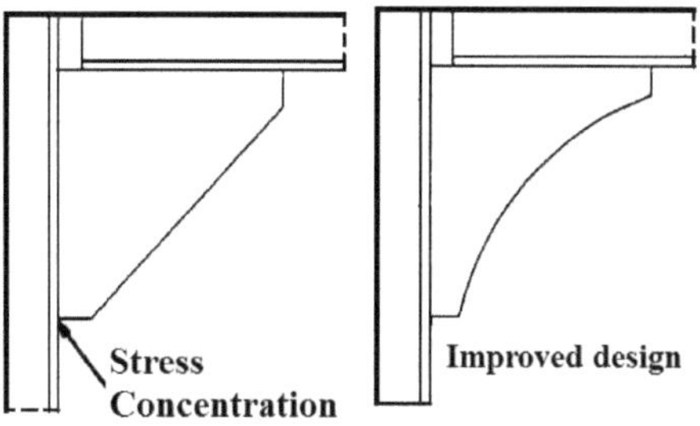

Fig. 2.34 Improved design of beam brackets

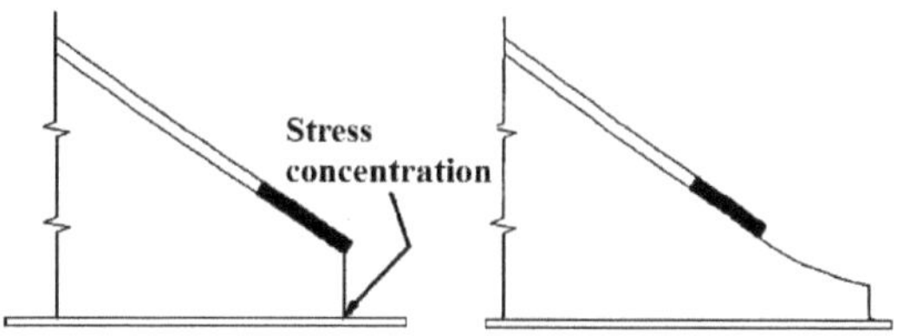

Fig. 2.35 Improved design of bracket toes

8.3 Tripping Brackets

Tripping brackets are used to connect deck beams with deck girders and also to prevent torsional deformations of the deck girder, see Fig. (2.36). Deck longitudinals are connected to deck transverses using tripping brackets, see Fig. (2.37).

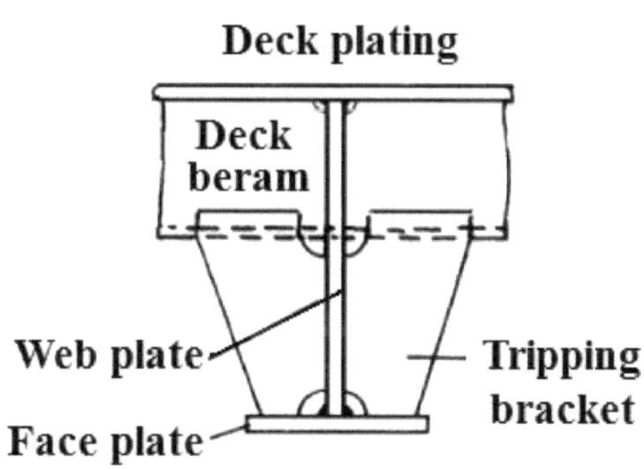

Fig. 2.36 Tripping bracket connecting a deck beam and a deck girder

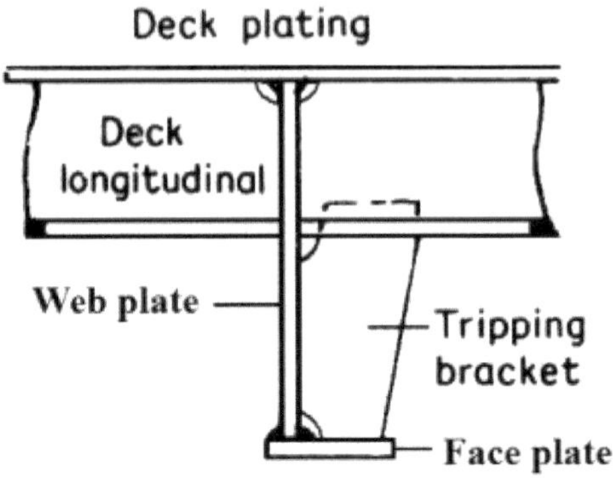

Fig. 2.37 Tripping bracket connecting a deck longitudinal and a deck transverse

8.4 Connection between Bottom Longitudinals and Bottom Transverses

Bottom longitudinals run through cut holes in the bottom transverses and is normally connected together by collars or lugs, see Fig. (2.38). A typical connection between a T-section bottom longitudinal, bottom plating and a T-section bottom transverse using a lug and a vertical stiffener is shown in Fig. (2.38). Using a bracket on the other side of the web plate improves the pattern of stress distribution and reduces significantly the stress concentration at the toe, see Fig. (2.39).

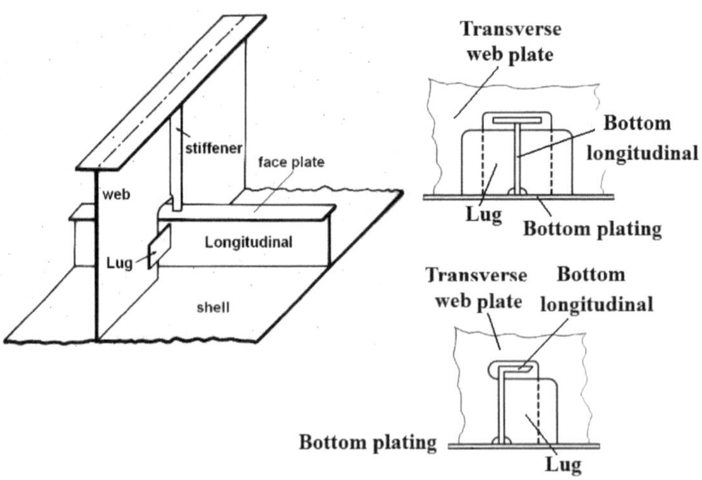

Fig. 2.38 Collars for cut outs for angle and T sections

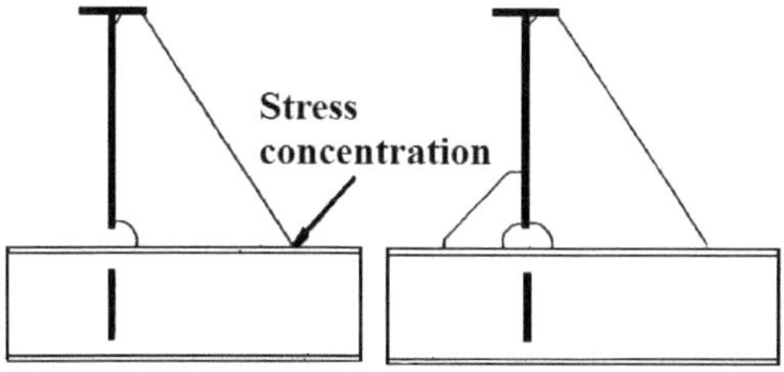

Fig. 2.39 Addition of a bracket on the other side of the web plate of the transverse member

Chapter 3
Configurations and Geometrical Properties of Ship Structure Members

1 Introduction

This chapter presents the main configurations and geometrical properties of ship structure members. The structural features of ship stiffened panels and frameworks are clarified. Standard and fabricated stiffening members commonly used in ship construction are presented. The geometrical properties of standard and symmetrical and asymmetrical fabricated sections with attached plating are determined. The effect of variation of thickness of attached plating on the magnitude of section modulus and second moment of area quantified. Geometrical and flexural properties of curved plates are given. Equivalent rolled and fabricated sections of commonly used ship strength members are given. Rational shapes of cross-sections of beams and columns are given. Procedures are given for the design of fabricated T-sections and the determination of optimum depth of I-section girders.

2 Structural Units of a Ship

There are two basic types of structure units used in the design and construction of ship hull girder:

(a) Stiffened panels
(b) Framework

2.1 *Stiffened Panels*

Stiffened panels represent the main structural elements used in the construction of ship hull girder and its internal structural assemblies and are generally subjected to lateral and in-plane loading. The lateral loads on the bottom plating come from hydrostatic and hydrodynamic water pressure. The in-plane loading on the bottom and deck plate panels comes mainly from longitudinal vertical and horizontal bending of ship hull girder in both conditions of still-water and among waves. These panels may be initially flat or curved. Unidirectional stiffened panels have several applications in ship construction, see Fig. (3.1). Cross-stiffened panels represent the most common structural units used in ship construction. These stiffened panels are composed of plate panels, primary and secondary stiffening structural members, see Figs. (3.2). The secondary stiffeners are supported by primary structural members spaced at ample distances apart in order to ensure that the plate panels shall have adequate strength and stiffness when loaded. The main function of the structural profiles is to serve as stiffeners on plate panels in bottom, side, deck and bulkhead plating.

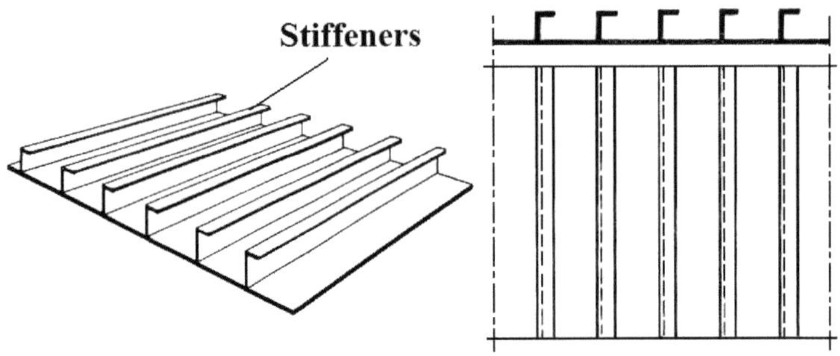

Fig. 3.1 Unidirectional stiffened panel

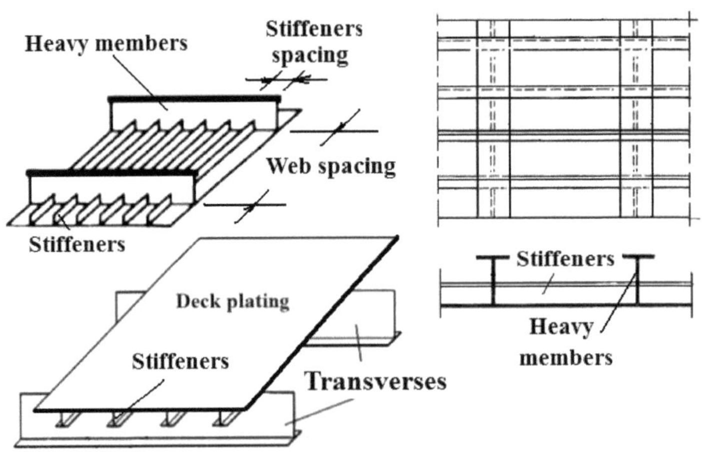

Fig. 3.2 Various configurations of stiffened panels showing the primary and secondary stiffeners

Three dimensional (3D) structural assemblies are constructed of two dimensional (2D) stiffened panels, see Fig. (3.3).

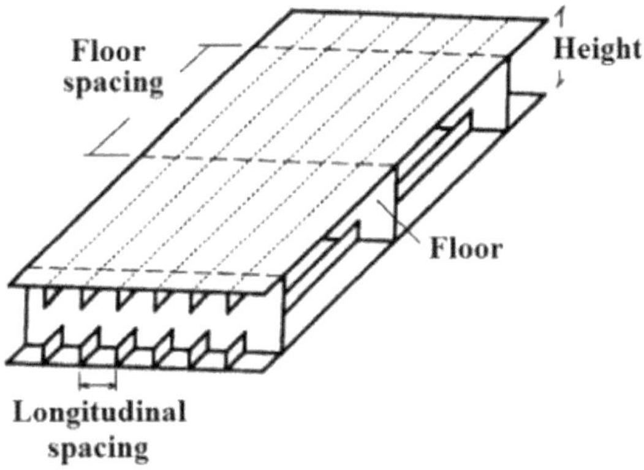

Fig. 3.3 A 3D stiffened structure assembly

2.2 Frameworks

Frameworks are transverse rings of deck beams, side frames and bottom floors or longitudinal rings composed of a deck girder, a vertical bulkhead stiffener and a

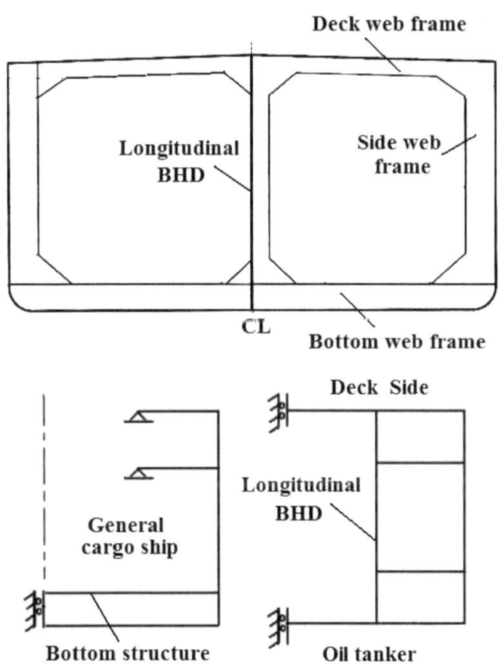

Fig. 3.4 A transverse ring of web frames in an oil tanker

longitudinal bottom girder. A transverse ring is composed of deck, bottom and side web frames. A typical web ring of an oil tanker and general cargo ship are shown in Fig. (3.4). Transverse frameworks are subjected either to uniformly or linearly distributed loads normal to their planes or to loads in their own planes. Other structural members or units are the hull fittings.

2.3 Hull Fittings

The various ship hull fittings must have adequate strength to withstand the static and dynamic loads imposed on them. These hull fittings include shaft brackets, masts, derricks, davits, machinery foundations and control surfaces such as rudders.

3 Configurations and Geometrical Properties of Ship Structure Members

The stiffening members are used as deck beams and side frames in transversely framed ships and as deck, bottom and side longitudinals in longitudinally framed ship structures

3.1 Standard Rolled Sections with Attached Plating

The main types of the standard rolled sections commonly used as frames, longitudinals and stiffeners in ship structure assemblies is shown in Fig. (3.5).

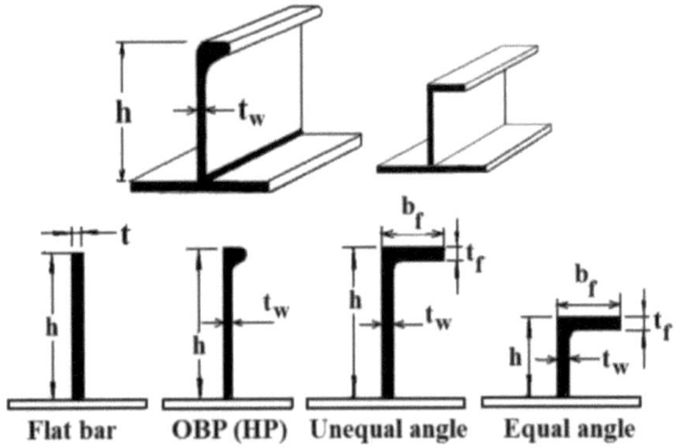

Fig. 3.5 Different types of standard rolled sections commonly used in ship structure assemblies as frames, longitudinals and stiffeners

The scantlings of the attached plating are the width and thickness. The width of the attached plating is the effective breadth which depends mainly on the thickness of the plating and the spacing between adjacent structural members.

3.2 *Fabricated Sections*

The primary stiffening members commonly used in ship structural assemblies are made of fabricated sections. The main types of the fabricated sections commonly used as web frames, girders and heavy transverses in ship structure assemblies is shown in Fig. (3.6, 3.7).

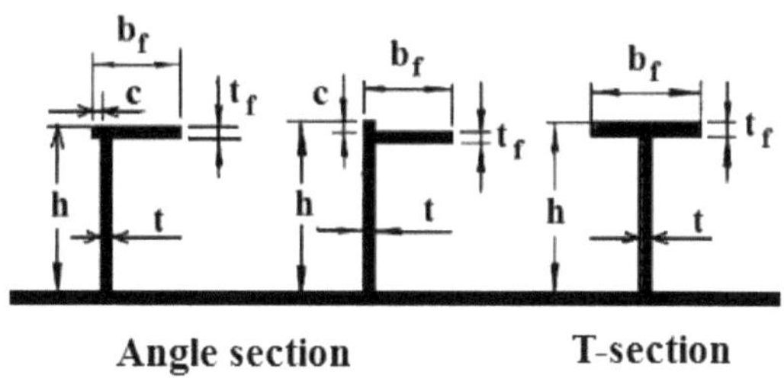

Fig. 3.6 Main types of the fabricated sections commonly in ship structure as web frames, girders and heavy transverses

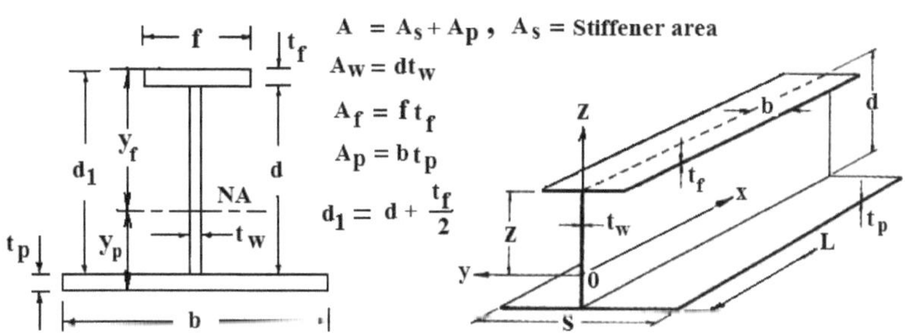

Fig. 3.7 Scantlings and geometrical properties of fabricated T-section

The main structural elements of a fabricated section are the web plate and the face plate. The fabricated section could be either a T-section or an angle section, see Figs. (3.7, 3.8). Tables 1 and 2 give scantlings, section area and section modulus for fabricated angle and T- sections. These tables are very useful for determining the scantlings of a fabricated section for a given section modulus. The values of the section modulus given in tables 3.1 and 3.2 are based on an attached plating of width 500 mm and thickness 9 mm.

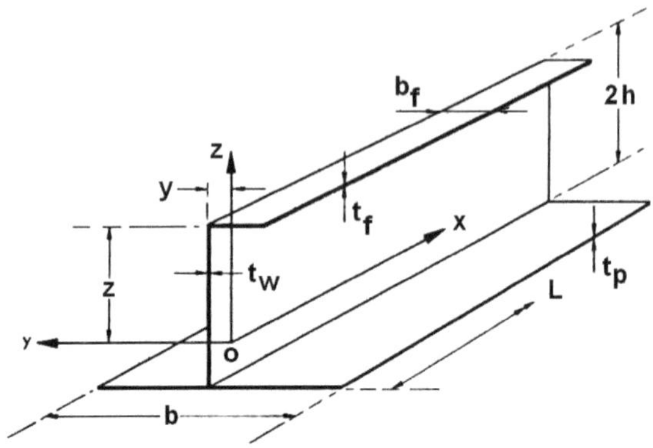

Fig. 3.8 Scantlings and geometrical properties of fabricated angle section

The scantlings of a fabricated T-section for a given value of the section modulus could be determined from Table 3.1. The scantlings of a fabricated angle section for a given value of the section modulus could be determined from Table 3.2.

For large fabricated sections, longitudinal stiffeners and tripping brackets may be required to stiffen the web plate, see Fig. (3.9).

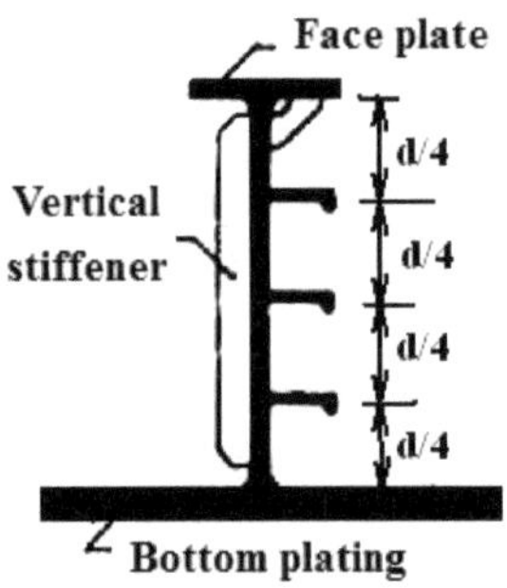

Fig. 3.9 Longitudinal and transverse stiffening of a large fabricated T-section

3.3 Geometrical Properties of Fabricated Symmetrical Sections with Attached Plating

The geometrical properties of sections with attached plating are given for the following commonly used symmetrical sections:

3.3.1 Flat-Bar

The geometrical properties for a flat bar stiffener with attached plating is shown in Fig. (3.10).

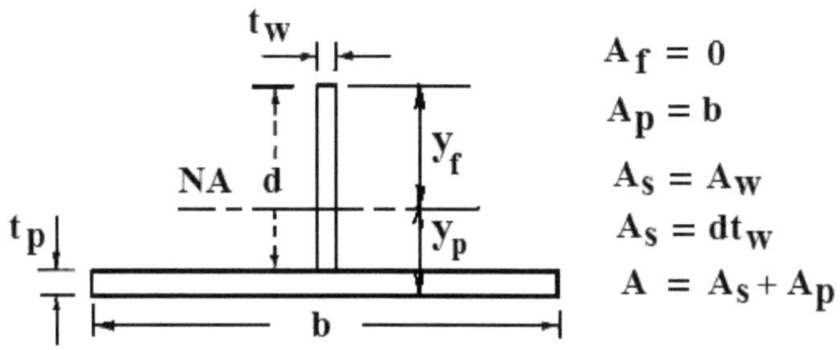

Fig. 3.10 Geometrical properties of a flat bar section with attached plating

Table 3.1 Scantlings, section area and section modulus of fabricated T-section with attached plating

W cm³	h × b × e 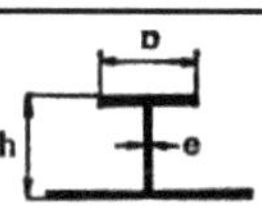		
	Scantlings	section area	section modulus
1583	**510 × 255 × 8,5**	65,02	1583,44
1672	**495 × 270 × 9**	68,85	1672,83
1736	**475 × 285 × 9,5**	72,20	1736,33
1867	**540 × 270 × 9**	72,90	1867,77
1944	**520 × 285 × 9,5**	76,47	1944,11
2014	**500 × 300 × 10**	80	2014,11
2183	**570 × 285 × 9,5**	81,22	2183,33
2268	**550 × 300 × 10**	85,0	2268,22
2319	**525 × 315 × 10,5**	88,20	2319,11
2531	**600 × 300 × 10**	90,0	2531,44
2625	**580 × 315 × 10,5**	93,97	2625,33
2652	**550 × 330 × 11**	96,80	2652,77
2913	**630 × 315 × 10,5**	99,22	2913,77
2986	**605 × 330 × 11**	102,85	2986,1
3331	**660 × 330 × 11**	108,90	3331,6

Table 3.2 Scantlings, section area and section modulus of fabricated angle section with attached plating

W cm³	h x b x e: scantlings	section area	section modulus
382	255 × 125 × 8,5	30,85	382,6
396	300 × 110 × 7,5	29,63	396,2
401	280 × 120 × 8	30,72	401,5
457	270 × 135 × 9	34,83	457,9
478	300 × 125 × 8,5	34,68	478,9
484	320 × 120 × 8	33,92	484,1
530	285 × 140 × 9,5	38,57	530,1
565	315 × 135 × 9	38,88	565,8
571	340 × 125 × 8,5	38,08	571
622	300 × 150 × 10	43,00	622,6
648	330 × 140 × 9,5	42,84	648,3
681	360 × 135 × 9	42,93	681,7
710	315 × 155 × 10,5	47,14	710,3
768	350 × 150 × 10	48	768,6
789	380 × 140 × 9,5	47,59	789,6
821	330 × 165 × 11	52,03	821,7
885	370 × 155 × 10,5	52,92	885,3
925	400 × 150 × 10	53,0	925,4
1013	385 × 165 × 11	58,07	1013,5
1056	420 × 155 × 10,5	58,17	1056,1
1058	360 × 180 × 12	61,92	1058,0
1219	440 × 165 × 11	64,13	1219,4
1304	420 × 180 × 12	69,12	1304,1
1334	390 × 195 × 13	72,67	1334,6
1568	480 × 180 × 12	76,32	1568,0
1644	455 × 195 × 13	81,12	1644,0
1654	420 × 210 × 14	84,28	1654,3
1975	520 × 195 × 13	89,57	1975,4
2019	450 × 225 × 15	96,75	2019,8
2036	490 × 210 × 14	94,08	2036,5
2433	480 × 240 × 16	110,08	2433,9
2445	560 × 210 × 14	103,88	2445,9
2485	525 × 225 × 15	108,0	2485,2
2983	600 × 225 × 15	119,25	2983,3
2993	560 × 240 × 16	122,88	2993,2

3.3.2 Standard Angle Sections

The geometrical properties of an unequal angle section with attached plating are shown in Fig. (3.11). Scantlings, section modulus and moment of inertia of unequal angle sections with attached plating are given in Tables 3.3a and 33b.

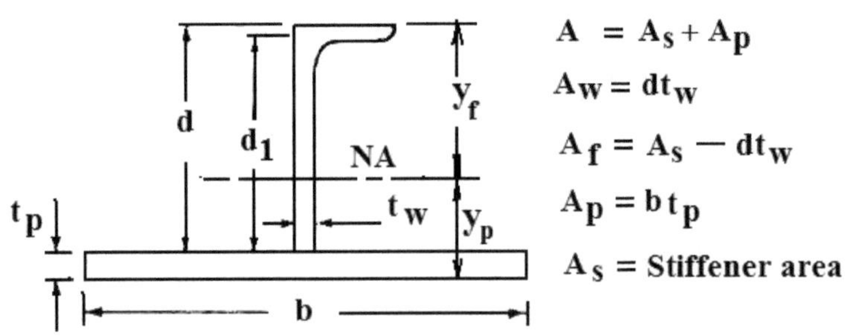

Fig. 3.11 Geometrical properties of an angle section with attached plating

Table 3.3a Flexural properties of angle sections with attached plating

W cm³	scantlings	section area	section modulus cm³
32	70 × 50 × 7	7,95	32,1
32	65 × 50 × 8	8,60	32,9
33	80 × 40 × 7	7,90	33,4
35	75 × 50 × 7	8,31	35,4
36	70 × 50 × 8	9,00	36,2
37	80 × 40 × 8	9,01	37,7
38	80 × 60 × 6	8,11	38,1
38	80 × 50 × 7	8,65	38,5
39	75 × 50 × 8	9,41	39,9
43	80 × 60 × 7	9,38	43,9
46	100 × 50 × 6	8,73	46,0
47	90 × 65 × 6	9,01	47,1
49	80 × 60 × 8	10,60	49,4
49	90 × 70 × 8	12,73	49,5
52	100 × 50 × 7	10,1	52,9
54	90 × 65 × 7	10,4	54,2
59	100 × 50 × 8	11,4	59,5
61	90 × 65 × 8	11,8	61,1
63	100 × 65 × 7	11,2	63,0
71	100 × 65 × 8	12,7	71,2
73	100 × 50 × 10	14,1	73,0
75	90 × 65 × 10	14,6	75,0
78	100 × 75 × 8	13,5	78,5
86	100 × 65 × 10	15,6	86,7
91	100 × 80 × 9	15,48	92,4
95	100 × 75 × 10	16,6	95,7
99	130 × 65 × 6	15,1	99,2

Table 3.3b Flexural properties of angle sections with attached plating

W cm³	scantlings	section area	section modulus
200	150 × 100 × 10	24,20	203,9
220	150 × 90 × 12	27,50	223,7
240	150 × 100 × 12	28,70	240,3
	180 × 90 × 10	26,20	244,3
270	150 × 100 × 14	33,20	276,5
280	180 × 90 × 12	31,20	289,1
300	200 × 100 × 10	29,20	301,8
350	200 × 100 × 12	34,80	356,9
390	250 × 90 × 10	33,20	388,7
460	250 × 90 × 12	39,60	460,7
530	250 × 90 × 14	45,90	530,9

3.3.3 Offset Bulb

The geometrical properties of an OBP with attached plating are shown in Fig. (3.12). The scantlings, section area, section modulus and moment of inertia of OBP sections with/without attached plating are given in Table 3.4. The thickness of the attached plating is assumed 10 mm and 15 mm. Table 3.5 gives scantlings and section modulus for some large OBP sections with attached plating.

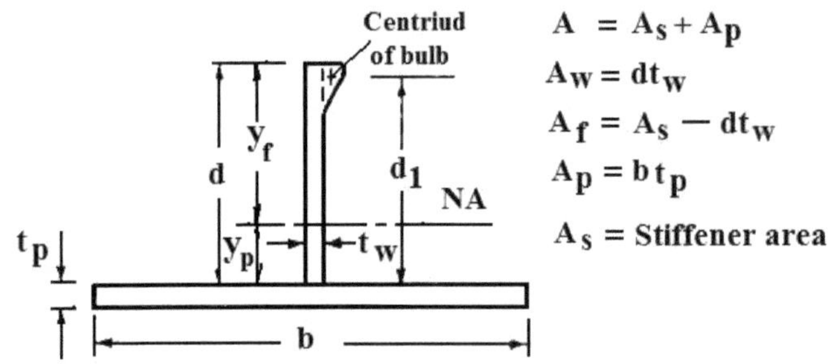

Fig. 3.12 Geometrical properties of an offset bulb plate with attached plating

Table 3.4 Flexural properties of OBP sections with attached plating

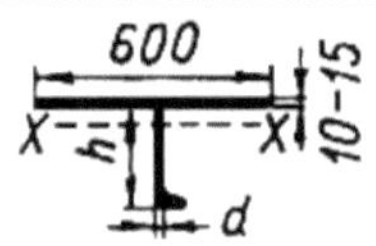

Geometric properties of section with plate					
depth, h	thickness, d	total area, F	moment of inertia about the axis xx, I_x	least section modulus about the axis xx, W_x	thickness of plate, t
mm	mm	cm²	cm⁴	cm³	mm
60	5	64,27	87	15	10
70	5	65,06	137	20	10
80	5	65,84	202	25	10
100	6	68,63	434	45	10
120	6,5	71,15	767	68	10
140	7	74,05	1274	100	10
160	8	107,96	2200/1980	147/140	15/10
180	9	112,20	3280/2860	200/188	15/10
200	10	117,36	4730	268	15
220	11	122,82	6500	343	15
240	12	128,75	8720	434	15
270	12	133,82	12180	552	15

Table 3.5 Scantlings, section area and section modulus of OBP with attached plating of 600 mm and thickness equal to the web thickness

W cm³	scantlings	section area	section modulus	W cm³	scantlings	section area	section modulus
210,0	200 × 9	23,60	212,5	490,0	260 × 13	43,90	499,2
225,0	200 × 10	25,60	225,6	530,0	280 × 11	42,60	543,7
240,0	200 × 11	27,60	240.7	560,0	280 × 12	45,50	571,7
250,0	200 × 12	29,60	255,4	590,0	280 × 13	48,30	597,3
282,0	220 × 10	29,00	288,7	635,0	300 × 11	46,70	647,5
298,0	220 × 11	31,20	304,3	665,0	300 × 12	49,70	679.8
314,0	220 × 12	33,40	318,5	710,0	300 × 13	52,80	711,2
346,0	240 × 10	32,40	351,3	740,0	300 × 14	55,80	740,7
370,0	240 × 11	34,90	375,7	780,0	320 × 12	54,20	793,6
394,0	240 × 12	37,30	394,0	820,0	320 × 13	57,40	829,4
450,0	260 × 11	(38,70	455,7	860,0	320 × 14	60,70	864,8
470,0	260 × 12	41,30	477,7	880,0	320 × 15	63,90	897,7

4 Flexural Properties of Fabricated Sections with Attached Plating

Consider the stiffened panel shown in Fig. (3.14). The stiffeners are fabricated T-sections. The scantlings of a fabricated section with attached plating are as follows, see Fig. (3.13):

Face plate: $A_f = b_f \times t_f$

Web plate: $A_w = d \times t_w$

Attached plating: $A_p = s \times t_p$

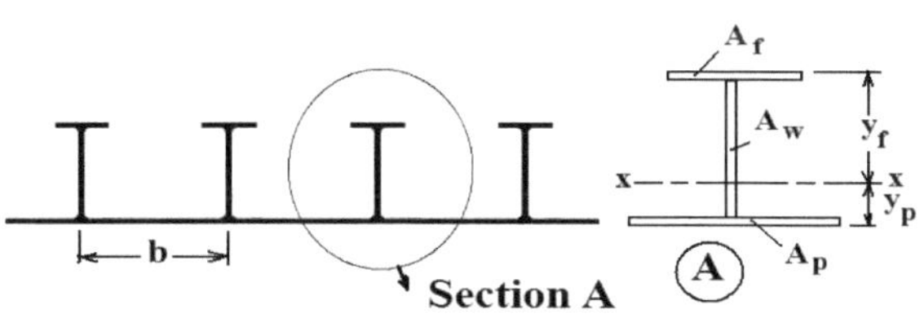

Fig. 3.13 A typical fabricated T-section with the attached plating

Where b_f, t_f = width and thickness of face plate
d, t_w = depth and thickness of web plate
b = stiffener spacing
b_e, t_p = effective breadth and thickness of attached plating.

The total section area of the girder is given by:

$$A_t = A_f + A_w + A_p$$

The position of the neutral axis from the outer side of the attached plating is given by:

$$y_p = \frac{A_f \cdot d + A_w \cdot d/2}{A_t} = \left(A_f + \frac{A_w}{2}\right).d/A_t$$

$$y_f = d - y_p$$

The second moment of area about the neutral axis of the section is given by:

$$I_x = A_f \cdot y_f^2 + A_p \cdot y_p^2 + \frac{t_w d^3}{12} + A_w\left(y_f - \frac{d}{2}\right)^2$$

Where I_x = second moment of area about the neutral axis of the section.
The section modulus of the section at the face plate is given by:

$$Z_f = \frac{I_x}{y_f}$$

The section modulus of the section at the attached plate is given by:

$$Z_p = \frac{I_x}{y_p}$$

5 Equivalent Section Modulus

In many occasions, because of the unavailability or shortage of certain profiles, equivalent sections having the same section modulus could be used. The selection of the equivalent section should satisfy, among several other requirements, the minimum weight requirement. Table 3.6 shows equivalent sections among unequal angle and OBP with attached plating.

Table 3.7 shows equivalent sections among equal angle and OBP with attached plating.

Table 3.8 shows equivalent sections among unequal angle and flat bar with attached plating. It is evident from table 3.7 that there is significant weight saving in the stiffening members when using OBP sections against unequal angle sections. As an example, for a section modulus of 670 cm^3, the section area of the unequal angle section is 60.0 cm^2 whereas the section area of the OBP section having the same section modulus is 49.1 cm^2 giving a weight saving of 18%.

Table 3.6 Equivalent sections between OBP and unequal angle sections

W						
cm^3	scantlings	area	section modulus	scantlings	area	section modulus
165	150 × 75 × 10	21,60	168,2	180 × 8	19,90	169,4
190	150 × 90 × 10	23,20	189,7	180 × 10	23,50	193,4
220	150 × 90 × 12	27,50	223,8	200 × 9	24,90	232,4
240	150 × 75 × 15	31,60	242,5	200 × 11	28,90	261,6
270	150 × 90 × 15	33,90	273,3	200 × 12	30,90	277,5
				220 × 9	28,30	296,5
460	200 × 150 × 12	40,80	466,3	260 × 10	37,90	470,3
570	200 × 150 × 15	50,50	572,3	280 × 11	44,60	588,9
670	200 × 150 × 18	60,00	675,9	300 × 11	49,10	701,1

As an example for a section modulus of 800 cm^3, the section area of the equal angle section in table 3.7 is 74.7 cm^2 whereas the section area of the OBP section having the same section modulus is 57.4 cm^2 giving a weight saving of 23%.

An example showing weight saving for a section modulus of 35 cm^3, Table 3.8 gives section area of the flat bar is 11.25 cm^2 whereas the section area of the unequal angle section having the same section modulus is 8.31 cm^2 giving a weight saving of 26%.

Table 3.9 shows equivalent sections among unequal angle and equal angle with attached plating. The attached plating has a width of 610 mm and a thickness equal to the thickness of the web of the attached section.

Table 3.7 Equivalent sections of equal angle and OBP with attached plating

W cm³						
	scantlings	area	section modulus	scantlings	area	section modulus
220,0	120 × 120 × 13 100 × 100 ×20	29,70 36,20	220,7 222,5	200 × 10	25,60	225,6
250,0	120 × 120 × 15	33,90	251,2	200 × 12	29,60	255,4
270,0	130 × 130 × 14	34,70	278,4	220 × 10	29,00	288,7
300,0	140 × 140 × 13	35,00	303,1	220 × 11	31,20	304,3
340,0	140 × 140 × 15	40,00	344,6	240 × 10	32,40	351,3
370,0	150 × 150 × 14	40,30	372,3	240 × 11	34,90	375,7
390,0	150 × 150 × 15	43,00	396,2	240 × 12	37,30	394,0
450,0	160 × 160 × 15	46,10	452,8	260 × 11	38,70	455,7
460,0	150 × 150 × 18	51,00	467,1	260 × 12	41,30	477,7
550,0	160 × 160 × 19	57,50	559,3	280 × 12	45,50	571,7
670,0	180 × 180 × 18	61,90	677,8	300 × 12	49,70	679,8
800,0	180 × 180 × 22	74,70	809,6	320 × 13	57,40	829,4
830,0	200 × 200 × 18	69,10	837,7	320 × 14	60,70	864,8

Table 3.8 Scantlings of flat bar and unequal angle sections having same section modulus

W cm³	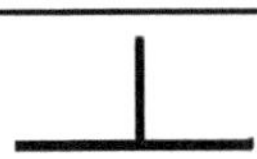			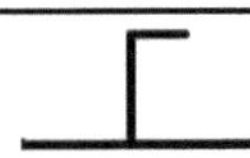		
	scantlings	area	section modulus	scantlings	area	section modulus
21,0	75 × 10	7,50	21,4	65 × 50 × 5	5,54	21,6
23,0	70 × 12	8,40	23,6	70 × 50 × 5	5,79	23,8
24,0	110 × 6	6,60	24,6	80 × 40 × 5	5,80	24,8
26,0	75 × 12	9,00	26,6	75 × 50 × 5	6,05	26,2
28,0	70 × 14	9,80	28,8	70 × 50 × 6	6,88	28,1
29,0	90 × 10	9,00	29,6	80 × 40 × 6	6,89	29,3
32,0	75 × 14	10,50	32,2	70 × 50 × 7	7,95	32,4
33,0	110 × 8 130 × 6	8,80 7,80	33,4 33,7	65 × 50 × 8 80 × 40 × 7	8,60 7,96	33,2 33,7
35,0	75 × 15 90 × 12	11,25 10,80	35,2 36,5	75 × 50 × 7 70 × 50 × 8	8,31 9,00	35,7 36,6

As an example for a section modulus of 195 cm^3, the section area of the equal angle section in table 3.9 is 29.0 cm^2 whereas the section area of the unequal angle section having the same section modulus is 24.2 cm^2 giving a weight saving of 16.5%.

Table 3.9 Scantlings, section area and section modulus of equal and unequal angle sections with attached plating

W cm^3						
	scantlings	section area	section modulus	scantlings	section area	section modulus
195	110 × 110 × 14	29,00	198,3	150 × 100 × 10	24,20	203,9
220	120 × 120 × 13 100 × 100 ×20	29,70 36,20	220,7 222,5	150 × 90 × 12	27,50	223,7
240	130 × 130 × 12	30,00	242,1	150 × 100 × 12 180 × 90 × 10	28,70 26,20	240,3 244,3
270	130 × 130 × 14	34,70	278,4	150 × 100 × 14 180 × 90 × 12	33,20 31,20	276,5 289,1
300	140 × 140 × 13	35,00	303,1	200 × 100 × 10	29,20	301,8
340	140 × 140 × 15	40,00	344,6	200 × 100 × 12	34,80	356,9
450	160 × 160 × 15	46,10	452,8	250 × 90 × 12	39,60	460,7
500	160 × 160 × 17	51,80	506,6	250 × 90 × 14	45,90	530,9

6 Effect of Variation in Thickness of Attached Plating on the Section Modulus and Second Moment of Area

The value of the section modulus depends on the width and thickness of the attached plating. An example showing the effect of the magnitude of the attached plating thickness can be seen clearly by considering an OBP section having a depth of 180 mm and a web thickness of 9 mm. For an attached plating of 15 mm the section modulus is 200 cm^3 and the moment of inertia is 3280 cm^4 whereas when the thickness of the attached plating is 10 mm, the section modulus is 188 cm^3 and the moment of inertia is 2860 cm^4. These values indicate a reduction of 6% in the value of the section modulus and a 12% in the value of the moment of inertia of the section with attached lating.

7 Geometrical and Flexural Properties of Curved Plates

The sectional area of the sector of the curved panel is given by, see Fig. (3.14):

$$A = \pi R_m . t / 2$$

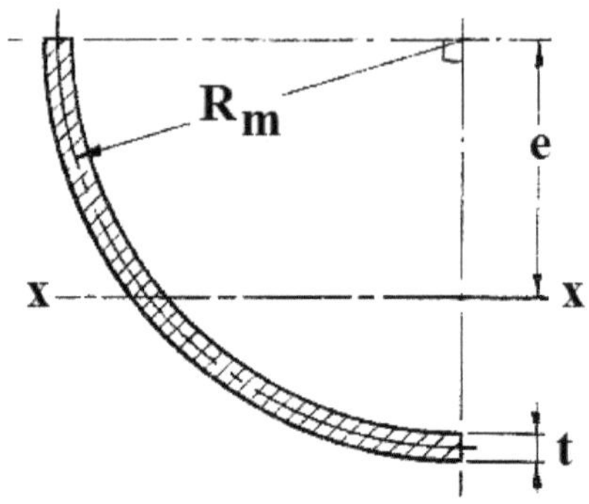

Fig. 3.14 Configuration and geometry of curved panels

The vertical position of the centroid from the top base line is given by:

$$e = 2R_m / \pi$$

the second moment of area about the x-axis is given by:

$$I_{xx} = (\pi / 4 - 2 / \pi) R_m^2 . t = 0.148 R_m^2$$

Example 3.2. Determine the geometrical and flexural properties of a circular bilge strake having the following data: R = 500 mm
t = 10 mm

Solution:

$$e = 318.3\, mm$$

$$A = 7.854 x 10^3\ mm^2 = 78.54\ cm^2$$

$$I_{xx} = 1.86 \times 10^8\ mm^4 = 1.86 \times 10^4\ cm^4\ \text{cm}^4$$

8 Rational Selection of Equivalent Rolled and Fabricated Sections of Ship Strength Members

The improper selection of geometry and scantlings of fabricated and rolled sections may lead to unnecessary increase in steel weight and higher weight/strength ratio and may also cause unfavorable stresses to be developed in these members. This could be easily illustrated by the lower structural efficiency and the higher stresses induced in the face plate of asymmetrical sections in comparison with symmetrical sections, see fig. (3.15).

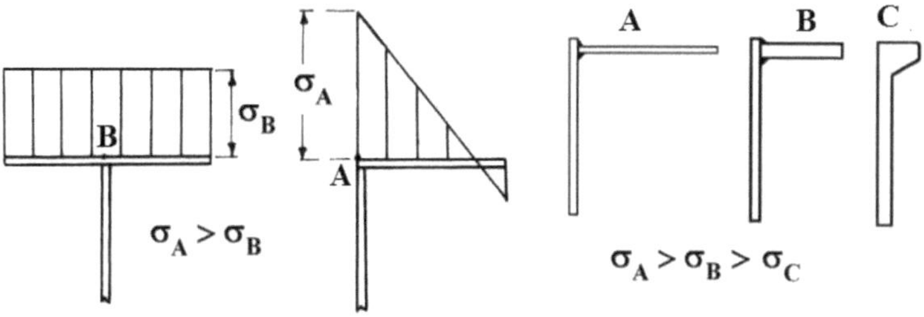

Fig. 3.15 Stress distribution in a symmetrical and asymmetrical face plate of fabricated sections

Table 3.10 Scantlings, section area and section modulus of angle section

Scantlings, mm	Section modulus, cm^3	Section area, cm^2
a-70 x 70 x 10.5	58.8	13.7
b- 90 x 60 x 8	59.5	11.4
a-130 x 90 x 13	191.5	27.05
b-150 x 90 x 10	190	23.2
a-200 x 90 x 12	343	33.6
b-225 x 90 x 10	343	30.7
a-200 x 100 x 13.	402	38.92
b-250 x 90 x 10	402	33.2

The irrational choice of scantlings for a given section modulus could be illustrated by the irrational selection of scantlings of a rolled angle section as shown in table (3.10). The scantlings given in table (3.10) are selected for a random sample of angle sections.

From table (3.10), it is obvious that a significant increase in steel weight will result from the irrational selection of the scantlings of rolled sections.

9 Scantlings of Ship Structural Members

The published rules of Classification societies are primarily simple semi-empirical in format and are calibrated to ensure successful past operational experience. These rules and other conventional codes and guides were used to determine the minimum required scantlings of ship structures, see Fig. (3.16).

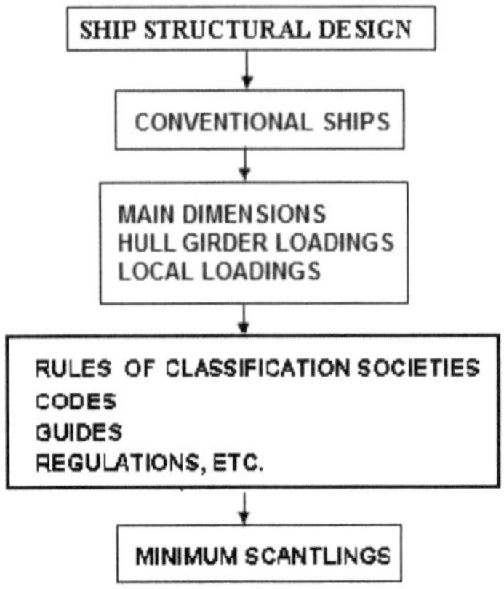

Fig. 3.16 Conventional procedure for determining ship minimum scantlings

However during the last 40 years, ship sizes and types have experienced drastic changes. These changes include the use of complex ship structures for the design of high speed crafts, large LNG carriers and complex structural assemblies with large deck openings for container ships, etc. The complete reliance on the empirical formulae given in the rules of classification societies cannot give economical, reliable and safe scantlings for these drastic changes in ship structures. The design of these new and complex ship structures require advanced methods of stress analysis using the finite element methods, FEM and rational design criteria to provide the required level of safety. The computational complexity has been greatly solved with the introduction of rational analysis procedures using highly sophisticated personal computers and the well-known packages of structural analysis software.

These advanced procedures include modeling of the complex ship structure, load modeling and application, modeling of boundary conditions a, selection of type and size of the finite elements.

The scantlings of the various structural members are calculated using the formulae specified in the rules of classification societies. These scantlings give plate thicknesses for the various plate panels of the hull shell plating, longitudinal and transverse bulkheads, etc. and the minimum section modulus for the primary and secondary structural members (frames and longitudinals). The ship designer selects the configuration and geometry of the primary/secondary sections that satisfies the minimum section modulus required by the rules of classification society.

10 Rational Shapes of Cross-Sections of Beams

The most efficient cross–section of a beam/girder is the most economical with regard to the use of material. These sections have most of the material in the top and bottom parts of the section where the stresses are highest. Thus the efficient section is the section that provides the required section modulus with the minimum section area. Therefore, the comparison between various sections could be based on the following non-dimensional quantity:

$$w = \frac{Z}{\sqrt{A^3}}$$

Where w = axial unit section modulus
Z = modulus of section, m^3
A = sectional area, m^2

Table 3.11 Values of "w" for different sections

Section	Circle ●	Square ■	Ring ○	Channel [	I-section I
w	0.14	0.167	0.73-0.84	0.57- 1.35	1.02-1.51

The magnitude of w depends only on the shape of the cross-section.

The higher the value of "w" the more efficient is the selected section. Table 3.11 gives numerical values of "w" for five different sections.

From table 3.11, it is obvious that significant weight saving could be achieved by proper selection of section configuration and geometry. However, this table is given only for illustration, as "w" is not the only parameter affecting the choice of section configuration. Instability, yielding, fatigue, fabrication, etc. are also important parameters affecting the choice of section.

11 Rational Shapes of Column Sections in Compression

For columns subjected to compressive loadings, the main parameters are the section area and moment of inertia. The radius of gyration k is given by:

$$k = \sqrt{I / A}$$

Let: $\rho = \frac{k}{\sqrt{A}} = \frac{\sqrt{I/A}}{\sqrt{A}} = \frac{\sqrt{I}}{A}$

Where: ρ = unit radius of gyration (non–dimensional)

Efficient sections have high ρ values, as given in Table 3.12:

Table 3.12 Unit radius of gyration for different sections

	Circle	Rectangle	Ring	Angel	I-section
Section					
ρ	0.283	0.204	1.6-2.25	0.3- 0.5	0.27-0.41
Geometry		h/b = 2	r/R=0.8-0.95		
Note		Least economical	Most economical		

12 Design of Girders Having Fabricated T-Sections

The geometry of a symmetrical fabricated T-section composed of a face plate, web plate and attached plating is shown in Fig. (3.17).

The scantlings of the face plate, web plate, attached plating and length of girder should satisfy the following conditions:

$$b/t_f \le 40, d/t_w \le 60, \ L/d \le 6 \text{ and } s/t_p \le 60$$

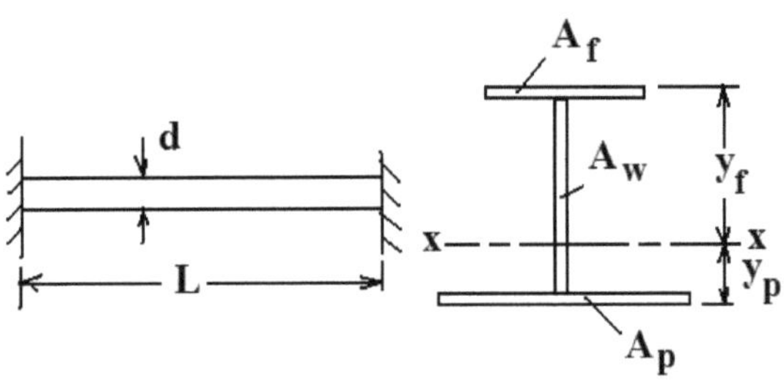

Fig. 3.17 Geometry and scantlings of a symmetrical fabricated girder

The maximum allowable stress at the face plate = $\sigma_f \le \sigma_y$

Where σ_y = yield stress of the material

The scantlings of the section are:

Face plate: $A_f = b \times t_f$

Web plate: $A_w = d \times t_w$

Attached plating: $A_p = s \times t_p$

The total section area of the girder is given by:

$$A_t = A_f + A_w + A_p$$

The second moment of area about the neutral axis of the section is given by:

$$I_x = A_f \cdot y_f^2 + A_p \cdot y_p^2 + \frac{t_w d^3}{12} + A_w \left(y_f - \frac{d}{2} \right)^2$$

The section modulus of the section at the face plate is given by:

$$Z_f = \frac{I_x}{y_f}$$

The position of the neutral axis from the outer side of the attached plating is given by:

$$y_p = \frac{A_f \cdot d + A_w \cdot d/2}{A_t} = \left(A_f + \frac{A_w}{2} \right).d/A_t$$

If the girder is subjected to a bending moment M, the minimum section modulus at the face plate of the section is given by:

$$Z_f = \frac{M}{\sigma_f}$$

The minimum depth of the girder section could be determined from the girder length as follows:

$$\mathrm{d} \geq L/6$$

The minimum thickness of the web plating is given by:

$$t_W \geq \frac{d}{60}$$

The minimum thickness of the attached plating is given by:

$$t_s \geq \frac{s}{60}$$

Assuming the width of face plate is b. the thickness of face plate is given by:

$$t_f \geq \frac{b}{40}$$

13 Determination of Optimum Depth of I-Section Girders

The rational depth of I-beam could be determined using the following assumptions, see Fig. (3.18):

a) The depth of section = h, i.e., the flange thickness is ignored

b) The moment of inertia of the flanges about own axes are not taken into account

$$I_x = 2A_f\left(\frac{h}{2}\right)^2 + \frac{t_w h^3}{12} = \frac{A_f h^2}{2} + \frac{t_w h^3}{12}$$

Where $A_f = b \cdot t_f$

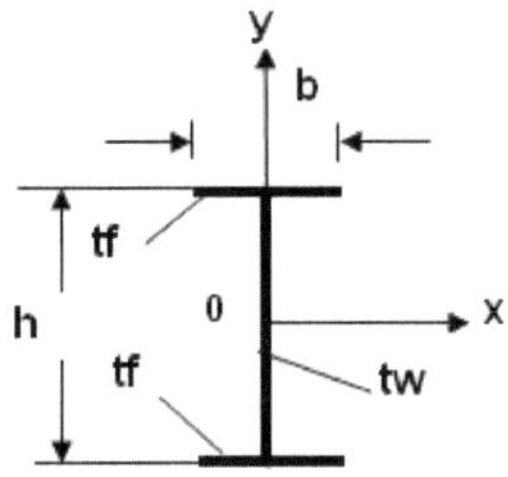

Fig. 3.18 Scantlings of the I-section girder

The section modulus at the flange is given by:

$$Z_f = \frac{I_X}{h/2} = A_f \cdot h + \frac{t_w h^2}{6}$$

Thus: $A_f = \frac{Z_f}{h} - \frac{t_w h}{6} = \frac{Z_f}{h} - \frac{A_w}{6}$

The total area is given by:

$$A_T = 2A_f + A_w = \frac{2Z_x}{h} - \frac{A_w}{3} + A_w$$

$$= \frac{2Z_f}{h} + \frac{2}{3}A_w = \frac{2Z_f}{h} + \frac{2}{3}t_w h$$

In order to find the depth "h" which makes the cross-sectional area a minimum, the relation between t_w and h must be established.

$$\text{Let: } \lambda = \frac{h}{t_w}, \quad 40 \le \lambda \le 60$$

$$A_T = \frac{2Z_f}{h} + \frac{2}{3}\frac{h^2}{\lambda}$$

The optimum height of a beam is obtained by setting:

$$\frac{dA_T}{dh} = 0$$

Thus: $-\frac{2Z_x}{h^2} + \frac{4}{3}\frac{h}{\lambda} = 0$

i.e. $h^3 = \frac{3}{2}\lambda Z_f$

$$h_{opt} = \sqrt[3]{\frac{3}{2}\lambda Z_f}$$

$$For\ \lambda = 60, h_{opt} \approx 4.49.\sqrt[3]{Z_f}$$

$$h_{opt} = \sqrt[3]{\frac{3}{2}\lambda Z_f} = \sqrt[3]{\frac{3}{2}\frac{h}{t_w}Z_f}$$

$$Hence: h_{opt} = \sqrt{\frac{3}{2}\frac{Z_f}{t_w}}$$

For: $Z_f = 300\,cm^3$

$$h_{opt} = 30.0 \quad cm$$

$$t_w = 300/60 = 5\ mm$$

This value of the web thickness is too small for a depth of 300 mm.

$$For\ t_f = 6.0 \quad mm$$

$$h_{opt} = \sqrt{\frac{3}{2}.\frac{300}{0.60}} = 27.4\ cm$$

$$For\ t_w = 6\ mm \quad \lambda = 27.4/0.6 = 45.6$$

Example 3.3. It is required to design a girder of length L = 3 m and subjected to end moments of 32 tm. The girder is assumed clamped at both ends and having a fabricated section as shown in Fig. (3.19). The section area of the attached plating $A_p = 100\ cm^2$.The girder should satisfy the following conditions:

L/d < 6 and the allowable stress at the face plate of the section is: $\sigma_f \leq 0.80 \quad t/cm^2$.

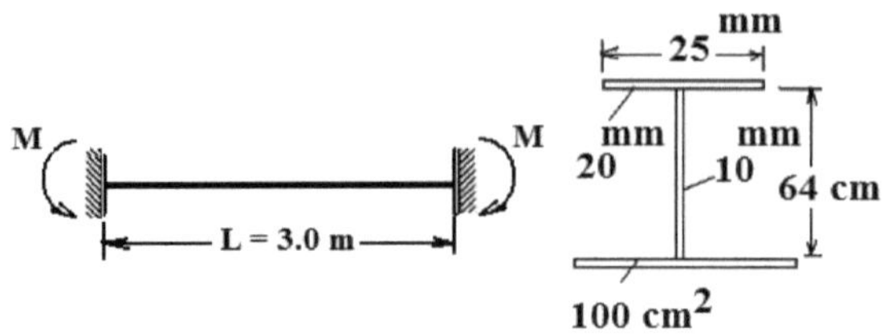

Fig. 3.19 Dimensions and scantlings of the girder section

Solution:

The given data of the girder is:

$$M = 32 \text{ tm}$$
$$L = 3\ m$$
$$A_p = 100 \quad cm^2$$
$$L/d < 6$$

The allowable stress at the face plate of the section is: $\sigma_f \leq 0.80 \quad t/cm^2$

The section modulus at the face plate is calculated as follows:

$$Z_f = \frac{M}{\sigma_f} = \frac{32 \times 100}{0.8} = 4 \times 10^3\ cm^3$$

The optimum depth of the girder section could be calculated as follows:

$$d \cong \sqrt{\frac{3Z_f}{2t_w}} = \sqrt{\frac{3 \times 4 \times 10^3}{2t_w}}$$

Assume: $t_w = 1.4$ cm

$$d \cong \sqrt{\frac{12000}{2.8}} = 65.46\ cm$$

$$L/d = \frac{3}{0.6546} = 4.58 < 6.0$$

For beams and girders commonly used in ship structure, see Fig. (3.20), the position of the N.A. of the section is approximately at a distance of 2/3 the depth of the section.

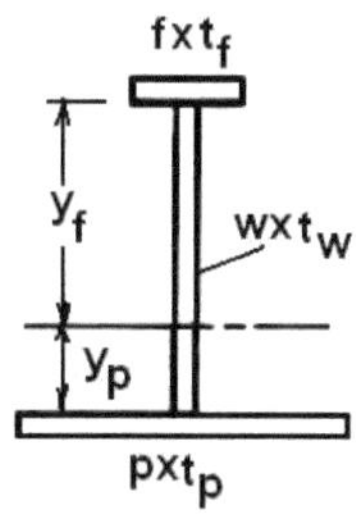

Fig. 3.20 Configuration and geometry of girder section

Hence, the section area of the face plate could be assumed as follows:

$$A_f \cong \frac{3}{4}\frac{Z_f}{d} = \frac{3\times4\times10^3}{4\times65.46} = 45.82 \quad cm^2$$

Take b =20 cm,

Then: $t_f = 22.9 \approx 23.0\ mm$

Chapter 4
Bending of Beams and Girders

1 Introduction

Chapter 4 introduces the application of the simple beam theory to ship structural members. The limitations of application of the simple beam theory to thin-walled asymmetrical sections are highlighted. A full explanation of the idealization of beam elements is illustrated. The various configurations of end supports commonly used in ship structure are given. The importance of determining the effective span of beams is stressed. The effective breadth concept is explained for uniform and curved structural members. The concept of modeling ship structure assemblies by beam elements is introduced. Several examples of 2D and 3D modeling of deck, bottom, side and bulkhead ship structure using beam and plate elements are illustrated. The various types of boundary end conditions commonly used in ship structure analysis are given. The concept of span points and effective span of a beam are clarified. A method is given to determine the optimum span length of a beam and the size of bracket to be fitted at both ends. The influence of the type of end support on the magnitude and distribution of the bending moment are presented. Bending stresses in beams constructed with high tensile steel. Flexural stresses in symmetrical sections, sections with one axis of symmetry and is explained fabricated asymmetrical sections are presented. The importance of calculating the equivalent stress is highlighted. A simple procedure for calculating flexural warping stresses is given. The main parameters affecting the magnitude and distribution of flexural warping stresses for asymmetrical sections are explained. The basic concept of effective breadth is introduced. The effective breadth of uniform symmetrical asymmetrical face plates are presented together with the effective flexural and properties of sections. Methods are given to determine the effective breadth of curved symmetrical and asymmetrical face plates.

2 Subdivision of Ship Structure into Members and Assemblies

The analysis of structural members aims at checking the strength, stability and rigidity of the preliminarily selected scantlings of the structure. The strength members of the structure are analyzed on the basis of the theory of the strength of materials and structural mechanics to determine the induced internal stresses under the action of the applied loads.

In order to materialize this aim, it is necessary first to prepare an assumed configuration of the general layout of the required structure. The analysis of ship structure requires idealization of the structure members and assemblies. The general structural configuration should consist of separate elements: plates, beams, frames, brackets, etc. and disregarding temporarily the actual shape of their cross sections. The most common categories of structure members and assemblies are:

1. Bar element (column)
2. Beam element
3. 2D- Frame structure
4. 2D-Grillage structure
5. 2D- Stiffened panel
6. 3D- Space frames
7. 3D structure assemblies.

3 Representation of Structure by Elements

Very many of the local strength problems in a ship can be solved adequately by the application of simple beam theory to a single stiffener with the attached plating. This is permissible if the end boundaries of the unit are truly represented by forces and moments that adjacent units apply to it. These edge constraints have a large influence and in many cases will not be so easily determined. It is important that the deflection of the supporting structure is negligible compared with the deflection of the isolated member.

Modern computer methods of structural analysis split the structure into a large number of beam or plate elements. In many cases, the use of beam elements gives results which are sufficiently accurate for design purposes. The beam elements are connected at joints or nodes and the computer calculates the deflection, forces and moments at each node.

4 Modeling of Structure

Modeling of ship structure could be carried out using beam elements, shell elements or hybrid elements. The chosen type and size of elements for modeling ship structure is to be carefully selected to reproduce the behaviour of the actual structure with a high degree of reliability. The modeling of a structure should include its surrounding members considered to have material influences on the behaviour of the structure members. Two or three dimensional structure idealization are commonly used for modeling ship structure assemblies.

4.1 Forces and Moments on a Beam Element

Consider a single beam element as shown in fig. (4.1). Forces and moments are developed at each end of the member to maintain the element in its constrained shape.

Using orthogonal Cartesian coordinates, there are three forces and three moments at each end, see Fig (4.1). Each force produces a corresponding deflection and end moment produces a corresponding rotation at the end of the member. The ability to move in a certain coordinate direction or rotate about a coordinate axis is called a degree of freedom. It follows that, in general, there are six degrees of freedom at each end of a member. The forces and moments in each degree of freedom cause a certain mode of deformation, see Fig. (4.2). The behavior in each mode can be calculated by using certain geometrical properties. By the principle of superposition, the final deflected shape may be found by compounding the deformations in each mode and the final stresses by adding the stresses in each mode. The validity of the principle of superposition relies mainly on the assumption that the structure is elastic and behaves linearly under the applied loading and also that the deformations and deflections of each member are small enough for secondary effects to be ignored..

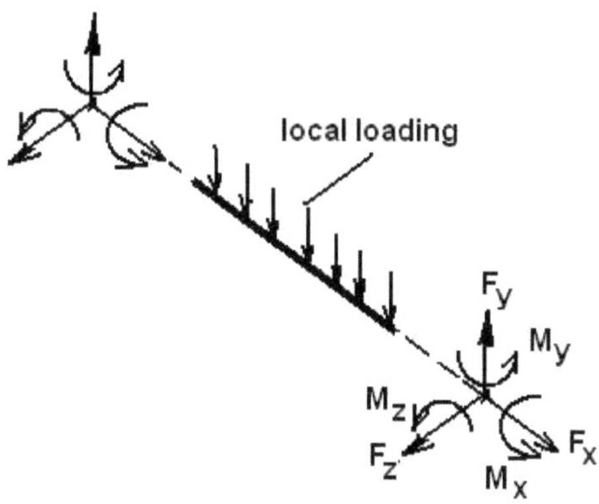

Fig. 4.1 Forces and moments on a beam element

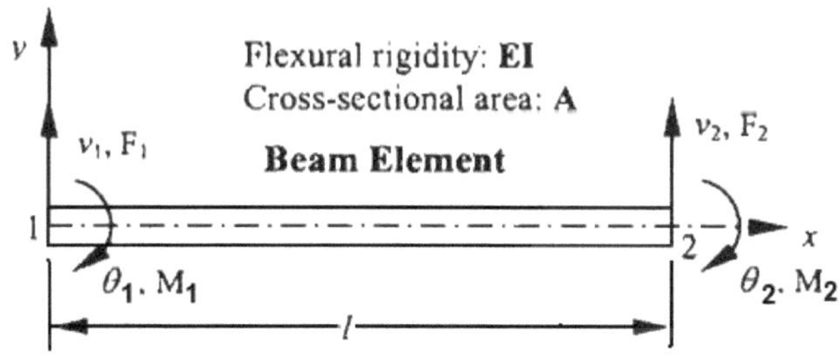

Fig. 4.2 Forces, moments and deformations at the two end nodes

4.2 *Modeling of Ship Structural Members*

Modeling of ship structural members is to be such that the chosen elements being beam elements, bar elements, plate bending elements, etc. can reproduce acceptable degree of reliability of the behaviour of the actual structure. To enable further study on the basis of the results of the analysis, re-meshing with fine meshes is normally used to obtain more detailed results.

4.3 Boundary Conditions of Idealized Beam Elements

Different types of structural idealizations using uniform beam elements for single and continuous beams and for various end support conditions are shown in fig. (4.3).

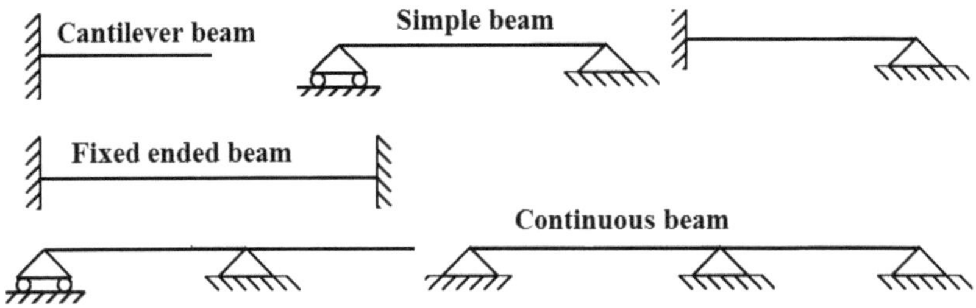

Fig. 4.3 Beam element idealizations for beams with different types of end support conditions

4.4 Modeling 2D Frame Structures Using Beam Elements

When modeling 2D frame structures using beam elements, the attached plating is idealized by its effective breadth. Beam elements of high rigidity are used at the corners of the 2D frame structure. An example of 2D structure idealization using beam elements is given for a transverse section of a conventional oil tanker, see Fig. (4.4). When modeling by using beam elements, attention is to be paid to the position of neutral axis of the members.

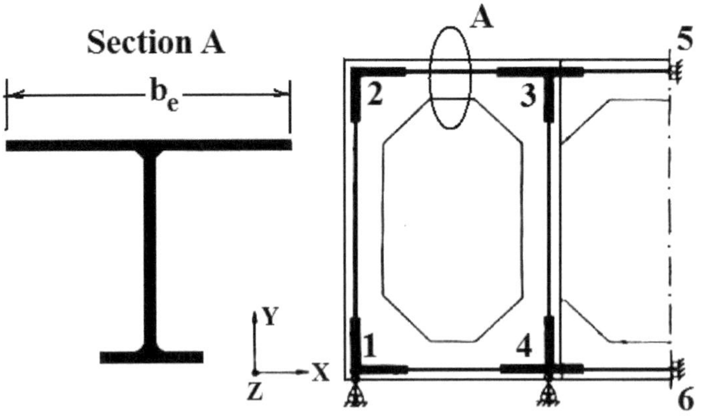

Fig. 4.4 A simple idealization of a web frame in an oil tanker

4.5 Modeling 2D Grillage Structure

Beam elements are used very effectively for the analysis of 2D plane structures using grillage idealization, see Fig. (4.5). Beam elements are used for the idealization of girders, transverses, longituydinals, beams, frames, etc.

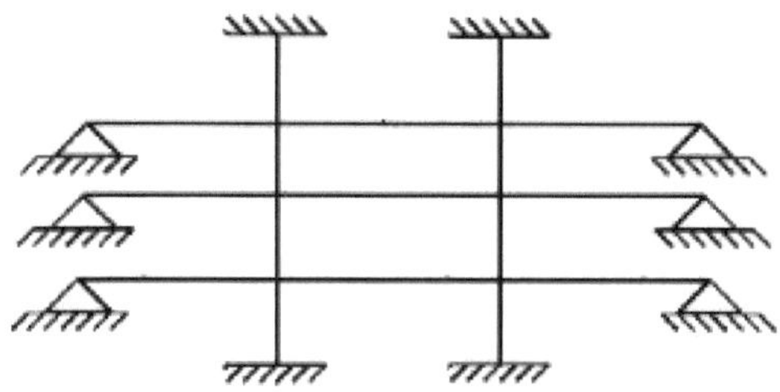

Fig. 4.5 A 2D beam element idealization of a grillage structure

4.6 Modeling 2D Deck Structure

Another example of using beam elements for an idealized deck structure is shown in Fig. (4.6). The idealized deck structure is modeled as a grillage structure composed of deck girders and deck beams.

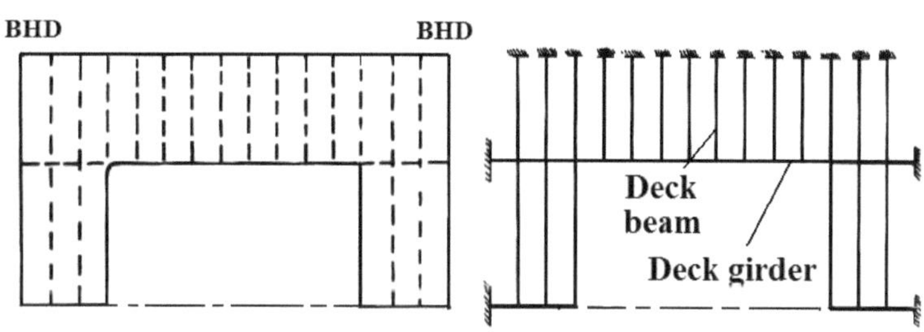

Fig. 4.6 Idealization of a transversely framed deck structure using beam elements

4.7 Modeling 2D Bottom Structure

The bottom structure of a ship could be idealized as a grillage structure composed of bottom girders and transverses. Beam elements are used for the modeling of the bottom structure, see Fig. (4.7).

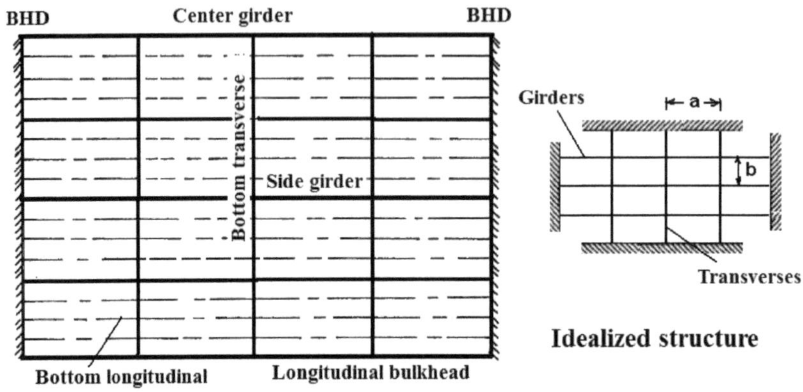

Fig. 4.7 Idealization of a longitudinally framed bottom structure using beam elements

4.8 Modeling 2D Side Structure

Side shell structure could be idealized as a grillage structure composed of the side stringer, web frames and ordinary frames. Beam elements are used for the modeling of the side structure assembly, see Fig. (4.8).

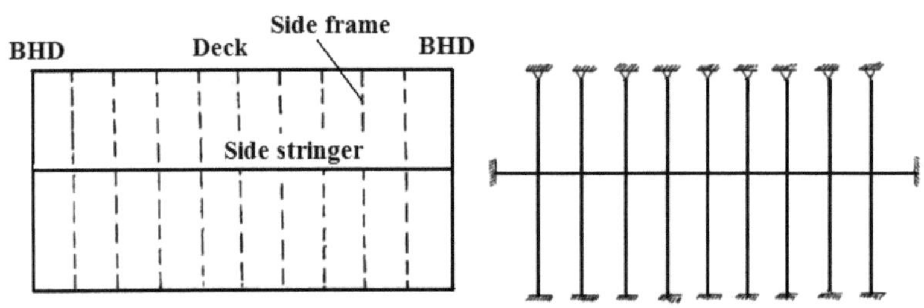

Fig. 4.8 Idealization of a side structure assembly using beam elements

4.9 Modeling 2D Transverse Bulkhead

Transverse bulkheads could be idealized by a 2D grillage structure composed of the heavy vertical and horizontal girders, see Fig. (4.9).

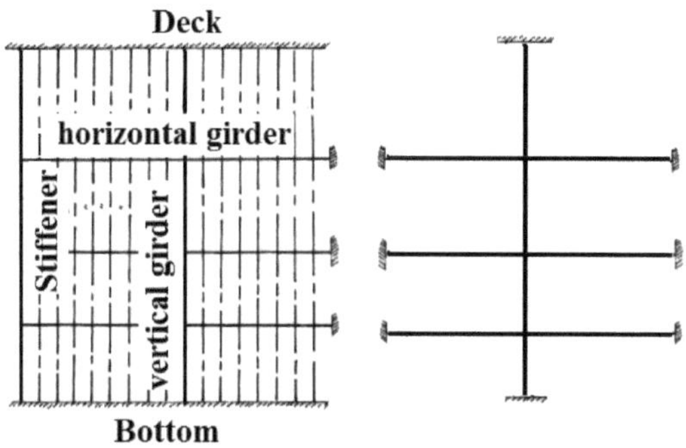

Fig. 4.9 Modeling 2D transverse bulkhead by a 2D grillage structure

4.10 Modeling 3D Space Frame Structures Using Beam Elements

Beam elements are commonly used for 3D space frame idealization of 3D ship structure assemblies, see Fig. (4.10).

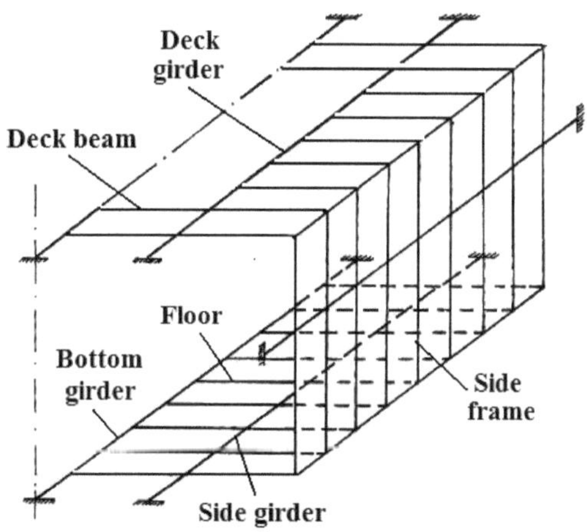

Fig. 4.10 Space frame idealization of a single deck cargo ship

5 Modeling by Using Plate Elements

Side shell, longitudinal bulkheads and other similar members are preferably to be modeled into two or three dimensional structure by using plate elements. Large aspect ratios of mesh are to be avoided.

5.1 FEM Idealization Using Plate Elements

The FEM could be also used for the analysis of ship structure using 2D and 3D idealizations. Fig. (4.11) shows a 2D idealization of a transverse web frame in a small oil tanker with a centerline longitudinal bulkhead. Fig. (4.12) shows a 3D idealization of ship hull girder of a container ship. Fig. (4.13) shows a 3D idealization of ship structure assembly of a bulk carrier. Fig. (4.14) shows a 3D idealization of a structure assembly of a small oil tanker.

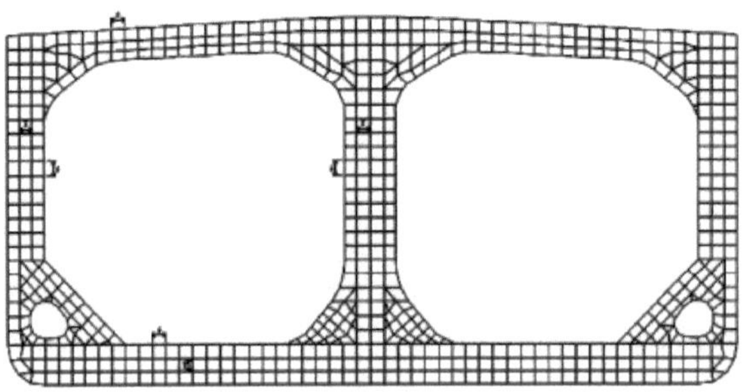

Fig. 4.11 A simple plate element 2D idealization of a web frame ring of a small oil tanker

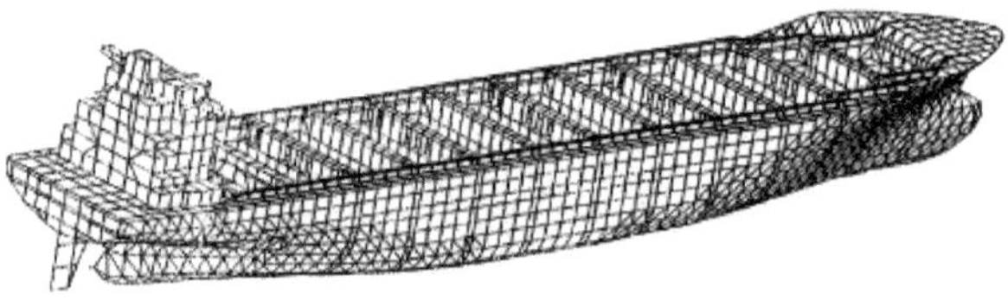

Fig. 4.12 A 3D idealization of the hull girder of a container ship

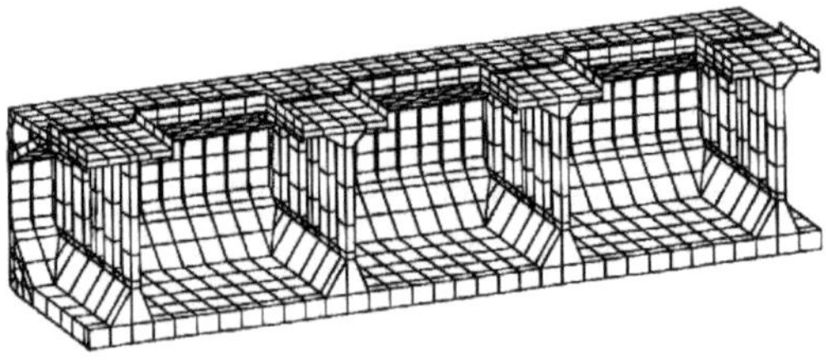

Fig. 4.13 A 3D idealization of a structure assembly of a bulk carrier

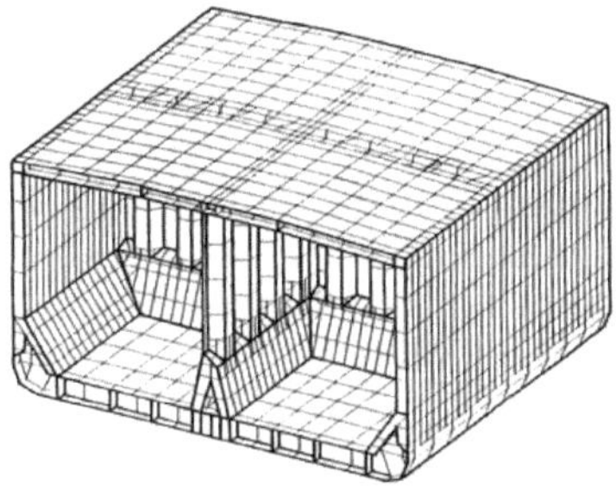

Fig. 4.14 A 3D idealization of a structure assembly of an oil tanker

Fig. (4.15) shows a 3D idealization of a girder of fabricated angle section with the attached plating. Proper sizes of meshes are to be selected in accordance with the stress distribution in the model and abnormally large aspect ratios of meshed are to be avoided. Girders and similar members having stress gradients along their depth are to be well meshed so as to obtain clear and reliable distribution of stresses and deformations. In order to cater for the cross bending effect of the flange of the girder, beam elements are used around the plate elements of the flange, see Fig. (4.16).

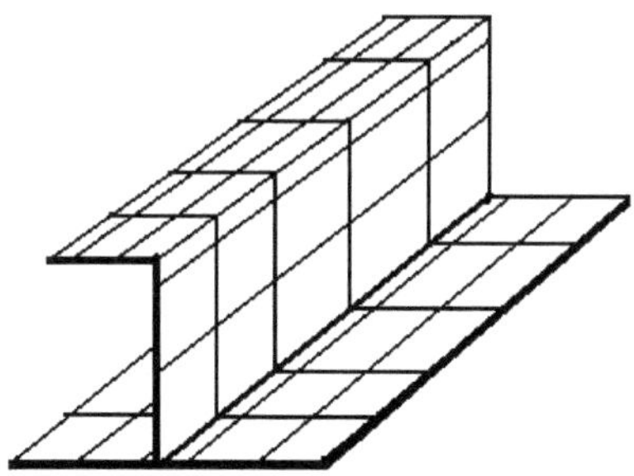

Fig. 4.15 A 3D plate element idealization of a fabricated angle section with the attached plating

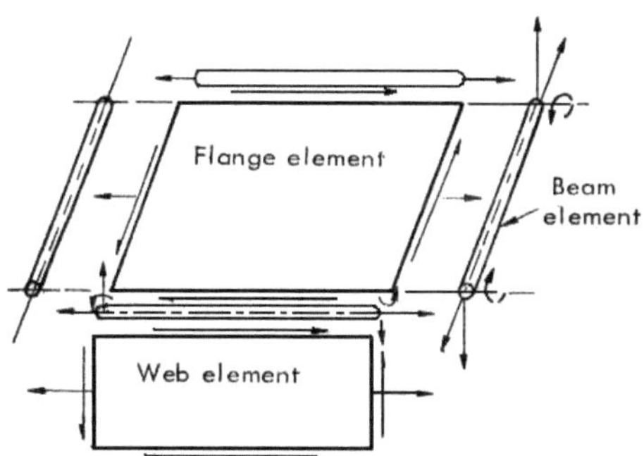

Fig. 4.16 A 3D plate element idealization of the flange with beam elements

6 Boundary Conditions of Beams and Columns

The boundary conditions of beams and columns could be assumed to be free, hinged, fixed or constrained, see fig. (4.17). The degree of rotational end constraint has more effect on deflections than it has on stresses. End constraints in practical ship structures approach the clamped condition for flexural behaviour, provided that the members are properly continuous at the joint.

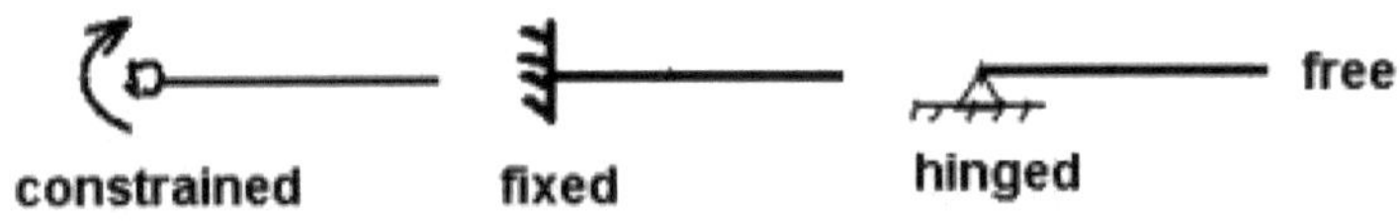

Fig. 4.17 Boundary support conditions

The degree of rotational end constraint of a member is due to the stiffness of the joint itself. It is relatively simple to produce a joint which can develop the full plastic moment of the strongest of the members entering the joint. The effects of the other members entering the joint can be calculated to give the actual rotational stiffness pertaining to the member. The degrees of freedom of the various types of end support conditions are shown in Fig. (4.18).

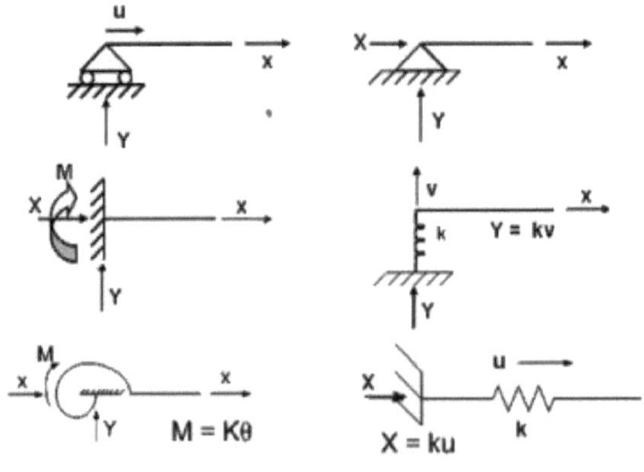

Fig. 4.18 Degrees of freedom of various types of end support conditions

7 Effective Span of a Beam

Before conducting any structural analysis using beam elements, it is essential to determine the span points and the effective length of the beam span, see

Fig. (4.19). For bracketed beam members at its supported ends, part of the bracket structure is commonly assumed to have very large rigidity. The location of the span points for bracketed beam elements is shown in Fig. (4.20). The location of the span points, the effective span length and the idealization of a bottom transverse member in an oil tanker is shown in Fig. (4.21).

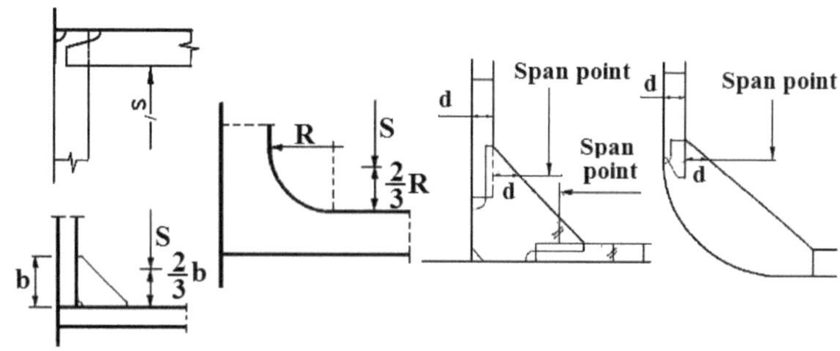

Fig. 4.19 Span points for various connections

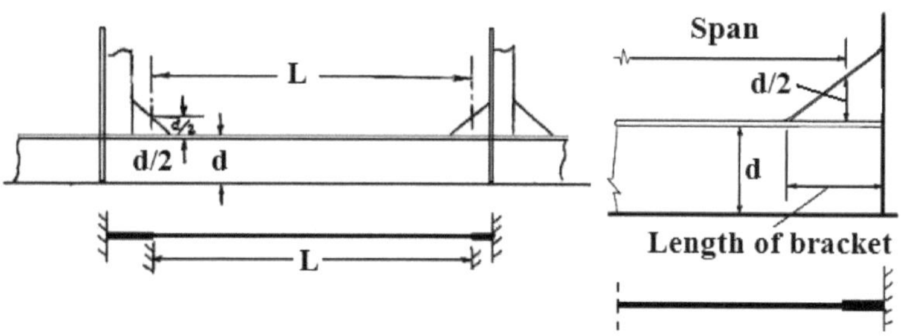

Fig. 4.20 Span points and effective span

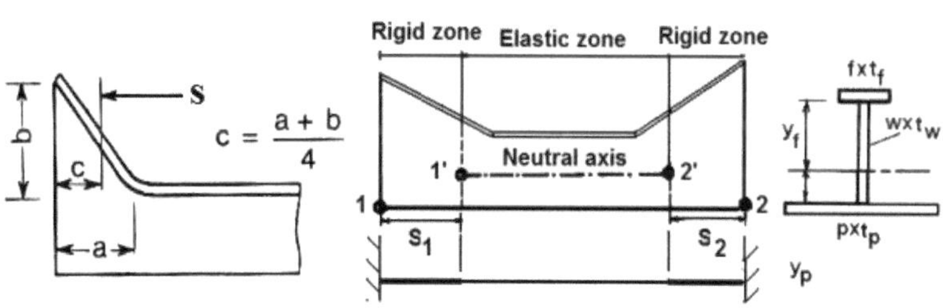

Fig. 4.21 Span points and effective span for a heavy bottom transverse member

8 Determination of the Optimum Span Length and Size of Bracket

For a beam clamed at both ends, the bending moment at the clamped end and at mid-span are given by, see Fig. (4.22):

$$M_0 = \frac{qL^2}{12}$$

$$M_{l/2} = -\frac{qL^2}{24}$$

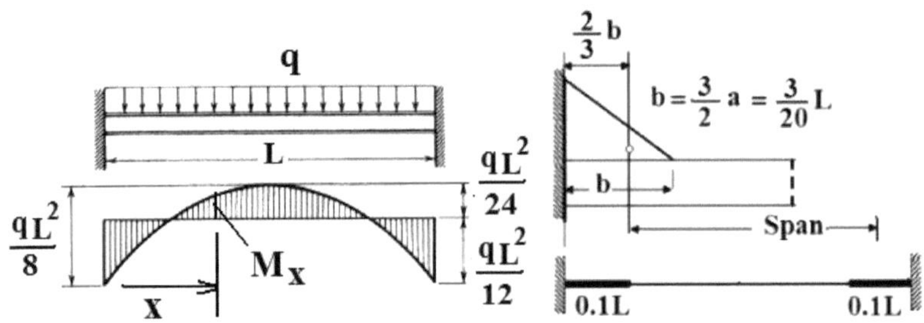

Fig. 4.22 Span points, effective span and size of bracket

It is clear that:

$$M_0 = 2M_{l/2}$$

The beam will be over-designed if the bending moment at the clamed end, M_0, is used for determining the section modulus and scantlings of the beam section. The optimum design of the beam could be achieved by using the effective span length. The latter could be obtained by determining the position on the beam length at which the bending moment is the same as at mid span.

At a distance x from the left clamped end of the beam, the bending moment is given by:

$$M_x = \frac{qL}{2}.x - \frac{qL^2}{12} - \frac{qx^2}{2}$$

Assuming the position at which $M_x = M_{l/2}$ is at as distance "a" from the clamped end, then we have:

$$-\frac{qL^2}{24} = \frac{qL}{2}.a - \frac{qL^2}{12} - \frac{qa^2}{2}$$

i.e. $$-\frac{L^2}{12} = La - \frac{L^2}{6} - a^2$$

$$a^2 - La + \frac{L^2}{12} = 0$$

Solving this equation for "a":

$$a = \frac{L \pm \sqrt{L^2 - 4L^2/12}}{2}$$

$$= \frac{L}{2}(1 \pm 0.81)$$

Hence $a \cong \frac{L}{2} \times 0.19 \; and \; \frac{L}{2} \times 1.81$

Then $a \cong 0.09L \cong L/10$

And $a \cong 0.9L$

Assuming the bracket length measured from the left end is "b", then the span point from the left end is at a distance "a" given by:

$$a \cong 0.66b \cong L/10$$

Because of symmetry of both ends of the beam, the optimum span length is given by:

$$L_{opt} \cong L - 0.2L = 0.8L$$

9 Simple Beam Theory

9.1 Beam Loading and Response

Bending moments and reactions at the end supports for different support conditions are given in Fig. (4.23) for the following loading condition:

- Uniform loading: Figs. a, b and d
- Bending and shear forces induced by deflection at the supports Fig. c,

Bending moments due to deflection of end supports

The bending moments at the end supports for a beam of length "L" clamped at both ends and uniformly loaded is shown in fig. (4.23). The end supports are assumed to be deflected as shown in Fig. (4.24).The magnitude of the induced bending moment at both ends due to the deflection of the supports is given by:

$$m = \frac{6EI}{L^2}(\delta_1 - \delta_2)$$

Where: δ1 = vertical deflection of end support 1
δ2 = vertical deflection of end support 2

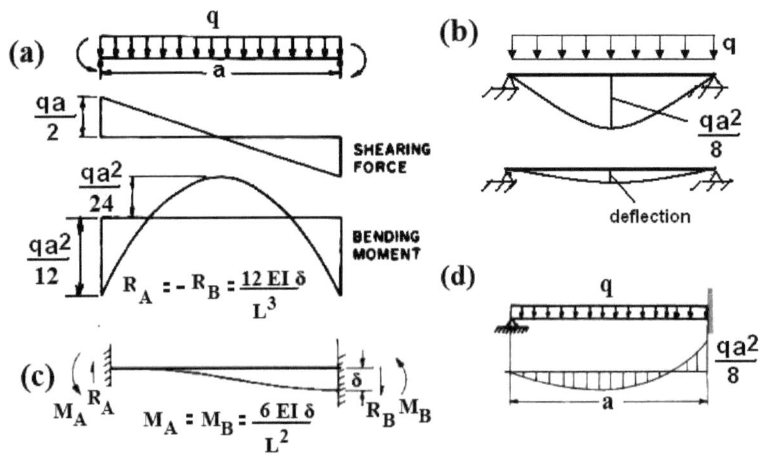

Fig. 4.23 End moments and reactions induced by transverse loading and end deflections for different support conditions

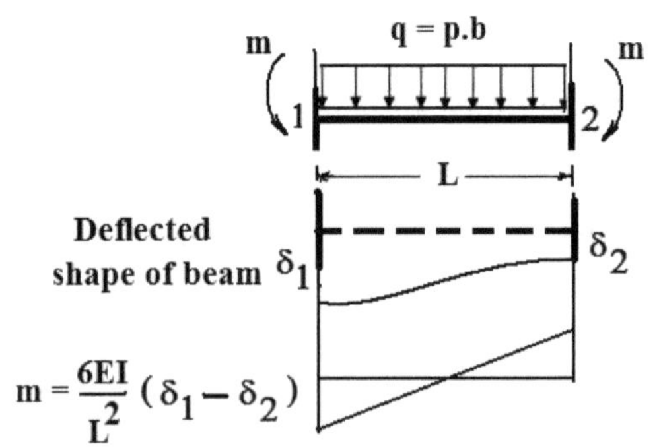

Fig. 4.24 End moments due to vertical deflection of supports

Simply supported beam carrying uniformly distributed load and end moments

Consider a simply supported beam carrying end couples and uniformly distributed load, see Fig. (4.25). The reactions at the end supports are given by:

$$R_B = \frac{wL}{2} - \frac{1}{L}\left(M_B - M_C\right)$$
$$R_C = \frac{wL}{2} + \frac{1}{L}\left(M_B - M_C\right)$$

The bending moment at a distance x from end support B is given by:

$$M = \frac{1}{2} w.x(L-x) + \frac{M_B}{L}(L-x) + \frac{M_C.x}{L}$$

The maximum bending moment occurs when:

$$\frac{dM}{dx} = \frac{1}{2} w(L-2x) - \frac{M_B}{L} + \frac{M_C}{L} = 0$$

$$x = \frac{1}{2}L - \frac{1}{wL}\left(M_B - M_C\right)$$

The maximum bending moment is given by:

$$M_{max} = \frac{1}{2} wL^2 + \frac{1}{2}\left(M_B + M_C\right) + \frac{1}{2wL^2}\left(M_B - M_C\right)^2$$

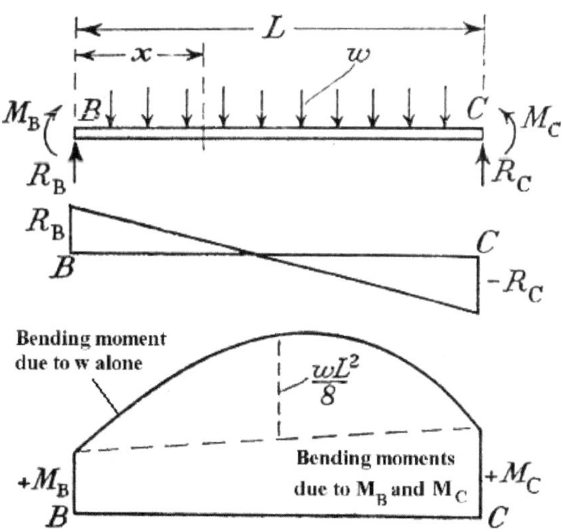

Fig. 4.25 Bending moment distribution due to end moments

Distribution of shear forces and bending moment along beam length for different loading conditions

The distribution of shear forces and bending moment along beam length for different loading conditions are given in Fig.(4.26, 4.27).

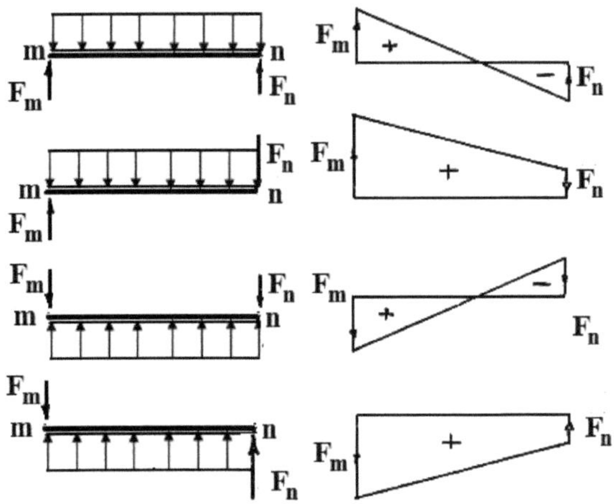

Fig. 4.26 Distribution of shear forces along beam length for different loading conditions

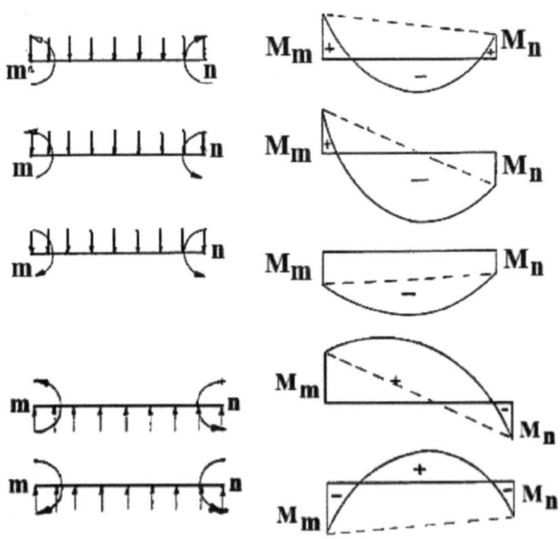

Fig. 4.27 Bending moment distribution along beam length for various loading conditions

9.2 Beam Deflections

Beam deflection could be determined by integrating the shearing force and bending moment as follows:

$$Shearing\ \ force = F = -\int q dx$$
$$Bending\ momentg = \int F dx = -\int \int q dx dx$$

The relationship between bending moment and deflection is given by:

$$-EI\frac{d^2y}{dx^2} = M$$

For a structural member of uniform section, we have:

$$-EI\frac{d^4y}{dx^4} = \frac{d^2M}{dx^2} = \frac{dF}{dx} = -q$$

For a constant rate of loading, we have:

$$EI\frac{d^3y}{dx^3} = qx + A$$

$$EI\frac{d^2y}{dx^2} = \frac{qx^2}{2} + Ax + B$$

$$EI\frac{dy}{dx} = \frac{qx^3}{6} + A\frac{x^2}{2} + Bx + C$$

$$EIy = \frac{qx^4}{24} + A\frac{x^3}{6} + B\frac{x^2}{2} + Cx + D$$

The values of the constants A, B, C, and D could be determined from the end conditions of the structural member.

For a simply supported beam, the end conditions are:

At x = 0 and L, $y = \frac{d^2y}{dx^2} = 0$

Substituting, we get: B = D = 0

Also, $0 = \frac{qL^4}{24} + A\frac{L^3}{6} + CL$

$$0 = \frac{qL^2}{2} + AL$$

$$A = -\frac{qL}{2}$$

$$C = -\left(\frac{qL^3}{24} - \frac{qL3}{12}\right) = \frac{qL^3}{24}$$

$$M = -EI\frac{d^2y}{dx^2} = \frac{qL}{2}x - \frac{qx^2}{2}$$

$$M_{L/2} = \frac{qL^2}{8}$$

$$y = \frac{1}{EI}\left(\frac{qx^4}{24} - \frac{qL}{12}x^3 + \frac{qL^3}{24}x\right)$$

The deflection of a beam clamped at both ends can be calculated using the same procedure and the boundary conditions at the clamped ends, see Fig, (4.28).

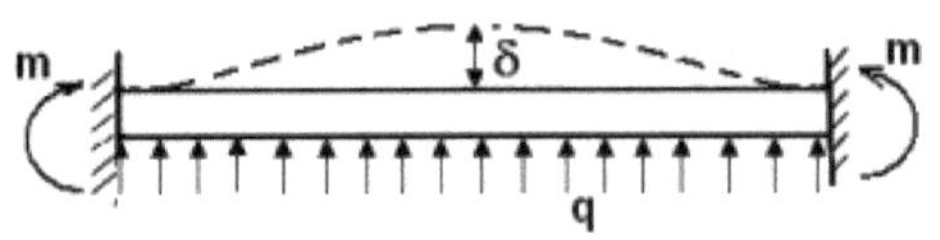

Fig. 4.28 Deflection shape of a beam uniformly loaded and clamped at both ends

The end deflections for different loadings and end conditions are given in Fig. (4.29).

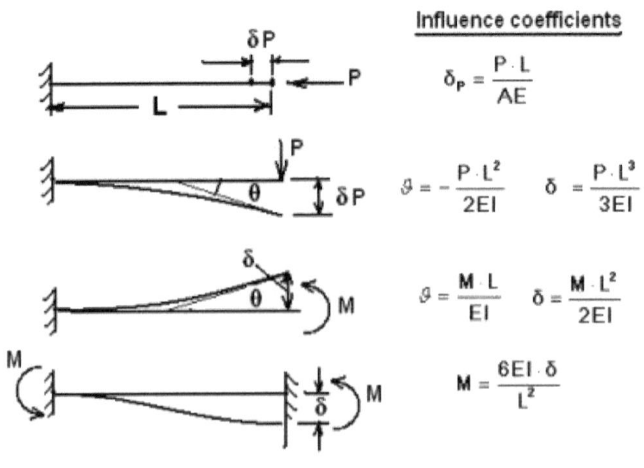

Fig. 4.29 End deflections for different loadings and end conditions

The reactions and deflection equation for different loading conditions and several types of end supports are given in Table (4.1).

Table 4.1 Reactions and deflection equation for different loading conditions and end supports

Type of Beam	Reactions	Deflections at Any Point x
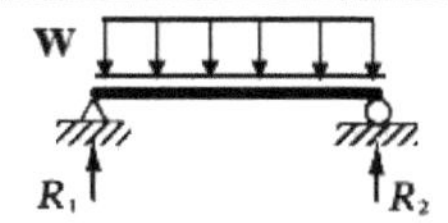	$R_1 = \frac{w.L}{2}$ $R_2 = \frac{w.L}{2}$	$\frac{w.x}{24EI}(L^3 - 2Lx^2 + x^3)$
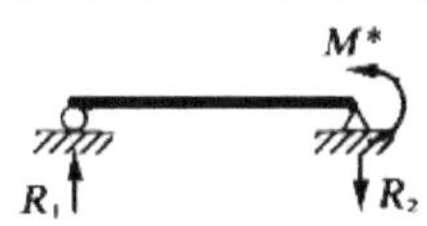	$R_1 = \frac{M^*}{L}$ $R_2 = \frac{M^*}{L}$	$\frac{M^*Lx}{6EI}\left(1 - \frac{x^2}{L^2}\right)$
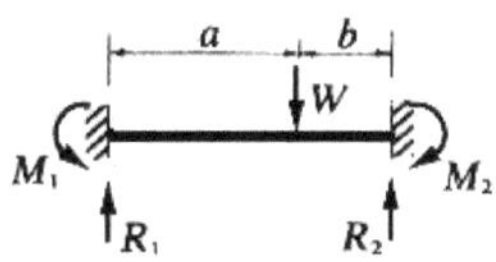	$R_1 = \frac{Wb^2}{L^3}(3a + b)$ $R_2 = \frac{Wa^2}{L^3}(3b + a)$ $M_1 = \frac{Wab^2}{L^2}$ $M_2 = \frac{Wa^2b}{L^2}$	For $x < a$, $\frac{Wb^2x^2}{6EIL^3}(3ax + bx - 3aL)$ For $x \geqslant a$, $\frac{Wa^2(L - x)^2}{6EIL^3}\psi$ $\psi = [(3b + a)(L - x) - 3bL]$
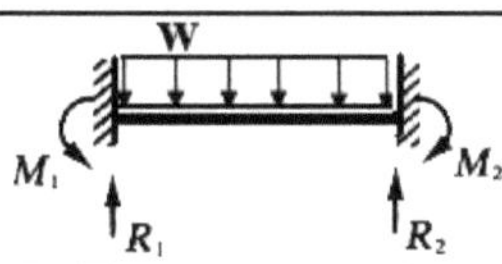	$R_1 = R_2 = \frac{w\,L}{2}$ $M_1 = M_2 = \frac{w\,L^2}{12}$	$\frac{w.x^2}{24EI}(2Lx - L^2 - x^2)$

10 The Influence of the Type of End Support on the Magnitude and Distribution of the Bending Moment

The influence of the type of end support on the magnitude and distribution of the response of a beam element to the applied loads could be very significant. The effect of the type of end support on the magnitude and distribution of the bending moment for a beam element under different patterns of loading is shown in Fig. (4.30).

Case (a) relates to pure bending. The load-deflection curve is linear up to first yield in the beam and remains flat after the plastic moment in the beam has been reached. In case (b) tensile membrane stresses are imposed in addition to the bending stresses. These tensile stresses have a stabilizing effect on the beam increasing its stiffness as deflection begins to develop. The counterpart to case (b) is

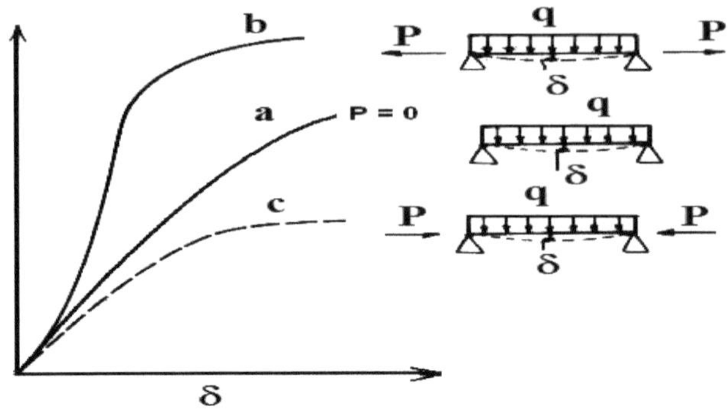

Fig. 4.30 Effect of boundary constraints from adjacent members

shown in case (c) where compressive membrane stresses are introduced. Such compressive stresses may have a violent effect on the beam strength and may also lead to a dramatic behaviour after the ultimate load has been reached.

10.1 Effect of Degree of Constraint at the End Support on the Magnitude and Distribution of the Bending Moment

The magnitude and distribution of bending moment along a beam of length ℓ m and uniformly loaded is shown in Fig. (4.31) for the following three cases of end support conditions:

- Simple support condition at both end supports, $f_0 = 0$
- Fixed end condition at both end supports, $f_1 = 1.0$
- Constrained condition at both end supports, $f_c = 0.75$

It is clear from Fig. (4.31) that the degree of constraint at both ends of support has a significant effect on the magnitude and distribution of the bending moment along the beam length.

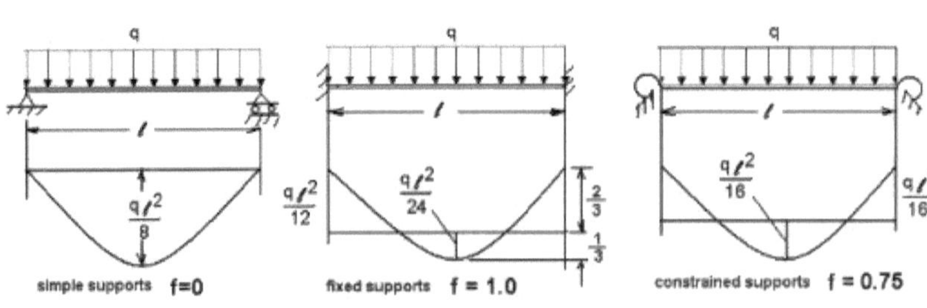

Fig. 4.31 Bending moment distribution along a uniform beam uniformly loaded and having different end constraints

10.2 General Case of Uniform Loading and Constrained End Supports

Consider the general case of a beam of length L and uniformly loaded, see Fig. (4.32). The end supports of the beam have the following degrees of constraint:

- The degree of constraint at the left end support = f_o
- The degree of constraint at the right end support = f_L

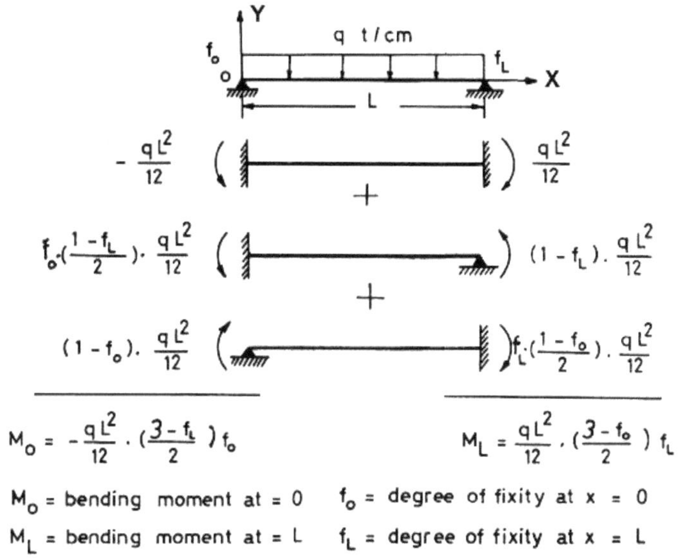

Fig. 4.32 The general case of a uniform beam uniformly loaded and having constrained end supports

Because of the presence of certain degree of constraint at each end of the beam, the bending moment at both ends of support are given by, see Fig. (4.32):

$$M_o = -\frac{qL^2}{12} \cdot \left(\frac{3 - f_L}{2} \right) f_o$$

$$M_L = \frac{qL^2}{12} \cdot \left(\frac{3 - f_o}{2} \right) f_L$$

Ex. Consider a deck beam of a cargo ship as shown in Fig. (4.33). Assume the following cases of end supports for the deck beam:

1- Free end at the hatch girder and Fixed at the connection with the side frame
2- Simple support at the hatch girder and Fixed at the connection with the side frame
3- Simple support at both ends of the deck beam
4- Clamped support at both ends of the deck beam.

The bending moments developed for these different support conditions are given in table 4.2:

Table 4.2 bending moments at both ends for different support conditions

Case 1	Left end	0
	Right end	4
Case 2	Left end	0
	Right end	4
Case 3	Left/right end	0
	Midspan	1
Case 4	Left end	0.666
	Right end	0.666

It is clear from table 4.2 that the assumed condition of end support has a significant effect on the magnitude of the bending moment.

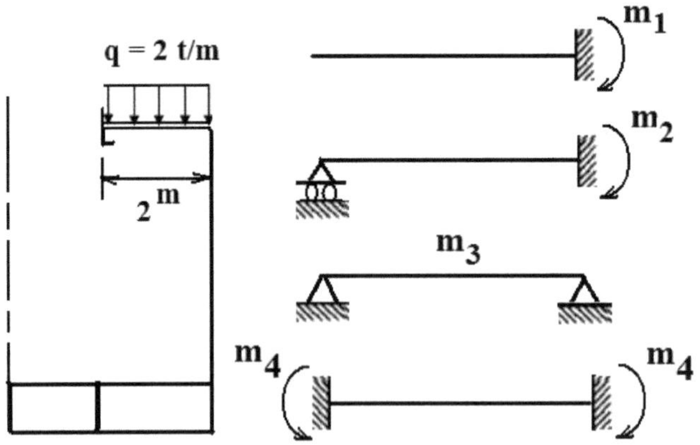

Fig. 4.33 Bending moments for assumed different conditions of end supports of a deck beam

11 Beam Stresses

11.1 Beam under Normal (Axial) Loading

The normal stress induced in a beam subjected to a normal force P is given by, see Fig. (4.34):

$$\sigma_N = \pm P/A$$

Where:
P = axial force with the sign indicating tension or compression
A = section area of the beam

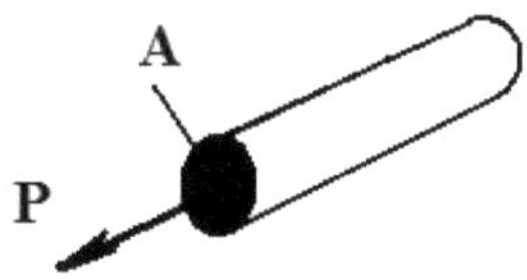

Fig. 4.34 A solid circular section beam under a tensile force

11.2 Beams Subjected to Bending Stresses

11.2.1 Bending of Symmetrical Sections

For fully symmetrical sections, the bending stress at the top and bottom flanges are equal and are given by, see Fig. (4.35).

$$\sigma_f = \sigma_p = M.\frac{h/2}{I}$$

Where: m = bending moment
h = depth of section
I = moment of inertia of the beam section
I/y = section modulus of the beam section

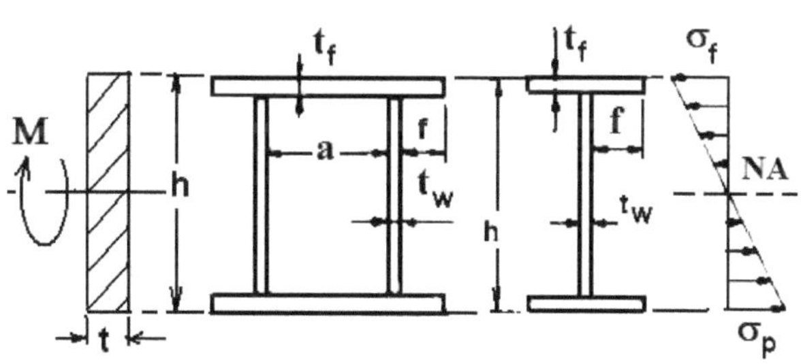

Fig. 4.35 Bending stress distribution over the depth of fully symmetrical sections of various geometrical shapes

The section modulus depends entirely on the geometry of the section. For symmetrical sections the centroidal axes lies at mid-depth of section and the extreme fiber stresses are equal.

The bending and shear stresses induced over a fully symmetrical section of an I-beam is shown in Fig. (4.36).

The average shear stress is given by:

$$\tau = F/a$$

Where:

$a = h.t$ = Section area of web plate only

The shear stress may always be taken as positive.

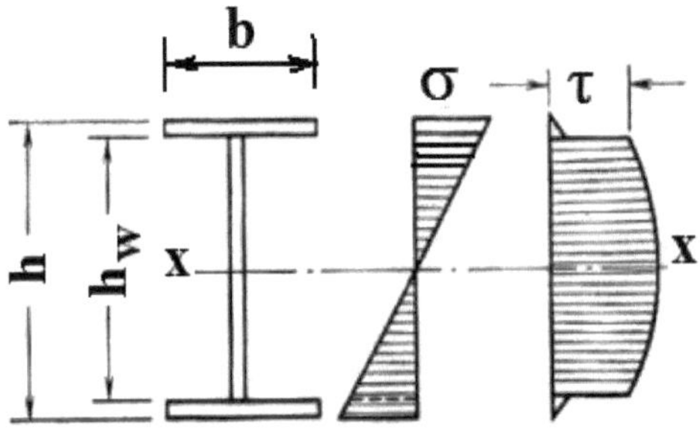

Fig. 4.36 Bending and shear stress distribution over the web of a fully symmetrical I-section

11.2.2 Bending of Sections with One Axis of Symmetry

Sections having one axis of symmetry, a principal centroidal axis will be the axis of symmetry. The other centroidal axis will be nearer to the flange having the larger area, see Fig. (4.37).

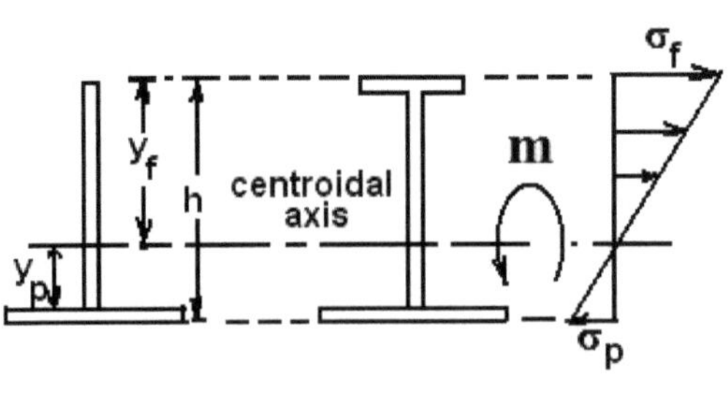

Fig. 4.37 Bending stress distribution over the depth of section

In this case the simple beam theory could be used for calculating the bending stresses. The bending stresses at the top and bottom flanges are given by:

$$\sigma_f = m.\frac{y_f}{I} \text{ and } \sigma_p = m.\frac{y_p}{I}$$

$$\sigma_f > \sigma_p$$

The section modulus at the flange and the plate are calculated as follows:

$$Z_f = I / y_f \quad and \quad Z_p = I / y_p$$

11.3 Bending of Asymmetrical Sections

Asymmetrical sections are those having no symmetry on both centroidal axes. In this case, neither of the centroidal principal axes coincides with the plane of bending, see Fig. (4.38). The bending stresses for the angle section subjected to a vertical bending moment is shown in Fig. (4.39). The bending stress at any point could be calculated using the simple beam theory by resolving the bending moment into its components acting about the principal axes of the section. The bending stress is given by:

$$\sigma = \pm\frac{M\cos\theta \cdot Y}{I_X} \pm \frac{M\sin\theta \cdot X}{I_Y}$$

The bending stress could be also calculated using the flexural characteristics of the asymmetrical section as follows:

$$\sigma = \frac{y(M_x \cdot I_y - M_y \cdot I_{yx}) + x(M_y I_x - M_x I_{yx})}{I_y I_x - I_{xy}^2}$$

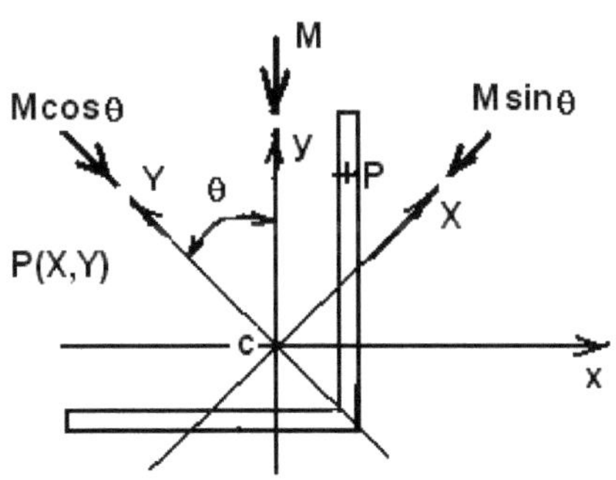

Fig. 4.38 An angle section subjected to a vertical bending moment in the vertical plane of the y-axis

Where: x, y = centroidal axes, see Fig. (4.38)
X, Y = the principal centroidal axes.
I_{xy} = product of inertia about x and y axes

$$I_{xy} = \int_0^A xydA$$

The distribution of the bending stress over an asymmetrical section is shown in Fig. (4.39).

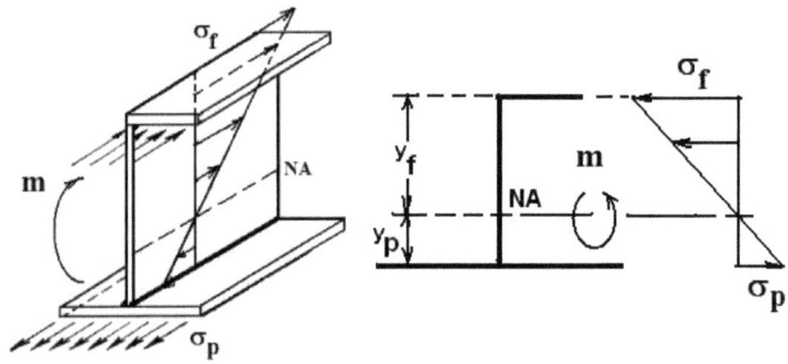

Fig. 4.39 Assumed bending stress distribution over an asymmetrical section

12 Bending Stresses in Beams Constructed with High Tensile Steel

Consider a girder of length "a" clamped at both ends and having an inverted T-section as shown in Fig. (4.40).

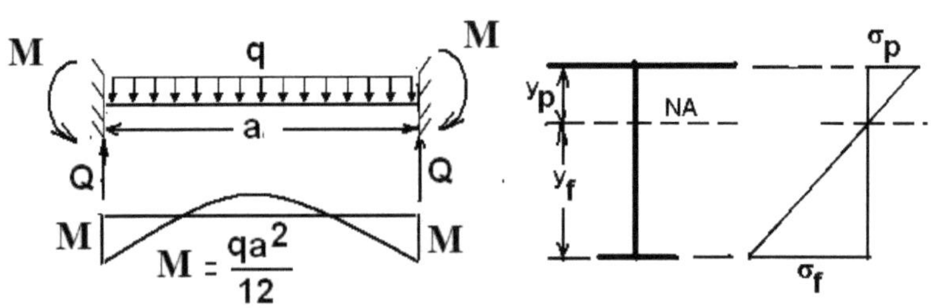

Fig. 4.40 Bending stress distribution over the section of a uniformly loaded girder

The bending moment at the clamped end of the girder is given by:

$$M = \frac{qa^2}{12}$$

Case 1: The girder is constructed of mild steel (MS)

Let σ_{Mf} = allowable bending stress at the face plate of the section

$$\sigma_{Mf} = \frac{M \cdot y_f}{I_M}$$

Hence: the section modulus of the MS girder section is given by:

$$I_M / y_{fM} = \frac{M}{\sigma_{Mf}}$$

Case 2: The girder is constructed of high tensile steel (HTS) having allowable bending stress σ_H

The section modulus at the face plate of the HTS girder is given by:

$$\frac{I_H}{y_{fH}} = \frac{M}{\sigma_{Hf}}$$

Where: H = high tensile steel

M = Mild steel

Assuming that the yield stresses for MS and HTS are given by:

$$\sigma_{y_M} = 2.4 \quad \text{t/cm}^2 \quad \sigma_{y_H} = 3.6 \quad \text{t/cm}^2$$

Hence: $\dfrac{\sigma_{y_H}}{\sigma_{y_M}} = \dfrac{3.6}{2.4} = 1.5$

$$\frac{I_H / y_{fH}}{I_M / y_{fM}} = \frac{\sigma_{Mf}}{\sigma_{Hf}} = \frac{2}{3}$$

i.e $$\left(\frac{I}{y_f}\right)_H = \frac{2}{3}\left(\frac{I}{y_f}\right)_M$$

13 Equivalent Stress

The general equation of equivalent stress for a plate element subjected to a biaxial system of stresses is given by, see fig. (4.41):

$$\sigma_{eq} = \sqrt{\sigma_x^2 + \sigma_y^2 - \sigma_x \cdot \sigma_y + 3\tau^2} \leq \sigma_y$$

The equivalent stress for an element subjected only to normal and shear stresses is given by:

$$\sigma_e = \sqrt{\sigma^2 + 3\tau^2}$$

Where: σ = normal stress
τ = shear stress

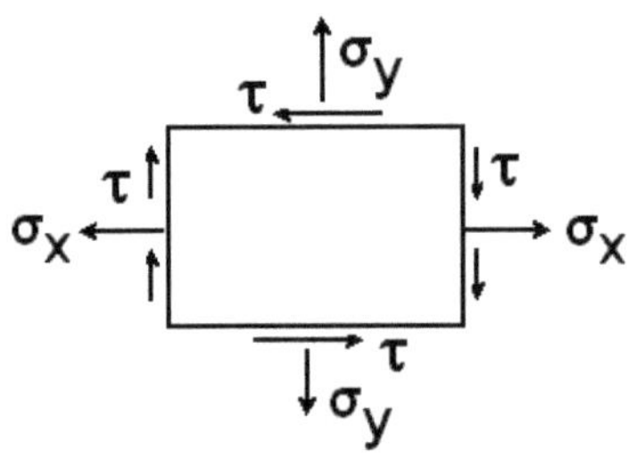

Fig. 4.41 A plate element subjected to in-plane normal and shear stresses

14 Flexural Stresses in Fabricated Asymmetrical Sections

Fabricated asymmetrical sections are widely used by shipbuilders as they are generally more economical to fabricate than symmetrical sections. Asymmetrical sections, however, are not structurally efficient because of the development of flexural warping stresses due to the torsional loading developed by the offset position of the shear center, see Fig. (4.42). The mode of deformation of a stiffened panel with asymmetrical angle sections results mainly from the offset position of the shear center C on the opposite side of the face plate, see Fig.(4.42).

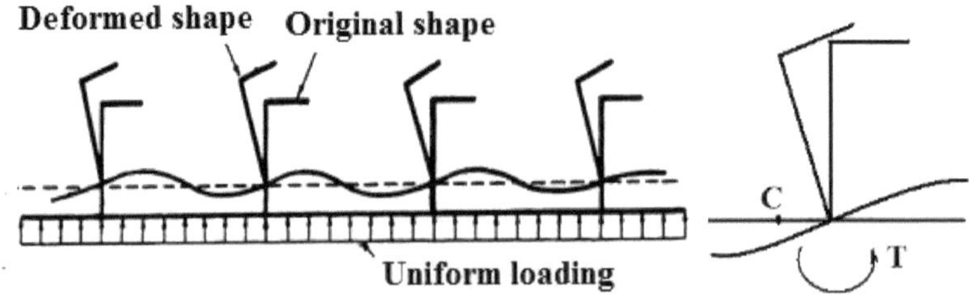

Fig. 4.42 Mode of deformation of a stiffened panel with asymmetrical angle sections

14.1 *A Simple Procedure for Calculating Flexural Warping Stresses*

For an isolated single member having an offset face plate and attached plating, the resultant of the external lateral pressure passes through point R, see fig. (4.43).

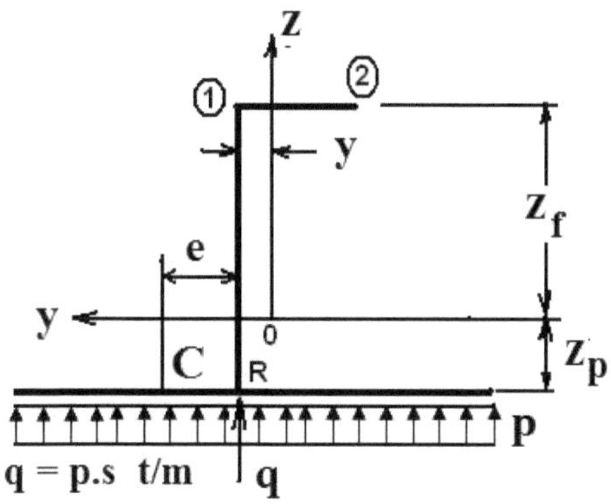

Fig. 4.43 An isolated single member having an offset face plate and attached plating

Since the shear center is at a distance "e" from R, torsional loading is created along the length of member. However, since the structural member is an integral part of a stiffened panel, the shear center should lie within the outer plating, as the latter cannot deform in its own plane. Therefore, the structural member will rotate about an enforced center of rotation C located at a distance "e" from the web plate, see Fig. (4.43). The location of C with respect to the web plate is given by:

$$e = -I_p \times I_{zy}.d / \left(I_y \times I_z - I_{zy}^2\right)$$

Because the load q does not pass through the enforced center of rotation ***C***, the torsional loading acting along the length of the member is given by, see Fig. (4.43):

$$m = q \times e \quad t \cdot m/m$$

Where: $q = p \times b \quad t/m$

p = pressure load t/m^2
b = spacing of stiffening members

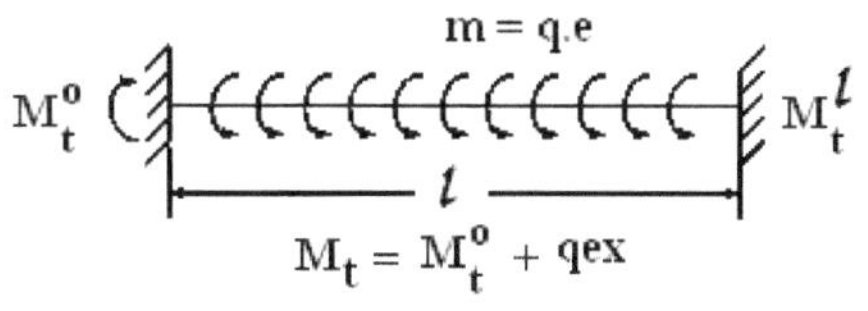

Fig. 4.44 Torque distribution along the length of member

The torque at any distance x along the length of member is given by, see Fig. (4.44):

$$M_t = M_t^0 + qex$$

Where: M_t^0 = torque at the left end support of the member.

The flexural warping stress σ_ω could be calculated using the general equation of a uniform member under a non-uniform torsion as given by:

$$M_t = C \cdot d\varphi/dx - C_1 \cdot d^3\varphi/dx^3$$

Where: $C = E \cdot J_t$ = torsional rigidity

$C_1 = E \cdot J_\omega$ Warping rigidity

For a member fully constrained at both ends against rotation and warping we have:

$$\text{At } x = 0, \text{ and } x = \ell: \ \varphi = \varphi' = 0$$

Hence:

$$\varphi = \frac{qe\ell}{kC} \cdot [\sinh(kx/2).\sinh k(\ell - x/2)/\sinh(k\ell/2) - kx \cdot (\ell - x)/2\ell]$$

And $\varphi'' = \dfrac{qe\ell}{2C} \cdot [2/\ell - k.\cosh(kx - k\ell/2)/\sinh(k\ell/2)]$

The flexural warping stresses induced by the lateral loading are given by:

$$\sigma_\omega = -E\omega \cdot d^2\varphi/dx^2$$

$$= -Eqe\ell/2C \cdot [2/\ell - k \cdot \cosh(kx - k\ell/2)/\sinh(k\ell/2)] \cdot \omega$$

Where: $k = \sqrt{\dfrac{C}{C_1}} = \sqrt{\dfrac{GJ_t}{EJ_W}}$

At the clamped ends of the member, the flexural warping stresses σ_ω attain the maximum values. Hence at points (i) and (o) on the offset face plate of the section, σ_ω are given by:

$$(\sigma_\omega)_i = -\frac{Eqe}{C} \cdot [1 - (k\ell/2)/\tanh(k\ell/2)] \cdot \omega_1$$

$$(\sigma_\omega)_o = -\frac{Eqe}{C} \cdot [1 - (k\ell/2)/\tanh(k\ell/2)] \cdot \omega_2$$

Where:

E = modulus of elasticity

φ = angle of twist,

$\omega_i = -(d \times e) + \omega_c$ = sectorial area at point i
$\omega_o = \omega_i + (b \times d)$ = sectorial area at point o
ω_c = correcting sectorial area

$$\omega_c = \frac{d}{2A_T}\left[e.A_w - 2A_f\left(\frac{b}{2} - e\right)\right]$$

$$e = -\frac{I_p.I_{yz}.d}{I_y.I_z - I_{yz}^2}$$

e = distance of shear center from the web plate of the section

$$I_p = t_p.\frac{b_e^3}{12}$$

b_e = effective breadth of the attached plating

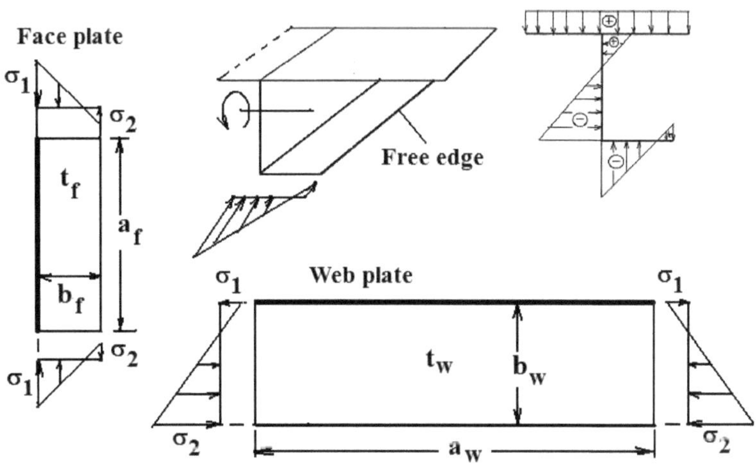

Fig. 4.45 Flexural warping stresses induced in the web and face plates of an asymmetrical section

The mechanism of development of flexural warping stresses in thin-walled asymmetrical sections is illustrated in Fig. (4.45). The distribution of the flexural stresses over the web and face plates of an asymmetrical section are shown in Fig. (4.45). The web plate is constrained along the bottom edge by virtue of the attachment with the attached plating and the top edge by the attachment with the face plate. The face plate is constrained along one edge by virtue of the attachment with the web plate and the other edge is free, see Fig. (4.45). It is evident that the simple beam theory cannot be used to predict the flexural stresses in asymmetrical sections.

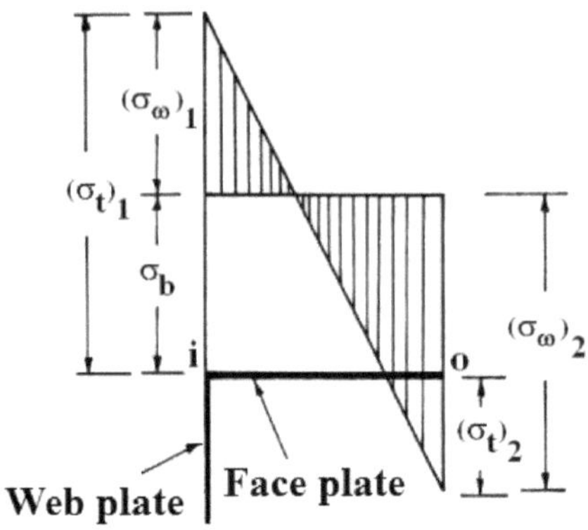

Fig. 4.46 Flexural and flexural warping stresses over an offset face plate

The total stress distribution over the asymmetrical face plate could be obtained by superimposing the flexural warping stresses over the bending stress, see Fig. (4.46).

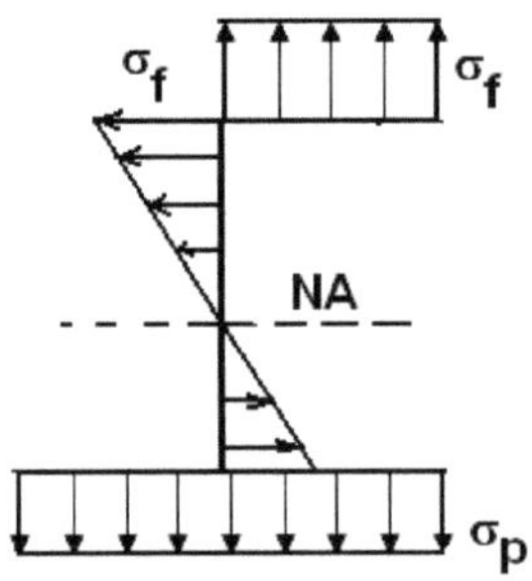

Fig. 4.47 Flexural stresses using simple beam theory

The bending stress at the flange of the section σ_f depends on the flexural properties of the section, intensity of loading and the degree of constraint at both ends of member. It could be calculated using the simple beam theory as follows, see fig. (4.47):

$$\sigma_f = \frac{M.y}{I}$$

$$M = \frac{q.L^2}{12}$$

Where: q = uniform load, tons/m
L = length of member

The total stress distribution over the asymmetrical section is shown in Fig. (4.46). It is evident from the stress distribution over an asymmetrical section that these sections are deficient in carrying lateral loading because of the high stresses developed at the junction between the web and face plate. Also the free edge of the face plate could be subjected to high in-plane stresses.

14.2 Main Parameters Affecting the Magnitude and Distribution of Flexural Warping Stresses

The main parameters affecting the magnitude and distribution of the flexural warping stresses developed in thin-walled asymmetrical sections are:

- Thickness of the web plate
- Width of face plate
- Thickness of face plate

In order to study the effect of these parameters on the magnitude of the flexural warping stresses developed over the asymmetrical face plate, consider the following data of a beam having an asymmetrical section and two values for each of the following main parameters:

- Thickness of web plate $t_w = 20$ mm and 30 mm
- Width of face plate b = 250 mm and 315 mm
- Thickness of the face plate $t_f = 25$ mm and 30 mm
- The beam is assumed clamped at both ends.

The results of calculations are given in terms of the applied uniform load q and are summarized as follows, see Table 4.3

The flexural warping stress at the inner edge of the face plate = $(\sigma_\omega)_i/q$
The flexural warping stress at the outer edge of the face plate = $(\sigma_\omega)_o/q$
The bending stress at the face plate using the simple beam theory = σ_b/q
The total stress at the inner edge of the face plate = $(\sigma_t)_i/q$.

It is clear from table 4.3 that the simple beam theory cannot be used to calculate flexural stresses for thin-walled asymmetrical sections. It is also clear that the main parameters affecting the magnitude and distribution of flexural warping stresses are the thicknesses of plating of the web and face plate.

Table 4.3 Effect of scantlings of section on σ_ω

Stress	t_w = 20 mm			t_w = 30 mm		
	b = 250		b = 315	b = 250		b = 315
	t_f = 25	t_f = 30	t_f = 30	t_f = 25	t_f = 30	t_f = 30
$(\sigma_\omega)i/q$	1.082	1.125	1.120	0.608	0.650	0.683
$(\sigma_\omega)o/q$	-3.615	-3.375	-3.305	-2.865	-2.720	-2.505
σ_h/q	2.085	1.893	1.842	1.740	1.608	1.437
$(\sigma_t)i/q$	3.167	3.018	2.962	2.348	2.258	2.120

The results given in Table 4.3 show that increasing the web thickness has a significant effect on the magnitude of the flexural warping stresses whereas increasing the thickness of face plate has a much less effect. Also, increasing the width of face plate has a negligible effect.

Ex. Consider the asymmetrical section beam shown in Fig. (4.48). The beam is uniformly loaded and is idealized by plate and beam elements as shown in Fig. (4.49).

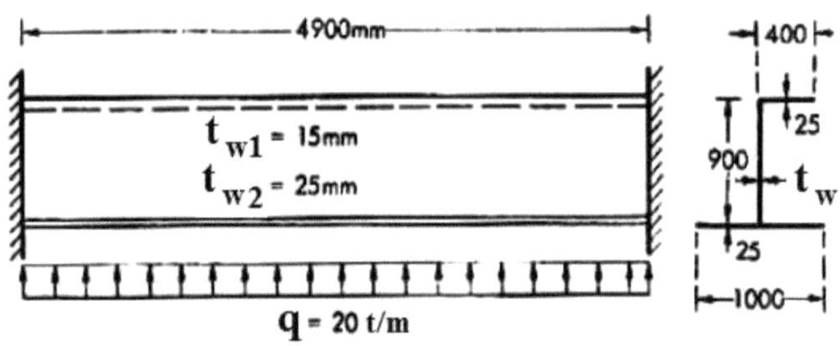

Fig. 4.48 Geometry and scantlings of an offset section beam

The beam is idealized using beam and plate elements, see Fig. (4.49). The analysis is carried out using the FEM for two values of the thickness of the web plate and the results are compared with the results using the given analytical method. Fig. (4.50) show the variation of the flexural stresses for the assumed two values of the thickness of the web plating.

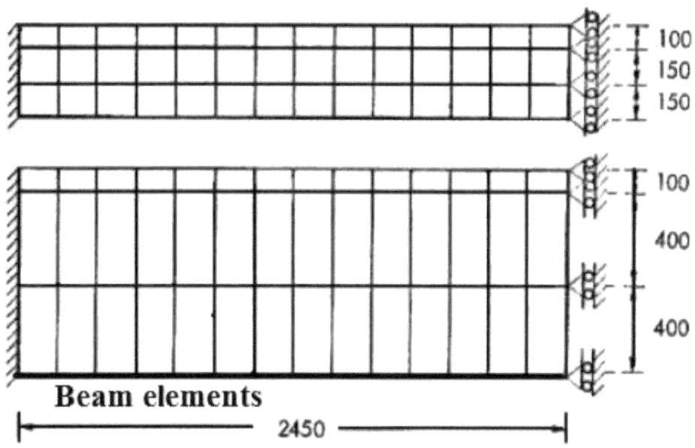

Fig. 4.49 Idealization of the asymmetrical beam using beam and plate elements

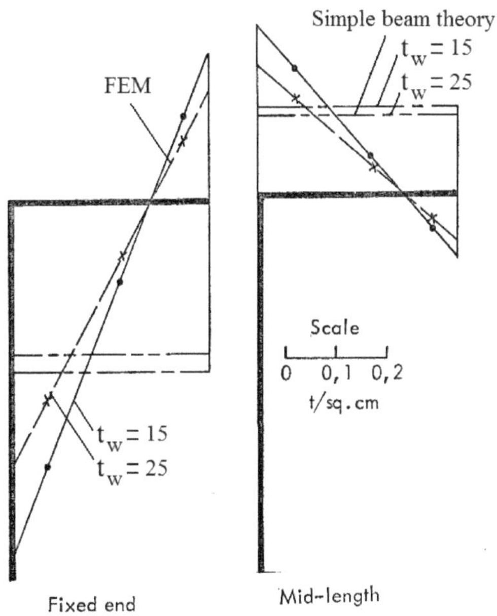

Fig. 4.50 Flexural bending and warping stresses over an offset face plate

15 Effective Breadth Concept

The concept of the effective breadth is clearly illustrated by Fig. (4.51). The analysis of structures using 2D and 3D idealization could be greatly simplified by using the concept of effective breadth for idealizing the face plates and attached plating of structural members.

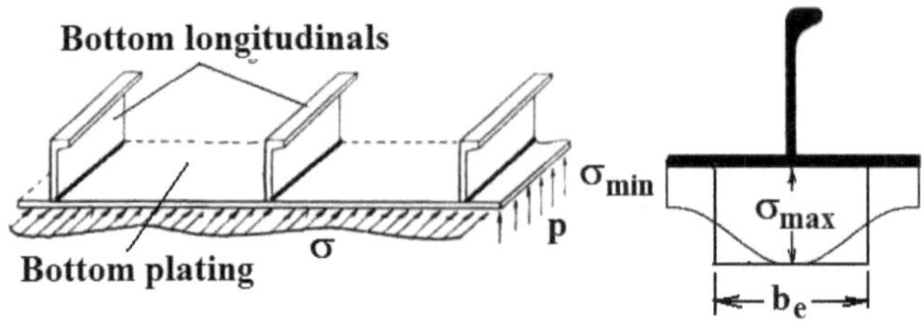

Fig. 4.51 The basic concept of the effective breadth

The effective breadth concept is considered for the following cases:

- Symmetrical sections
- Asymmetrical sections (Offset face plates)
- Curved sections
  - Symmetrical face plates
  - Offset face plates

15.1 Effective Breadth of Uniform Symmetrical Sections

The fall off of normal stresses across the width of the symmetrical face plate and attached plating result from shear lag effect, see Fig. (4.52).

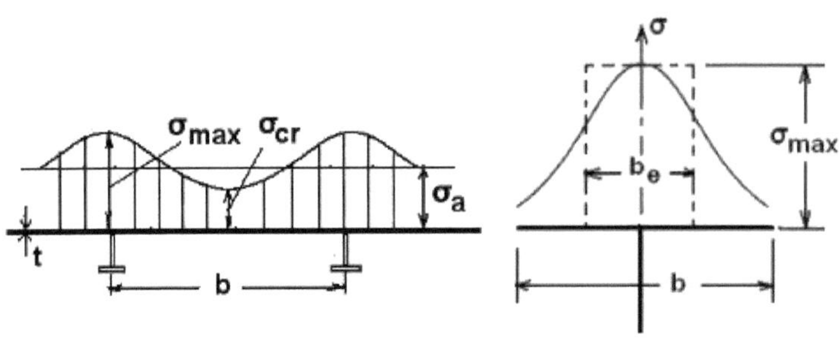

Fig. 4.52 Basic elements of the concept of effective breadth

The effective breadth b_e of the face plates and attached plating of girders depends on the span of the beam or girder and the width of the attached plating and could be determined using Fig. (4.53) or Table 4.4.

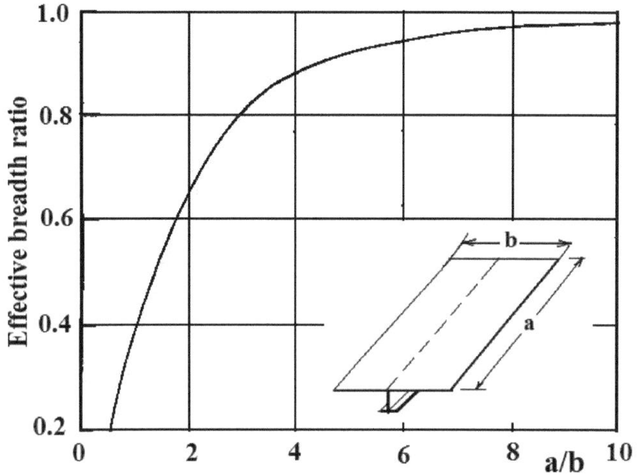

Fig. 4.53 Effective breadth ratio (uniform loading

Table 4.4 Effective breadth ratio of the face plates and attached plating

a/b	0	1	2	3	4	5	6	7
b_e/b	0	0.36	0.64	0.82	0.91	0.96	0.98	1

Where:

b = width of plating, measured from center to center

a = length between zero-points of bending moment curve, i.e. unsupported span in case of simply supported girders

a = 0.6 x unsupported span for clamped ends of girder

b_e = the effective breadth of plating

λ = effective breadth ratio

15.2 Effective Flexural Properties of Sections

Using the concept of effective breadth, the distribution of the flexural stresses over the modified geometry of the section is shown in Fig. (4.54). Because of the fall-off of normal stresses across the face plate and attached plating, the flexural properties of uniform symmetrical sections are greatly reduced.

The change in position of the neutral axis, geometry and flexural properties of a T-section is shown in Fig. (4.54). The geometry and flexural properties of the effective section are as follows, see Fig. (4.55):

a_o, I_o and Z_o are the original geometrical and flexural properties of the section

a_e, I_e and Z_e are the effective geometrical and flexural properties of the section

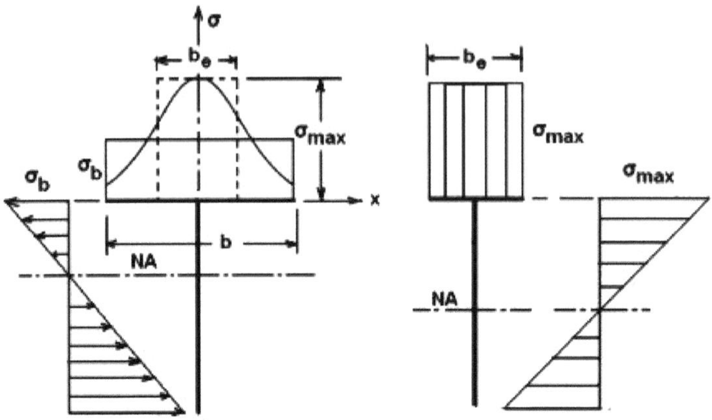

Fig. 4.54 Modified stress distribution over the effective section

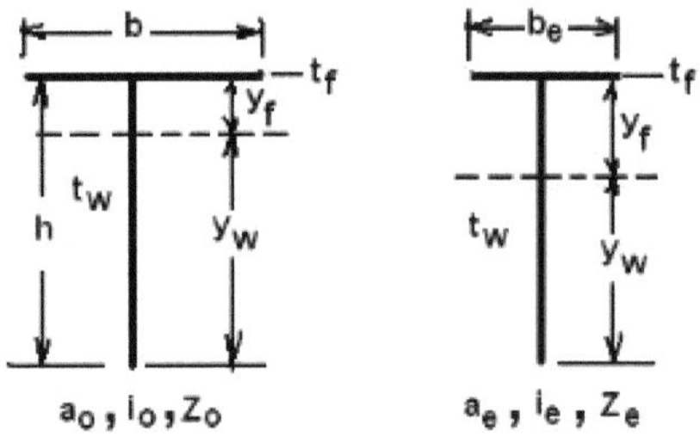

Fig. 4.55 Effective section of girder

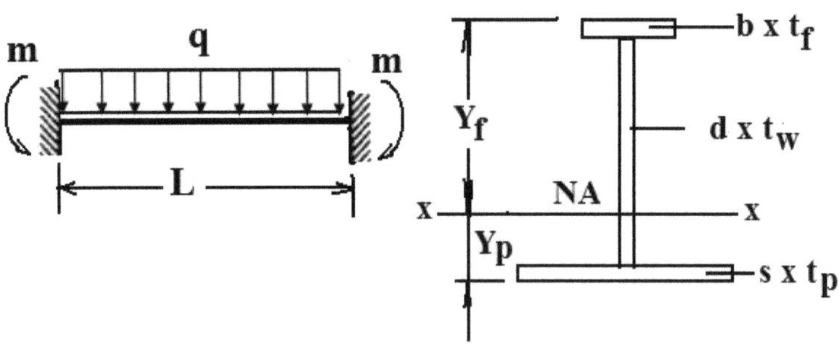

Fig. 4.56 T-section beam with attached plating

Example 3.1 Determine the geometrical and flexural properties of the T-section beam with attached plating shown in Fig. 4.56. The beam is assumed clamped at both ends and subjected to uniform loading.

Given:
Length of beam, L = 2.74 m
Spacing of beam, s = 3.05 m
Uniform Pressure, p = 107 kN/m^2
Thickness of Plate, t_p = 15.88 mm
Web depth, d = 39.7 cm
Web thickness, t_w = 9.65 mm
Flange width, $b_£$ = 18.0 cm
Flange thickness t_f = 16.0 mm
Yield stress, σ_y = 2.35 kN/cm^2
Allowable stress, σ_b = 1.95 kN/cm^2
Proportional Limit, σ_p = 1.76 kN/cm^2
Young's Modulus, E = 2068.50 kN/cm^2
Poisson's' ratio, v = 0.3
Effective breadth of plating = b_e
Shear lag approach: b_e = L/4 = 2.74/4 = 0.685 m
Post buckling approach b_e = 60t = 60 x 15.88 x 10^{-3} = 0.953 m
Spacing of beam, s = 3.05 m
The effective breadth is the least value of the above 3 values.
Therefore: b_e = 0.685 m

Calculation of the geometrical properties of the T-section
ΣA = 175.89 cm^2
$\Sigma A.y$ = 2141.232 cm3
Yp = $\Sigma A.y/\Sigma A$ = 12.174 cm
Y_f = 30.714 cm
$I_x = \Sigma A.y^2 + \Sigma I_o$ = 74.2119605 cm4
I_{NA} = 48.1446021 cm^4
$Z_p = I_{NA}/y_p$ = 3.954707 cm^3
$Z_f = I_{NA}/y_f$ = 1.567513 cm^3
Shear area A_{sh} = 41.39 cm^2
Radius of gyration r =√481446.021/17589 =16.54 cm
Loading and stresses
Local load q = b.p = 3.05 x 107 = 326.25 kN/m
Bending moment at end support $M_1 = qL^2/12$ = 0.2041 kNm
Bending moment at midspan M_2 =qL2/24 = 0.10205 kNm
Shear force F = 0.447 kN

Bending stresses at the end supports
Bending stress at the face plate = $\sigma_f = M_1/Z_f$ = 13.03 kN/cm^2
Bending stress at the attached plate = 5.16 kN/cm^2

Bending stresses at midspan
Bending stress at the face plate = $\sigma_f = M_2/Z_f$ = 6.51 kN/cm^2
Bending stress at the attached plate = 2.58 kN/cm^2

Ultimate strength of plating F_u

$F_u = \sigma_y$ for $\beta < 1.25$

$F_u = \sigma_y\,(2.25/\beta - 1.25/\beta^2)$ for $\beta > 1.25$

Where $\beta = b/t\,\sqrt{\sigma_y/E} = 3050/15.88\sqrt{2.35/2068.5} = 6.47$

Since $\beta > 1.25$, $F_u = 7.471$ kN/cm^2

15.3 Effective Breadth of Asymmetrical Face Plates

In the elastic stress analysis of ship structures using the finite element method (FEM), the face plates of fabricated girders, transverses, brackets, etc., are normally idealized by bar or beam elements. For narrow symmetrical face plates, the sectional properties of the idealized member could be computed directly from the scantlings of the symmetrical face plate, assuming that it is fully effective. On the other hand, when dealing with offset or curved face plates attached to curved brackets, the sectional properties of the idealized member should be based on its actual contribution to the load carrying capacity of these face plates. This could be achieved using the concept of "effective breadth". Using the full area of the faceplate for calculating the sectional properties of the idealized member could lead to misleading values of stresses, particularly in the curved regions of the structure assembly. The deficiency of the face plates of asymmetrical sections could be easily explained by considering the face plate as a long narrow plate loaded by shear forces along one edge while the other edge is free, see fig.(4.57).

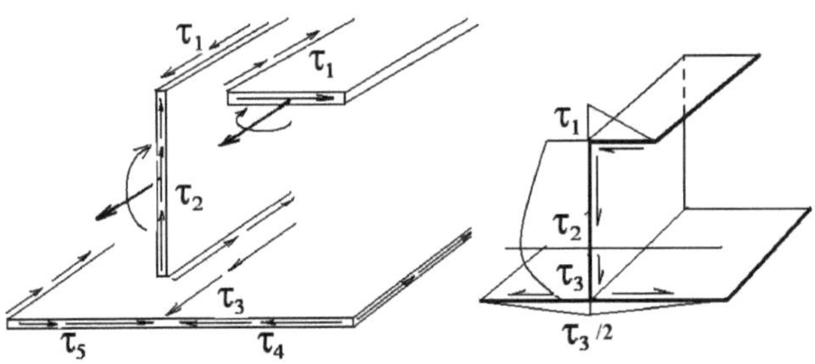

Fig. 4.57 Main stress components of asymmetrical sections fitted with offset face plate

The normal stress over the section of the face plate is given by:

$$\sigma_N = \frac{N}{b.t}$$

And the bending stress is given by, see Fig. (4.58):

$$\sigma_M = \frac{M \cdot b/2}{I} = \frac{M.b}{2 \cdot tb^3/12} = \frac{6M}{tb^2}$$

$$= \frac{6}{tb^2} \cdot N \cdot \frac{b}{2}$$

The maximum stress at the loaded edge is therefore given by:

$$\sigma_{\max} = (\sigma_t)_i = \sigma_N + \sigma_M = \frac{N}{bt} + \frac{3N}{bt} = \frac{4N}{bt}$$

Hence: $N = \sigma_N bt = b_e \cdot t \cdot \sigma_{\max} = b_e \cdot \frac{4N}{b}$

Thus: $b_e = \frac{b}{4}$

This means that the effective breadth of offset face plates is about 25% of the width of the face plate. The change in the position of the neutral axis and geometry of the section is shown in Fig. (4.59).

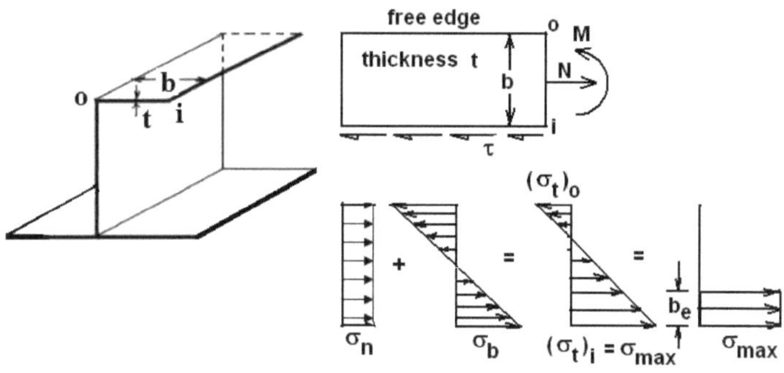

Fig. 4.58 Normal and flexural stresses over the offset face plate

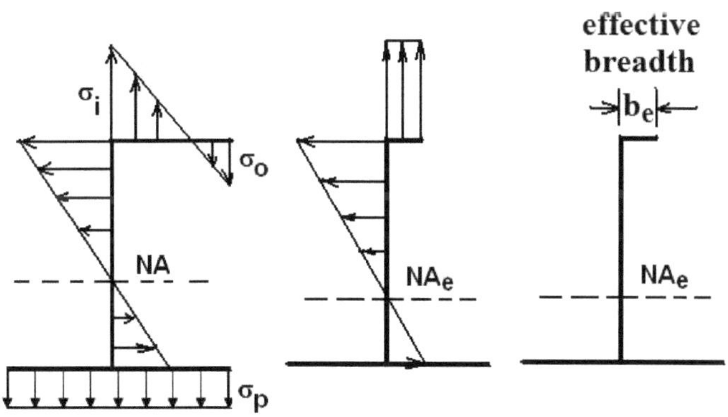

Fig. 4.59 Effective breadth of the face plate

15.4 Effective Breadth of Curved Face Plates

15.4.1 Symmetrical Face Plates

The circumferential stresses in a symmetrical face plate induce radial forces which cause the outer edges of the face plate to deflect radially. This effect is sometimes called "cross bending", see Fig. (4. 60). Cross bending has an adverse effect on the effectiveness of these curved face plates. As a result, high stresses may be induced in both the curved plate and the web near the face plate, see Fig. (4.60). The effectiveness of curved symmetrical face plates depends on:

- The width of the curved face plate
- Thickness of the face plate
- Radius of curvature of the face plate

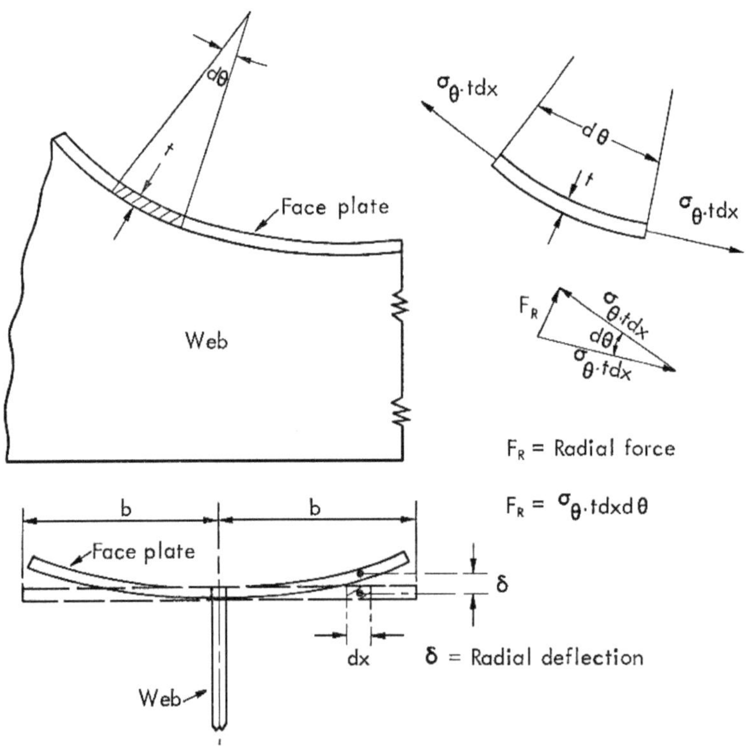

Fig. 4.60 Cross bending of symmetrical face plates and the fall-off of bending stress over the width of the face plate

The main parameter combining these factors is b/√ Rt. The variation of the of effective breadth ratio with the main parameters of the curved face plate is shown in fig. (4.61).

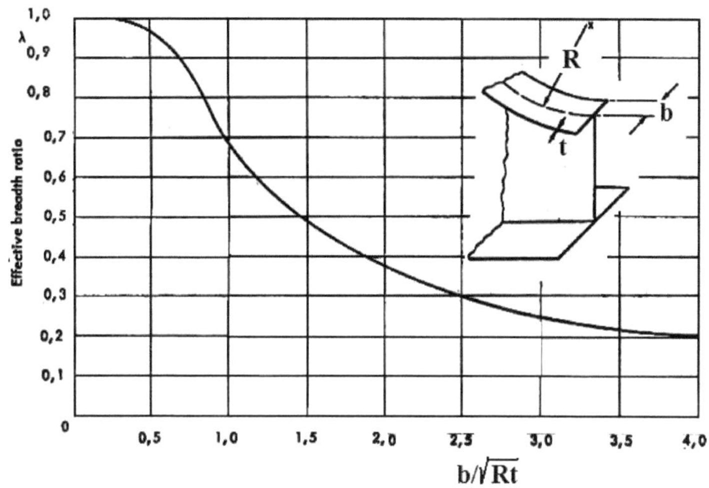

Fig. 4.61 Effective breadth of curved symmetrical face plates

15.4.2 Asymmetrical Face Plate

Offset curved face plates are less efficient than symmetrical curved face plates. This could be realized by the pattern of stress distribution at different locations over the structural member of the asymmetrical face plate shown in fig. (4.62).

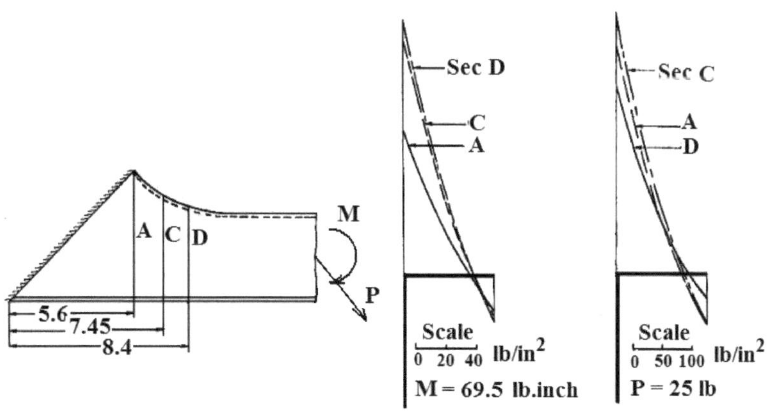

Fig. 4.62 Stress distribution over an asymmetrical face plate at different locations over the structural member

The variation of the of effective breadth ratio with the main parameters of the asymmetrical curved face plate is shown in fig. (4.63).

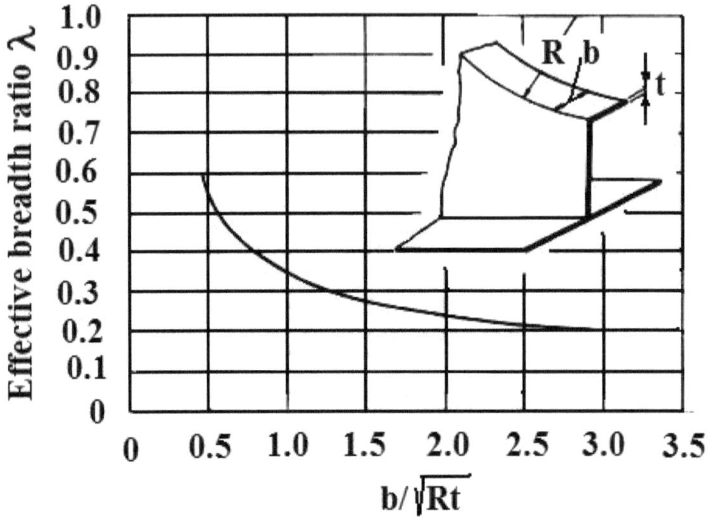

Fig. 4.63 Effective breadth of face plate of curved asymmetrical face plate

Part II
Chapter 5 – Chapter 9

Chapter 5
Hull Girder Loading

1 Introduction

Ship hull girder loads consist of static and dynamic components. The most significant of these components are the still-water bending moments and shear forces. The still-water component results from the difference between the distributions of the various weight items and the distribution of the supporting buoyancy forces along the ship length. The weight items include light weight of the ship, cargo weight and weight of consumables. The dynamic loads include wave induced hydrodynamic loads, sloshing, slamming, inertia loads due to vessel motion and impact loads. The dynamic wave-induced loads include the vertical and horizontal shear forces, bending moments and torsional moments. Wave loads are normally evaluated by means of one of the two methods: design wave method or spectral analysis method. The latter method is based on either short or long term predictions, transfer functions, and the assumption that loads are linearly dependent on wave height. Depending on the loading condition of the vessel, the still-water bending moments and shear forces are determined using the published rules of classification societies. A more detailed analysis is required when determining the dynamic aspects of the hull girder loads for extreme sea conditions that the vessel is bound to encounter over its lifetime.

The loadings on ship hull girder are not deterministic quantities but are subjected to several sources of uncertainties and should be treated statistically. The design load is the maximum load likely to occur over a specified period of time. Wave Loads are generally predicted either over short periods (for a particular sea condition) or over long periods (taking account of all sea conditions).

2 The Nature of Hull Girder Loads

A ship floating at rest in still-water is subject to gravitational forces and hydrostatic pressures over the immersed volume of the hull. The intensity of this hydrostatic pressure varies directly with the depth of immersion below the water surface. When the ship is subjected to a wave system, the pressures surrounding the hull are modified due to the hydrodynamic effects of water particle movement and the simple hydrostatic law no longer applies. This complicated loading system comprising pressures and mass acceleration forces can be represented more simply by considering their overall effect on the hull. These effects can be defined in terms of various load parameters such as longitudinal bending moments, either about a horizontal transverse axis or about a vertical axis through the ship's centerline. Other loads include the vertical or horizontal shearing forces at a transverse section and the external torque moment about a fore and aft axis.

For offshore structures, the design load refers normally to the maximum value likely to occur over the expected service life of the structure and is normally based on a recurrence period of 50-100 years. Wind loads are evaluated by assuming sustained and gusty wind speeds.

3 Classification of Hull Girder Loads

Hull girder loads could be categorized as follows:

- Hull girder bending moment
- Hull girder shear loading
- Hull girder torsion loading
- Local loadings

The main components of hull girder bending moment are:

- Stillwater bending moment (M_s)
- Wave induced bending moment (M_w)
- Dynamic bending moment (whipping, slamming, springing)

The still-water loads contribute an important part of the total shear and bending moment in most ships, to which wave-induced effects must be added.

4 Hull Girder Longitudinal Vertical Bending Moments

4.1 Stillwater Shear Force and Bending Moment

When dealing with the longitudinal strength of a ship, the ship hull girder may be treated as a non-uniform beam subjected to varying load throughout its length. This load is obtained from the difference between the non-uniform distribution of weight of the structure and its contents, which are acting downwards, and the non-uniform distribution of the supporting buoyancy forces which is acting upwards, see Fig. (5.1).

There is a balance between the total downward forces of weight and mass acceleration, and the total upward forces of buoyancy and hydrodynamic reaction. For equilibrium conditions, the lines of action of these two forces should coincide. Because of the non-uniformity of the upward and downward forces, shear and bending moments are created in the same manner as in ordinary beams, see Fig. (5.1). The still water loads of a ship experience little variations during a voyage.

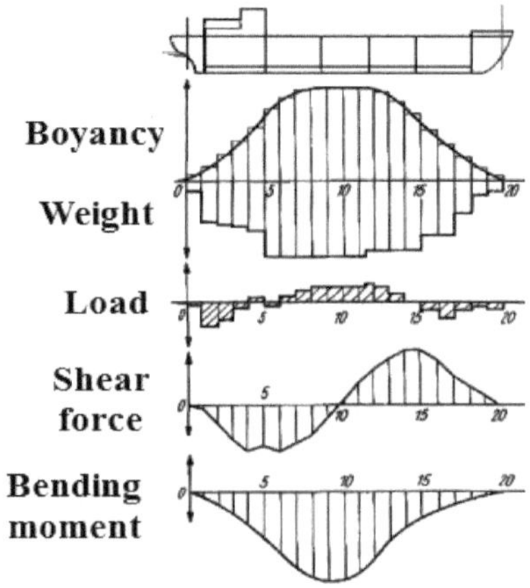

Fig. 5.1 Load, shear force and bending moment distribution along ship length

These variations result mainly from the changes in the distribution of consumables on board. The Stillwater loads vary significantly only during loading and unloading operations. The still water bending moment and shear forces are obtained from the buoyancy and weight distributions along ship length, see Fig. (5.2).

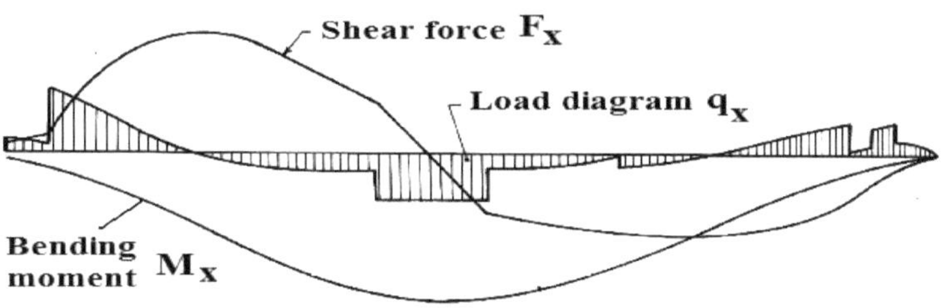

Fig. 5.2 Load, shear force and bending moment distribution along ship length

Assuming the weight and buoyancy distributions are w_x and b_x, then the load distribution is given by:

$$q_x = b_x - w_x$$

The distributions of the shear force and bending moment along ship length are given by:

$$F_x = \int_0^x (b_x - w_x)dx$$

$$M_x = \int_0^x F_x dx$$

Because the distribution of load along ship length is never a continuous function, the above integrations are carried out using numerical methods.

4.2 Wave-Induced Component

The wave induced shear force and bending moment components result from the distribution of the forces of support throughout the length of a ship during her passage among waves, see Fig. (5.3). The operational factors influencing the magnitude and distribution of the wave induced bending moment are the characteristics of sea waves, draft, displacement, trim, ship speed and heading to the waves,. The wave bending moments and shear forces are determined assuming the vessel either poised on a wave of length equal to ship length with a crest at amidships, see Fig. (5.4), or with trough at amidships, see Fig. (5.5).

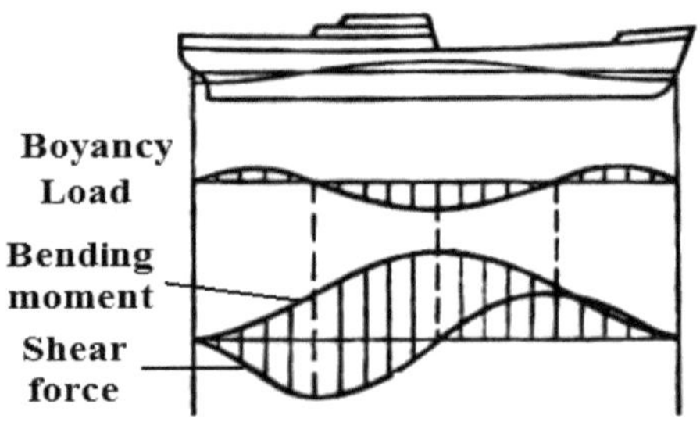

Fig. 5.3 Wave induced bending moment and shear force

Fig. 5.4 The deformed hull shape of a ship in a hogging condition

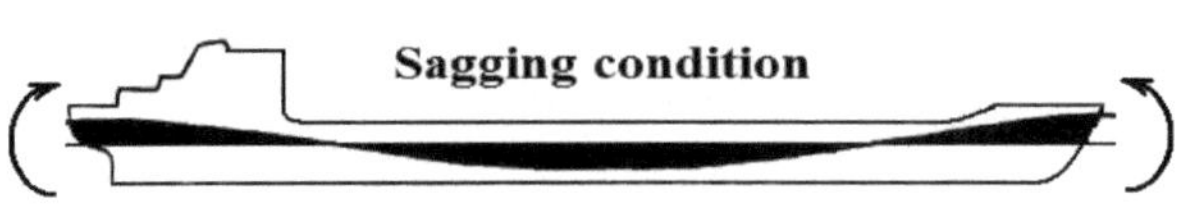

Fig. 5.5 The deformed hull shape a ship in a sagging condition

5 Effect of Hull Girder Vertical Deflection on the Distribution of Shear Force and Bending Moment along Ship Length

The general deflection theory of beams is given in detail in several text books. The total deflection curve of a floating free-free beam of variable cross-section under the action an arbitrary loading system is composed of two parts, bending deflection and shear deflection, see Fig. (5.6). The bending deflection is calculated from the following general differential equation:

$$\frac{d^2}{dx^2}\left(EI_x \cdot \frac{d^2 w}{dx^2}\right) = q_x \tag{1}$$

The shear deflection is calculated from the following equation:

$$w_s = \int_0^x \frac{\lambda F}{GA} dx \tag{2}$$

Where:

$q_x = f(x) - p(x)$

f(x) = downward forces, see Fig. (5.6)

p(x) = upward supporting forces

λ = a constant depending on the geometry of the cross-section and is given by:

$$\lambda = \frac{1}{AF^2} \int_0^A \tau^2 dA$$

Where: A = shear area of the section
F = shear force
τ = shear stress

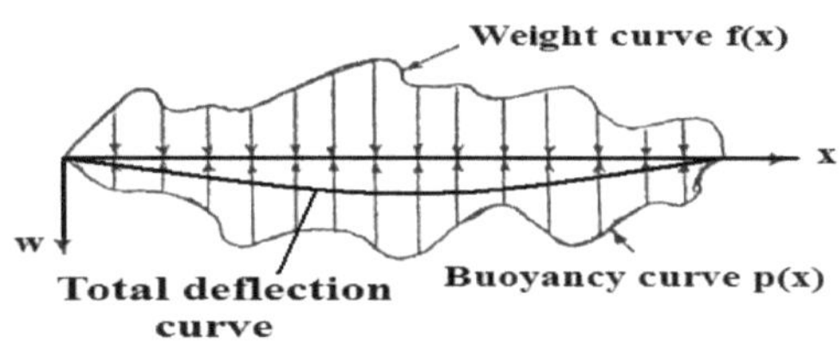

Fig. 5.6 Arbitrary buoyancy and weight curves

The total curvature of the elastic line at any position x along the ship length is therefore given by:

$$\left(\frac{d^2 w}{dx^2}\right)_t = \left(\frac{d^2 w}{dx^2}\right)_b + \left(\frac{d^2 w}{dx^2}\right)_s \tag{3}$$

$$\text{Thus: } \left(\frac{d^2 w}{dx^2}\right)_t = M'_x / EI_x \tag{4}$$

Where: M'_x = bending moment corrected for shear force

$$M'_x = M_x + \alpha EI_x . q_x \tag{5}$$

$$M_x = \int_0^x \int_0^x q_x . dx^2 \tag{6}$$

Where: $\alpha = \lambda / AG$

The equation to the elastic line is therefore given by:

$$(w_t)_x = \int_0^x \int_0^x \frac{M'_x}{EI_x} . dx^2 + \theta_o x + w_o \tag{7}$$

Where θ_o and w_o are arbitrary constants depending on the chosen reference line and the end conditions, see Fig. (5.7).

Assuming the line joining the two end points of the elastic line is the reference line, then we have:

$$(w_t)_o = (w_t)_L = 0$$

The equation to the elastic line is therefore given by:

$$(w_t)_x = \frac{1}{E}\left[\int_0^x \int_0^x \frac{M'_x}{I_x} . dx^2 - \frac{x}{L} \int_0^L \int_0^L \frac{M'_x}{I_x} . dx^2 \right] \tag{8}$$

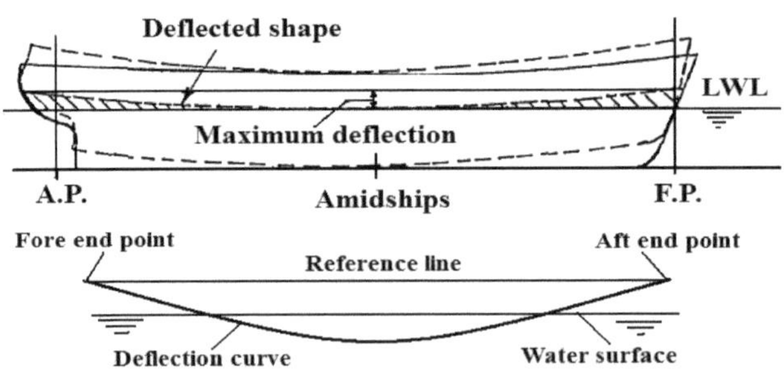

Fig. 5.7 Deflected shape of ship hull girder in a sagging condition

On the other hand, if the stillwater surface is assumed to be the reference line, see Fig. (5.7), the two constants θ_o and w_o are determined from the following equilibrium conditions:

$$i - \int (\Delta q) dx = 0 \qquad (a)$$

$$ii - \int(\Delta q)x.dx = 0 \quad (b)$$

Where Δq = change in load due to hull girder deflection

$$= \gamma.y_x.(w_t)_x$$

y_x = breadth of the ship at the waterline and is assumed to be constant over the ship deflection range

Substituting $(w_t)_x$ from equation (7) into (a) and (b), we get:

$$\int_0^L \gamma.y_x\left[\frac{1}{E.}\int_0^x\int_0^x \frac{M_x'}{I_x}.dx^2 + \theta_o.x + w_o\right]dx = 0 \tag{9}$$

$$\int_0^L \gamma.y_x\left[\frac{1}{E.}\int_0^x\int_0^x \frac{M_x'}{I_x}.dx^2 + \theta_o.x + w_o\right].x.dx = 0 \tag{10}$$

Solving equations (9) and (10) for θ_o and w_o and then substituting in equation (7), the deflection curve of the elastic line of the hull girder could be determined with reference to the stillwater surface.

5.1 Shear Force and Bending Moment Correction due to Ship Deflection

The deflection of a hull girder alters the assumed distribution of buoyancy along ship length. This variation in buoyancy affects the magnitude and distribution of the shearing force and bending moment as well as the shape of the deflection curve.

The calculation of the corrected deflection curve as well as the distribution of shearing force and bending moment could be carried out by an iterative process. The mathematical procedure is given by:

$$q(x) = q_x + \sum_{i=1}^{n} \delta q_{xi}$$

$$F(x) = F_x + \sum_{i=1}^{n} \delta F_{xi}$$

$$M(x) = M_x + \sum_{i=1}^{n} \delta M_{xi}$$

$$w_t(x) = (w_t)_x + \sum_{i=1}^{n} \delta(w_t)_{xi}$$

Where: i = 1, 2, .is the number of iterations

$$\delta q_x = \gamma.y_x.(w_t)_x$$

$$\delta F_x = \int_0^x \delta q_x .dx$$

$$\delta M_x = \int_0^x \int_0^x \delta q_x .dx^2$$

$$\delta (w_t)_x = \int_0^x \int_0^x \frac{\delta M_x}{EI_x} .dx^2 + \theta_o x + w_o$$

6 Hydrodynamic Loads

The hydrodynamic loadings are the periodic loads generated by the wave actions in particular situations (springing) or by mechanical excitation (main engine/propeller). Included in the same category are the transient impulsive loads that excite free structural vibrations. The cargo distribution along ship length is an important factor affecting the dynamic component as it determines the moment of inertia and the natural frequency of pitching and heaving. The most severe response depends on the actual distribution of weights along ship length and the sea condition.

Ships navigating in rough seas experience various types of impacts from the waves which induce elastic vibration throughout the hull. The impact loadings on the external surface of a ship involves impulsive pressures induced by the hydrodynamic loadings acting over the outside ship surface during very short time periods relative to the natural response rates of the ship. The dynamic response of a ship structure to extreme hydrodynamic loadings is a highly transient and non-linear process and therefore cannot be modeled as a quasi-static phenomenon. These hydrodynamic loadings are associated with the short time exchange of momentum between the ship and the sea water. Such change of momentum can only take place in the vicinity of the free surface of the sea water. The terms "slamming" and "pounding" have been associated with wave impact. The term pounding has been used to mean a wave slap resulting in a shuddering load which is abrupt but not severe. Slamming has been used to mean an impact at the ship bow, an impact on the bow flare, an impact on the bottom of a displacement vessel.

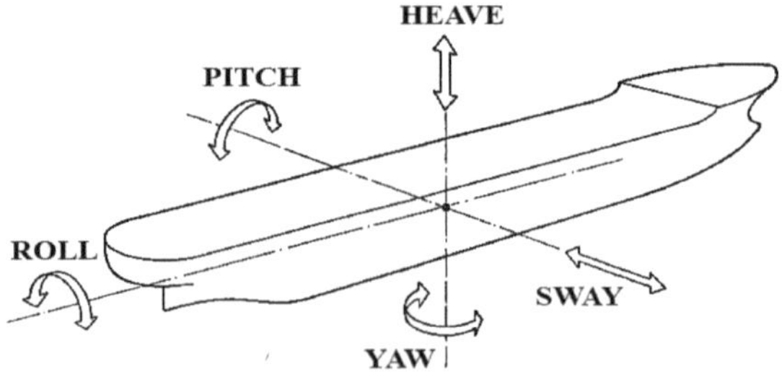

Fig. 5.8 Types of ship motions

The term slamming is generally used to describe forward bottom impact. When a vessel experiences large pitch and heave motions during heavy sea states, see Fig. (5.8), the forefoot of a vessel can rise above the water surface, see Fig. (5.9). As the vessel re-enters the water, large impulse pressures are imparted to the hull structure due to the relative motions of the sea water and the ship. The hull slams into the water surface and the vessel experiences heavy impulse pressures to the local forefoot structure and subsequent whipping forces to the entire hull structure. It is these large impulse pressures and whipping forces that cause high frequency stresses and extensive local damage to ship's structure.

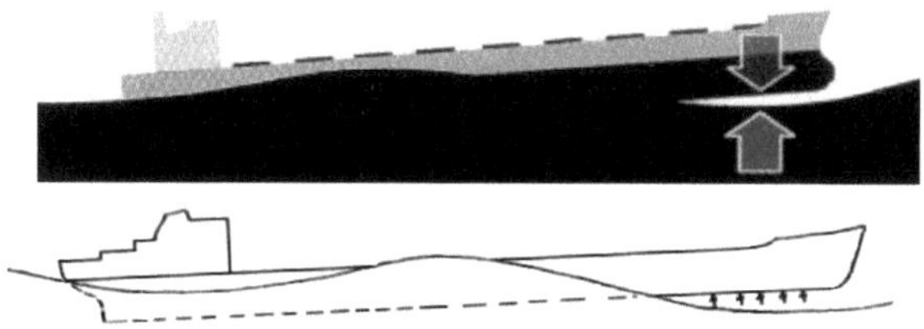

Fig. 5.9 Bottom slamming

The duration of slamming pressure is in the order of milliseconds and is generally of local nature. The variation of wave induced and slam induced bending moment with time is shown in Fig. (5.10).

It is worth noting here that the term flare slamming describes the impact forces applied to the bow flare of a vessel. As a result of large ship motions, an impact force is generated on the bow flare as it enters an oncoming wave system. This impact produces not only high forces, but also high frequency vibrations associated with bottom slamming.

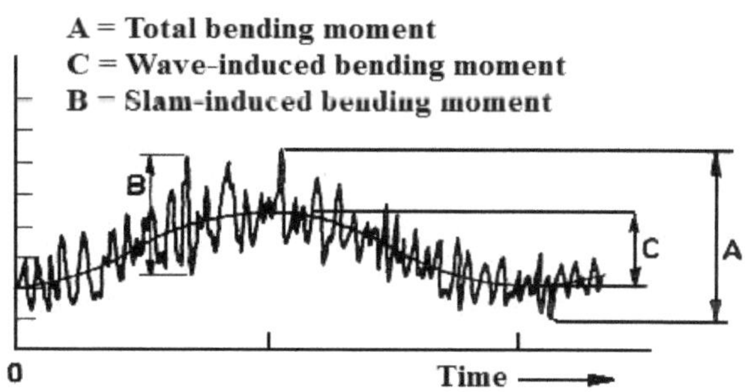

Fig. 5.10 Increase in range of wave bending stresses due to slam-induced stresses

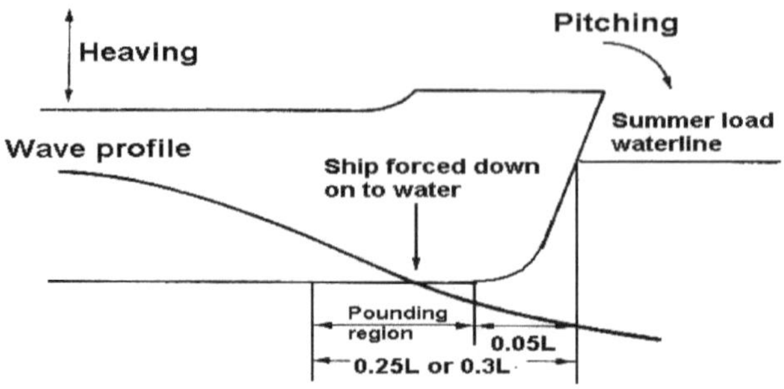

Fig. 5.11 Specified length of bottom forward to be strengthened to sustain the excessive hydrodynamic pressure due to bottom pounding

Some major differences exist between bow flare slamming and bottom slamming. The speed of impact is slower with flare slamming than with bottom slamming. The impulse pressure is spread across the rapidly increasing area of the bow flare causing larger total forces on the structure. The forefoot emergence is a characteristic of bottom slamming while it is not for flare slamming. The duration of the impact force is relatively long for the flare impact as compared to that for forward bottom impact.

Classification society rules require that when the forward draft is less than a specified percentage of ship length, a certain length of the bottom structure forward should be strengthened to sustain the severe hydrodynamic pressure induced by bottom pounding, see Fig. (5.11). The damage sustained by a vessel due to hydrodynamic impact loading can manifest itself in many forms, from deformed shell plating, to distorted and buckled longitudinals and frames, and to fatigue cracking in local structure connections. Damage is generally sustained by the vessel's tertiary structure at the location of impact. The secondary structure can be damaged by the direct action of impulse forces or by the high frequency whipping forces that accompany hydrodynamic impact. The primary structure is usually only affected by the whipping forces.

6.1 Dynamic Loadings due to Shipping Green Seas

Under severe weather conditions and excessive pitching and heaving particularly in the fully loaded conditions of the ship, the bow of a vessel can plunge into the water surface of an oncoming wave system causing the water to break over the bulwark and onto the deck of the vessel, see Fig. (5.12). This could cause large pressures to be applied and may cause local damage to the deck structure.

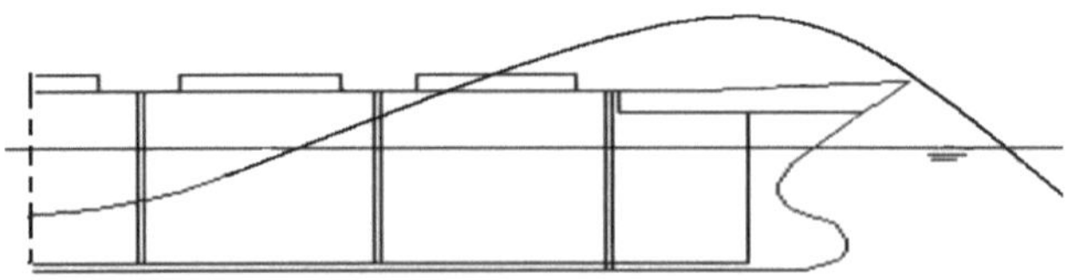

Fig. 5.12 Shipping of green seas on the fore deck of a ship

7 Hull Girder Dynamic Shear Force and Bending Moment

The hull girder dynamic loadings could be either impulsive, resulting mainly from ship motions among waves, slamming, or non-impulsive, resulting from the vibration of the hull girder. The magnitude of this component is influenced by ship's speed, heading, length, draft, trim, sea condition, and weight distribution. These factors affect the amplitudes of pitching and heaving, which affect the magnitude of the dynamic component. The impulsive component increases directly as the area under the bow that appears above the water surface increases. Therefore, for ships having a full form in the bow and running at a shallow draft in a severe sea condition, the impulsive component may be rather significant. This is the case of a bulk carrier, or an oil tanker, in the ballast condition.

An approximate estimate of the maximum impulsive dynamic shear force component could be obtained from the dynamic bending moment component. It is shown that the maximum impulsive dynamic shearing force component could be estimated by:

$$\left(F_{Dyn}\right)_{\max} \cong 3\left(M_{dyn}\right)_{\max}/L$$

Where: L = ship length

The maximum dynamic bending moment is given by:

$$\left(\mathrm{M_{Dyn}}\right)_{\max} = \mathrm{K.B^2.L.h/\lambda}$$

Where:

$$K = 0.2 \times [a + b \cdot (C_b - 0.6) +]C \cdot T/0.08$$

Where: A, b and c are coefficients depending on d/L ratio and Froude No.

L, B and d are length of load line, breadth and draft of the ship

h and λ are the height and length of the resonant wave, (wave period = natural pitching period of the ship in still-water)

T = trim

C_b = block coefficient

Therefore, for bulk carriers having zero trim and a Froude No. 0.15, the maximum impulsive bending moment is approximately given by:

$$\left(M_{Dyn}\right)_{\max} \cong 0.1.B^2.L.h/\lambda$$

The maximum impulsive dynamic shearing force is approximately given by:

$$\left(F_{Dyn}\right)_{\max} = 0.3.B^2 \cdot h/\lambda$$

8 Hull Girder Design Vertical Bending Moment

The hull girder design vertical bending moment is composed of the vertical standard stillwater bending moment M_s and the vertical wave bending moment M_w, see Fig. (5.13).

The formulae of the vertical wave induced bending moment given in the published Rules of classification societies include the dynamic component. The total vertical bending moment is given by:

$$M_V = M_S + M_W$$

Where: M_s = standard stillwater bending moment

M_w = Wave bending moment

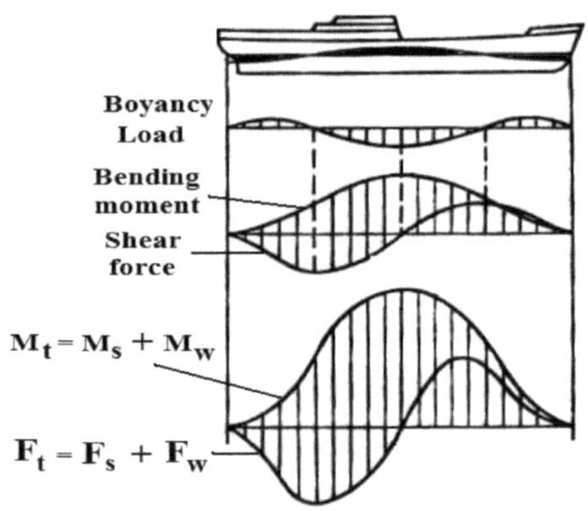

Fig. 5.13 Total shear force and bending moment

8.1 Standard Still Water Bending Moments

Published rules of Classification Societies provide approximate design values for still water bending moment over a specified length at the mid-ship region of the ship. Classification societies also give values for the longitudinal distributions of shear forces and bending moments along ship length, see Fig. (5.14). Fig. (5.15) shows the distribution of stillwater bending moment over ship length of oil tankers, bulk carriers and ore carriers.

Classification Societies defines the main loading conditions of the ship. These loading conditions include homogenous loading condition at maximum draught, ballast conditions, non-homogeneous loading at maximum draught, light load at less than maximum draught, short voyage or harbor condition, ballast exchange at sea, docking conditions, etc

It should be realized that the Stillwater shearing forces and bending moments are subject to various sources of uncertainties. A significant contribution to uncertainties comes from the quantification and location of the various cargo and consumable load items on board ship.

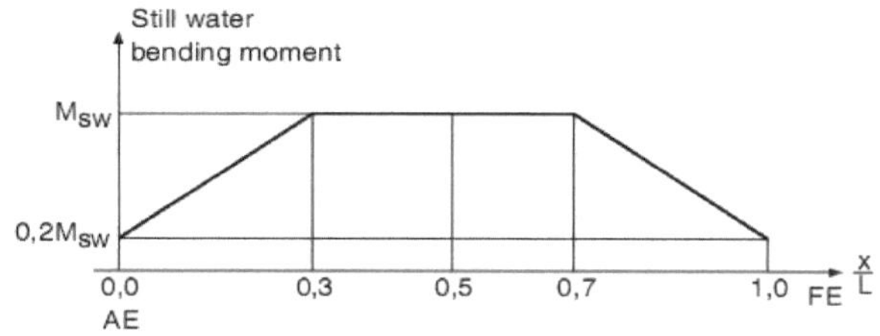

Fig. 5.14 Distribution of Stillwater bending moment over ship length as given by the rules of Classification societies

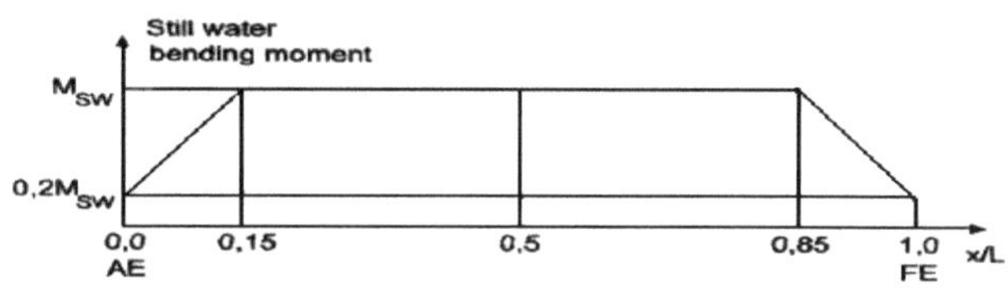

Fig. 5.15 Distribution of Stillwater MSW over ship length of oil tankers, bulk carriers and ore carriers as given by the rules of Classification societies

The actual distribution of cargo weight in bulk carriers and container ships are subject to more significant uncertainties than some other ship types. In order to confine these uncertainties, container ships should have close control during their operation on the actual weights of all single containers on board.

8.2 Vertical Wave Bending Moment

The vertical wave bending moment given by the rules of classification societies, expressed in kN-m may be obtained from the following equations:

M_{ws} = $-110\, C_1 L^2 B(C_b + 0.7) \times 10^3$ kNm Sagging Moment, Negative

^{M}wh = $+\,190\, C_1 U B C_b \times 10^3$ kNm Hogging Moment, Positive

C_1 =10.75-[(300-10/100]" 190<L<300m

=10.75 300 < L< 350m

= 10.75-[(L-350)/150F 350< L< 500m

L = length of vessel as defined in ABS Rules for Building and Classing Steel Vessels in m

B = the greatest molded breadth in m

C_b = block coefficient at summer load waterline, based on L as defined above.

8.3 An Approximate Estimate of the Maximum Value of the Wave Induced Bending Moment M_W

An approximate estimate of the maximum value of the wave induced bending moment M_W could be obtained from the maximum values of the wave bending moments of box-shaped and diamond-shaped vessels as follows:

Let: $M_b = M_W$ for box shaped vessel

$$M_b = \gamma BhL^2/4\pi^2$$

$M_d = M_W$ for diamond shaped vessel

$$M_d = \gamma BhL^2/12\pi^2 = 1/3 \cdot M_b$$

$M_{sh} = M_W$ for ship shaped vessel

Hence the wave induced bending moment for a ship is given by:

$$M_{sh} = \alpha M_b + (1-\alpha)M_d \qquad \text{For } 0 \le \alpha \le 1.0$$

Where: α is a factor depending on the block coefficient of the vessel and could be determined as follows:

$$\alpha = 2(C_b - 0.5)$$

C_b = block coefficient of the ship

Hence: $M_{sh} = 2M_b \cdot (C_b - 0.5) + [1 - 2(C_b - 0.5)] \cdot M_d$

$$= 2[M_b \cdot (C_b - 0.5) + (1 - C_b) \cdot M_d]$$

Substituting for M_b and M_d we get:

$$M_{sh} = 2[M_b \cdot (C_b - 0.5) + (1 - C_b) \cdot M_b/3]$$

$$= M_b/3 \cdot (4C_b - 1)$$

$$= \gamma BhL^2/12\pi^2 \cdot (4C_b - 1)$$

For $h = 0.607\sqrt{L}$

$$M_{sh} = 0.61/(3 \times 40) \cdot B \cdot L^{2.5} \cdot (4C_b - 1)$$

$$\cong 5 \times 10^{-3} \cdot B \cdot L^{2.5} \cdot (4C_b - 1)$$

9 Horizontal Bending Moment

The longitudinal horizontal wave bending moment amidships may be obtained from the following equation, as given by the rules published by classification societies:

$M_H = K_3C_1L^2DC_b \times 10^{-3}$ kN.m

Where: $K_3 = 180$

D = depth of ship at side, m

L, C_1 and C_b are as specified before.

The distribution of the horizontal bending moment along ship length could be obtained using the non-dimensional factors given in Fig. (5.16).

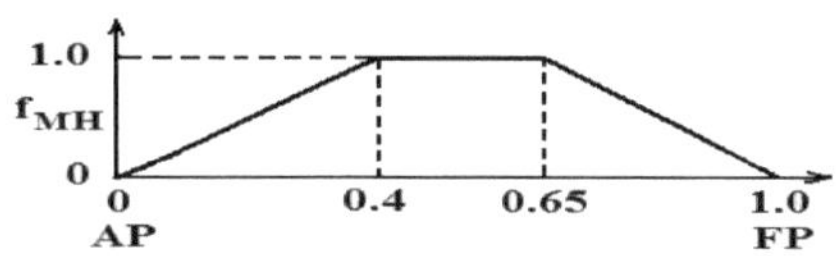

Fig. 5.16 Distribution factor over ship length of the wave-induced horizontal bending moment

10 Hull Girder Shearing Forces

High values of stresses may be induced in the shear carrying-members of ships by longitudinal vertical hull girder shear forces. Ships are also subjected to horizontal shearing forces induced by wave actions when the ship travels at an oblique direction to wave direction. For small ships, the horizontal component of shear force, FT_H, is generally much smaller than the vertical component, FT_V. For large ships, FT_H may attain high values, especially when the ship breadth is exceptionally large, which is the practice in the design of modern tankers and bulk carriers.

10.1 Total Vertical Shearing Force F_V

The total vertical shear force at any section along a ship steaming among waves is the sum of the still-water component, wave-induced component, and the dynamic component.

$$F_V = F_S + F_W + F_d$$

Where:

F_s = vertical still water component

F_W = vertical wave induced component (Low frequency component)

F_d = Dynamic component

The variation of the various components of shear stresses with time is shown diagrammatically in fig. (5.17). The relative magnitude of each component depends upon several factors such as ship type, size, etc.

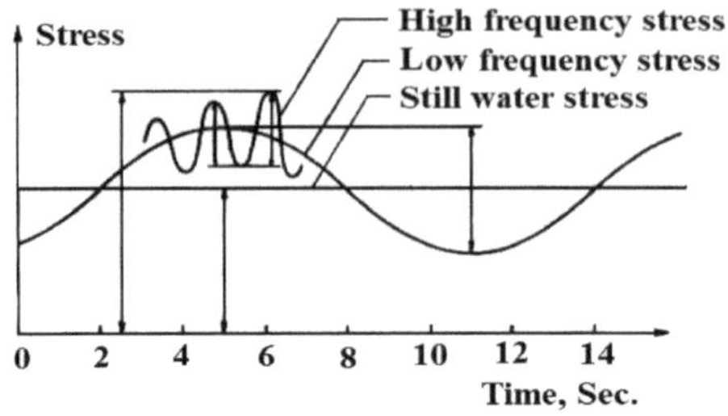

Fig. 5.17 Variation of Stillwater, wave induced and dynamic shear force with time

10.2 Stillwater Shear Force Component F_S

The conventional calculation of longitudinal vertical shearing force and bending moment gives results pertinent only to the general behaviour of the idealized ship girder, without due regard to the effect of local loading. The hull girder shear force and bending moment distributions along the length of a bulk carrier is shown in Fig. (5.18).

The effect of local loadings, however, may become rather significant for bulk carriers, especially when loading is in alternate holds and when the bottom and transverse bulkheads are of double skin.

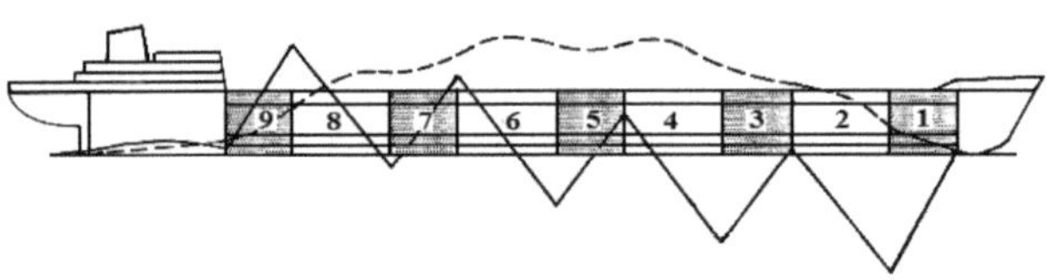

Fig. 5.18 Hull girder shear force and bending moment

The net local loading on the bottom structure of a ship is the difference between the downward load of cargo, or ballast, and the upward water pressure, see Fig. (5.19).

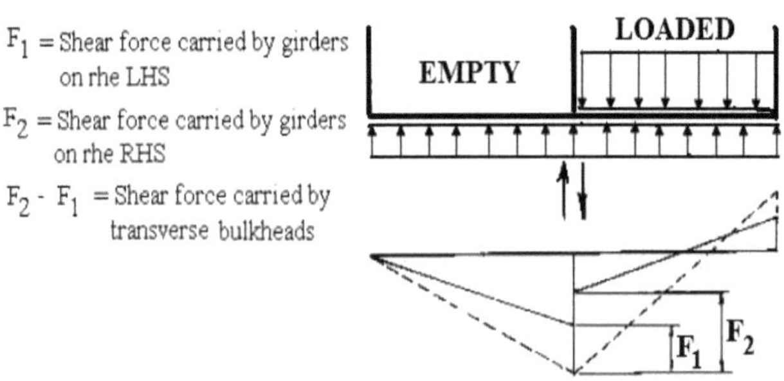

Fig. 5.19 Participation of bottom structure to hull girder shear loading

This net loading is transmitted to the side shell and transverse bulkheads in the form of shear forces and bending moments. The mechanism of load transmission depends mainly on the construction of the bottom structure, i.e. the stiffness and rigidity in the longitudinal and transverse directions, the construction of the attachments between the bottom members and the side shell and transverse bulkheads, the shear area of transverse bulkheads, the hold aspect ratio and upon the loading condition (uniform or in alternate holds).

10.3 The Distribution of the Vertical Wave-Induced Shearing Force

The wave induced shear forces are given in the published Rules of Classification societies for both sagging and hogging conditions as follows, see Fig. (5.20)

$$F_{hog} = +0.3 \cdot F_1 \cdot C \cdot L \cdot B \cdot (C_b + 0.7) \quad \text{kN}$$
$$F_{sag} = -0.3 \cdot F_2 \cdot C \cdot L \cdot B \cdot (C_b + 0.7) \quad \text{kN}$$

Where: L and B are in meters

$$C = 10.75 \times [(L-300)/150]^{1.5} \quad \text{For } 90 \le L \le 300$$
$$C = 10.75 \; C = 10.75 \quad \text{For } 300 \le L \le 350$$
$$C = 10.75 - [(L-350)/150]^{1.5} \quad \text{For } 350 \le L$$

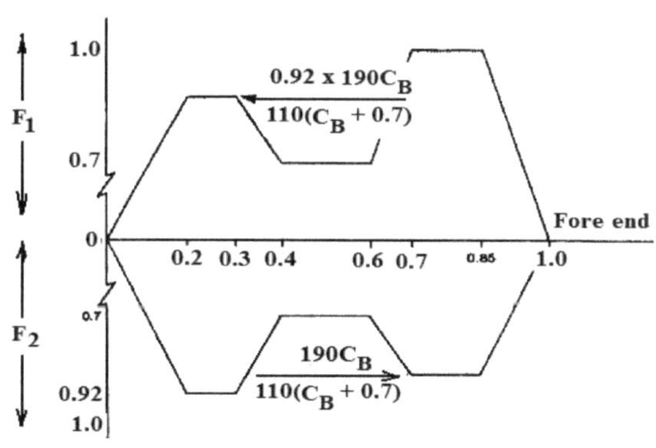

Fig. 5.20 Shear load coefficients

10.4 Approximate Value of the Maximum Vertical Shear Force

Assuming that the distribution of the wave induced bending moment is given by, see Fig. (5.21):

$$M = M_{L/2}(1-\cos(2\pi x/L))/2$$

Fig. 5.21 Assumed distribution of the wave induced bending moment along ship length

Since at $x = 0$ and $x = L$

$$M = dM/dx = F = 0$$

Then; $F = dM/dx = M_x[2\pi/L \cdot \sin(2\pi x/L)]/2$

$$= \pi/L \cdot M_x \cdot \sin(2\pi x/L)$$

F_{max} occurs at approximately $x = L/4$ *and* $x = 3L/4$

Then: $F_{max} = \pi M/L \cong C.M/L$

Where C = factor depending on ship type as given in table 5.1

Table 5.1 Values of C for some ship types

Ship type	Passenger ships	Cargo ships	Oil tankers
C	3.5	3.75	4.3

11 Wave Induced Torsion Loading

All ships are subjected to torsional moments which tend to twist the hull girder along its length, see Fig. (5.22). Torsional loading induces additional stresses, normally called warping stresses, near hatch corners. Normally, torsional stiffness is more than adequate to prevent the development of high warping stresses and undue distortion of the structure.

Wave induced torsion loading results from the motion of a ship in oblique waves, see Fig. (5.23). Severe rolling motion induces large dynamic forces and torsional loadings. When the ship is sailing obliquely into the predominant waves, the vertical wave bending moments are reduced, but the horizontal and torsional moments are increased. In a sea-way, torsional moments are set up by both hydrostatic and hydrodynamic forces including slamming, and by the mass-acceleration forces as a result of the ship's motions, see Fig. (5.24).

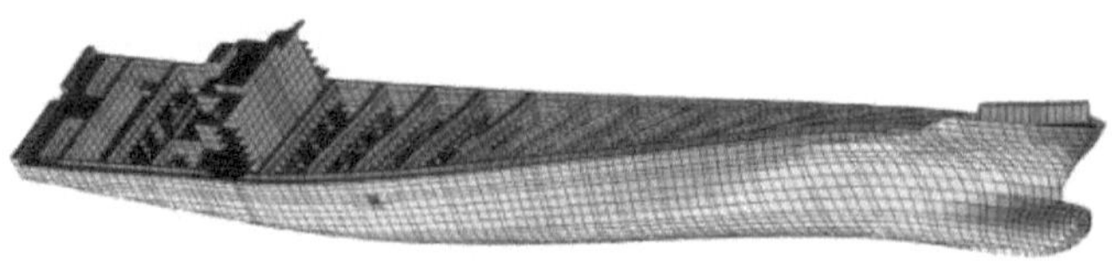

Fig. 5.22 Open deck container ship hull girder torsional deformation

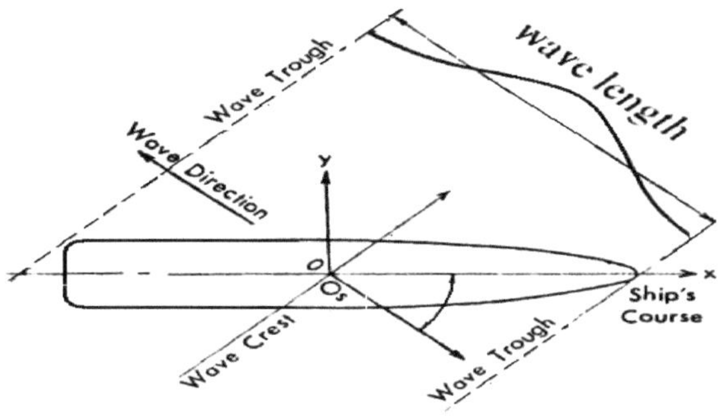

Fig. 5.23 Ship traveling in oblique waves

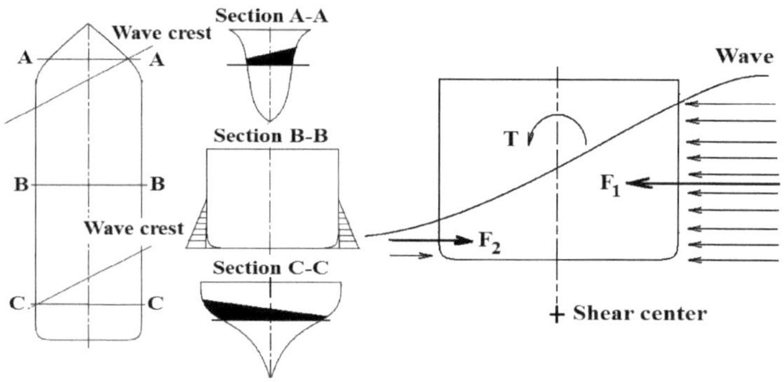

Fig. 5.24 Hull girder torsion loading at several ship sections along her length for a ship traveling obliquely among waves

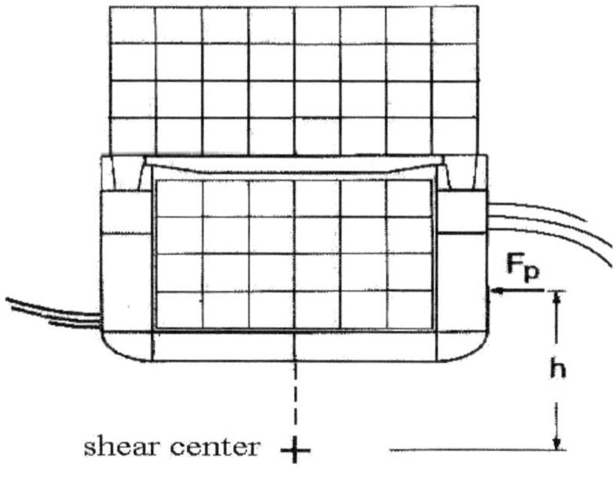

Fig. 5.25 Torsion load resulting from the effect of wave forces, cargo loading and the offset location of the shear center

Improper distribution of cargo loading and fuel also induces still-water torsional loading. Stillwater torsional moments could be determined from the published Rulers of Classification Societies. An approximate value of the Stillwater torque is given by

$$T_S = k \cdot B \cdot W_T \quad kN \cdot m$$

Where:: k = 0.004

B = breadth of the vessel, m

W_T = Maximum total container weight, kN

All external forces acting on the hull girder which do not pass through the axis of shear center, will produce torsion. Open ships, therefore, are also subjected to additional torsional loading induced by the horizontal component of the shearing force when acting at a distance from the shear center of the ship section see Fig. (5.25).

The values of the wave torques, calculated with respect to the section center of torsion, are obtained, in kN.m, from the following formula:

$$M_{WT} = \frac{H \cdot L}{4} n\left(F_{TM} C_M + F_{TQ} C_Q d\right)$$

Where: $\mathbf{F_{TM}}$, $\mathbf{F_{TQ}}$: Distribution factors defined in the Table 5. 2 for conditions 1, 2

C_M = wave torque coefficient

$$C_M = 0.45 B^2 C_W^2$$

C_Q = Horizontal wave shear coefficient

$$C_Q = 5TC_B$$

C_w = Water-plane coefficient, to be taken not greater than the value obtained from the following :

$$C_W = 0.165 + 0.95 C_B$$

Where C_B is to be assumed not less than 0.6.

d = vertical distance, in m, from the center of torsion to a point located 0.6T above the base line

Table 5.2 Distribution factors F_{TM} and F_{TQ}

Ship condition	Distribution factor F_{TM}	Distribution factor F_{TQ}
1	$1 - \cos\frac{2\pi x}{L}$	$\sin\frac{2\pi x}{L}$
2	$1 - \cos\frac{2\pi(L-x)}{L}$	$\sin\frac{2\pi(L-x)}{L}$

Condition 1: ship direction forming an angle of 60° with the prevailing sea condition
Condition 2: ship direction forming an angle of 120° with the prevailing sea condition

H= wave parameter

$$H = 8.13 - \left(\frac{250 - 0.7L}{125}\right)^3$$

12 Probabilistic Prediction of Hull Girder Loading

12.1 Short Term Prediction of Loading

The main properties of any wave system are the heights, lengths, and periods. These properties are commonly expressed by the power or energy spectrum. The power spectrum is transformed into a "spectrum of encounter" using the speed and heading of the ship. The structural response to waves used in the global structural analysis of a ship is calculated based on the ship's Response Amplitude Operators (RAOs).

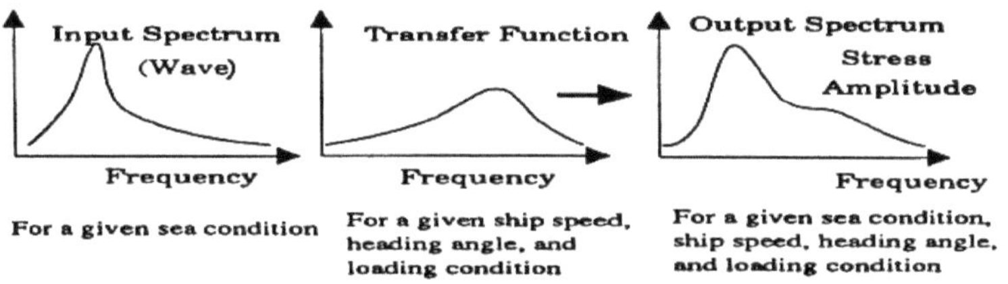

Fig. 5.26 Input-Output procedure for the stress amplitude spectrum

The RAOs are obtained for regular sinusoidal waves, different wave headings and frequencies. The bending moment response spectrum is obtained by multiplying the encounter spectrum by the appropriate transfer function, see Fig. (5.26). The largest response amplitude occurs at the largest value of area under the response spectrum.

The most severe operating conditions would then be defined as the power spectrum which gives the response spectrum with the largest area. The wave spectrum giving maximum response is assumed to have its energy largely concentrated at wave components of approximately the same length as the ship.

The short-term response of a ship is obtained from the evaluation of the seaway spectrum, which is assumed to be stationary in a period of a few hours. The long-term response and the probability of exceedance of the load are evaluated from the short-term prediction. Full scale measurements have verified that short term (about 30 to 45 minutes) statistical distribution of ship responses as well as wave heights and length can be characterized by the Rayleigh probability distribution which is given by, see Fig. (5.27):

$$f_X(x) = \frac{x}{m_O} \cdot \exp \cdot \left(-x^2/2m_O\right), \quad x \geq 0$$

Where: $m_o = \left(\sum_{i=1}^{n} x^2{}_i\right) / 2n$ = area under the wave spectrum,

n = total number of peak values

x = amplitude of response

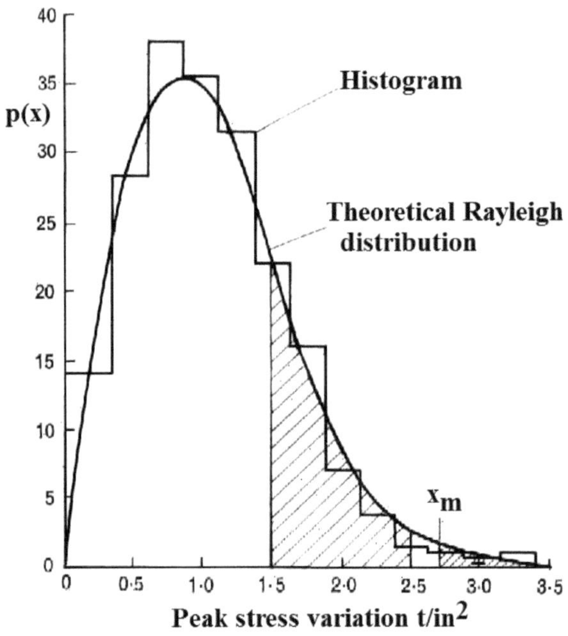

Fig. 5.27 Histogram and theoretical Rayleigh distribution

Within this short period, it is assumed that external conditions (ship speed, weather, etc.) remain unchanged. Hence, the probability that ship response (or wave height) takes on value equal to or greater than certain value x_1, i.e. probability of ascendance is given by:

$$P(X \geq x_1) = \int_{x_1}^{\infty} f(x)dx = \exp\left\{-\frac{x_1^{\,2}}{2m_0}\right\}$$

The cumulative probability is given by:

$$P(X \leq x_1) = \int_0^{x_1} f(x)dx = 1 - \exp\left\{ -\frac{x_1^{\,2}}{2m_0} \right\}$$

Therefore, the prediction of any particular ship response (or wave height) within the stipulated short period is entirely dependent upon the estimate of the parameter m_O.

12.2 Long Term Predictions

The long-term predictions of the load are normally based on the Weibull distribution function:

$$f_X(x) = 1 \cdot \exp \cdot \left\{ -\left(\frac{x}{w} \right)^k \right\}$$

Where: k = Shape factor

w = scale parameter

The Extreme values of the wave load could be determined as follows:

$$P[X > x_\alpha] = 1 - \left[1 - \exp\left(-\frac{x_\alpha}{w} \right)^k \right]^n$$

Where: n = total number of wave encounters

x_α = the extreme value of the wave load

Hence: $x_\alpha - w \left\{ \log\left[1 - (1-\alpha)^{1/n}\right]^{-1} \right\}^{1/k}$

12.3 Extreme Value Distributions

Given n observations of the variable X , see Fig. (5.28), and assuming X to be normally distributed, then we have,:

$$X \equiv N(u_X, \sigma_X)$$

Let: X_n = extreme high values of the load X

Then: $X_n \equiv N\left(u_{Xn}, \sigma_n\right)$

The mean values of X_n is given by:

$$u_{X_n} \cong u_X + \sigma\sqrt{\ell_n n} = u_X + k_n \sigma_x$$

The standard deviation of X_n is given by:

$$\sigma_n = \frac{\pi\sigma}{\sqrt{6\ell_n n}} = \alpha\sigma$$

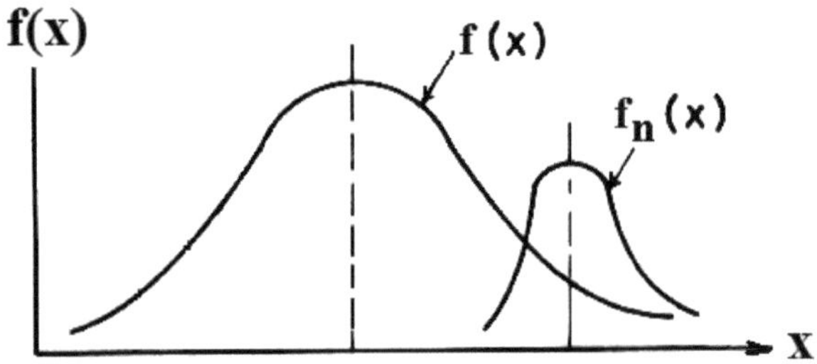

Fig. 5.28 Probability density function

The variation of the standard deviation with n is given in table 5.3.

Table 5.3 Variation of the standard deviation with n

n	$\ell_n n$	$k=\sqrt{l_n n}$	σ_n
100	4.605	2.146	0.5975
1000	6.9077	2.628	0.488
10000	9.21	3.0348	0.422
10^5	110513	3.393	0.3778
10^6	13.815	3.717	0.345

Chapter 6
Hull Girder Bending Stresses

1 Introduction

This chapter presents the methods commonly used to calculate the primary stresses induced by the vertical and horizontal hull girder bending moments. The strength members of ship hull girder sustaining the primary hull girder stresses are specified. The common procedures used for the calculation of ship section geometrical and flexural properties are given. The method of calculating hull girder bending stresses when ship is in the inclined condition is presented.

2 Hull Girder Bending Stress Components

Hull girder primary stresses are induced by the following bending moment components:

1. Stresses due to longitudinal vertical bending moments:
   a. Stillwater bending stresses
   b. Wave induced bending stresses (sagging/hogging conditions)
   c. Dynamic stresses
2. Stresses due to horizontal bending moment

The variation of the stillwater bending stresses with voyage time for certain ship types is negligible. The variation of wave induced bending stresses with time follows the same variation of the low frequency wave system. The induced dynamic stresses is of high frequency and results from whipping and springing of ship hull girder. The total stress at any point on the ship hull girder due to the total vertical bending moments is the algebraic sum of the induced three stress components, see Fig. (6.1)

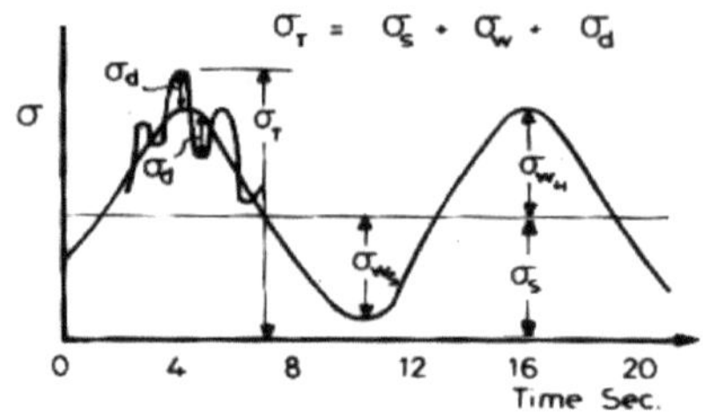

Fig. 6.1 Stillwater, wave induced and dynamic hull girder bending stresses

2.1 Hull Girder Primary Stresses Induced by Longitudinal Vertical Bending Moments

The hull girder primary stresses at the deck and bottom plating due to a longitudinal vertical bending moment are given by, see Fig. (6.2):

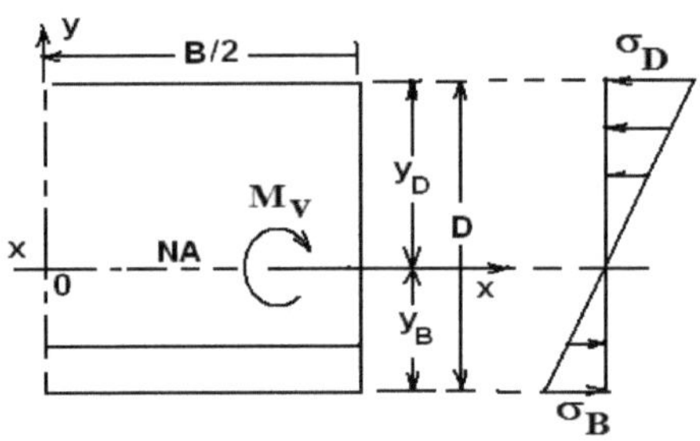

Fig. 6.2 Idealised section of a general cargo ship subjected to a longitudinal vertical sagging bendin moment

$$\sigma_D = M_v \cdot \frac{y_D}{I_x}$$

$$\sigma_B = M_v \cdot \frac{y_B}{I_x}$$

Where: $M_v = M_s \pm M_w$

M_V = Total vertical bending moment
M_s = Stillwater vertical bending moment
M_w = Vertical wave bending moment
I_x = Second moment of area of ship section about x-x axis
σ_D = Hull girder stress at deck due to Mv
σ_B = Hull girder stress at bottom plating due to Mv

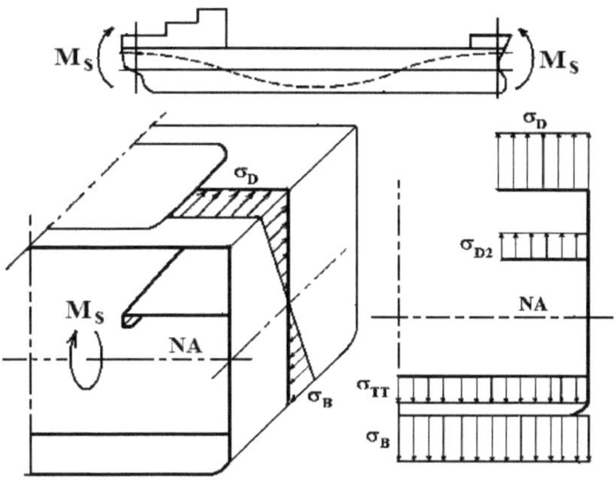

Fig. 6.3 Hull girder stresses over a ship section of a twin decker general cargo ship due to a sagging bending moment

For the sagging bending moment M_V, the hull girder stress at the deck is compressive and at the bottom is tensile. The magnitude of the deck and bottom stresses depend on the magnitude of the section modulus at the deck and bottom respectively.

$$Z_D = \frac{I_x}{y_D}$$

The section modulus at the bottom of the ship section is given by:

$$Z_B = \frac{I_x}{y_B}$$

The section modulus at the side of the ship section is given by:

$$Z_s = 2Iy/B$$

The second moment of area of ship section about the x-x axis is given by:

$$I_x = A_D \cdot y_D^2 + A_B \cdot y_B^2 + 2A_S \cdot \frac{D^2}{12} + 2A_s\left(y_D - \frac{D}{2}\right)^2$$

Where:

A_D = sectional area of the strength members of the deck structure
A_B = sectional area of the strength members of the bottom structure
A_S = sectional area of the strength members of the side structure
y_D = distance of the deck plating from the neutral axis of the ship section
y_B = distance of the bottom plating from the neutral axis of the ship section

The distribution of the primary hull girder bending stresses over the upper deck, second deck, side shell, tank top and bottom plating of a twin decker general cargo ship subjected to a sagging bending moment is shown in Fig. (6.3).

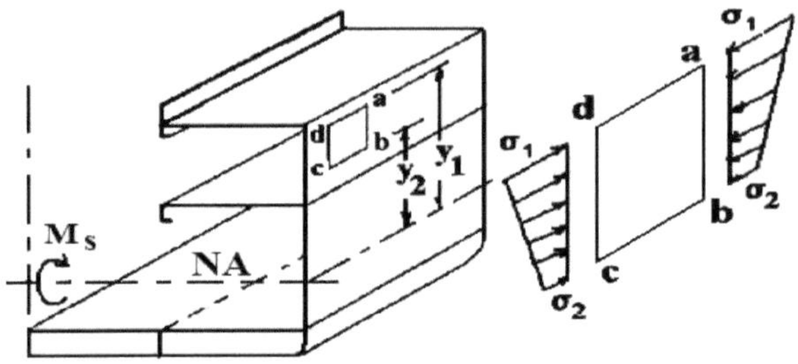

Fig. 6.4 Hull girder compressive bending stresses over a plate panel of a twin decker general cargo ship due to a sagging bending moment

The primary hull girder compressive stresses over a plate panel in the side shell plating above the neutral axis of the ship section of a twin decker general cargo ship is shown in Fig. (6.4). The compressive bending stresses at the upper and lower edges of the plate panel are given by, see Fig. (6.4):

$$\sigma_1 = M_s \cdot \frac{y_1}{I_x}$$

$$\sigma_2 = M_s \cdot \frac{y_2}{I_x}$$

Where: I_x = moment of inertia of ship section about its neutral axis

The hull girder primary compressive bending stress over a deck plate panel in a conventional oil tanker due to a sagging bending moment is shown in Fig. (6.5).

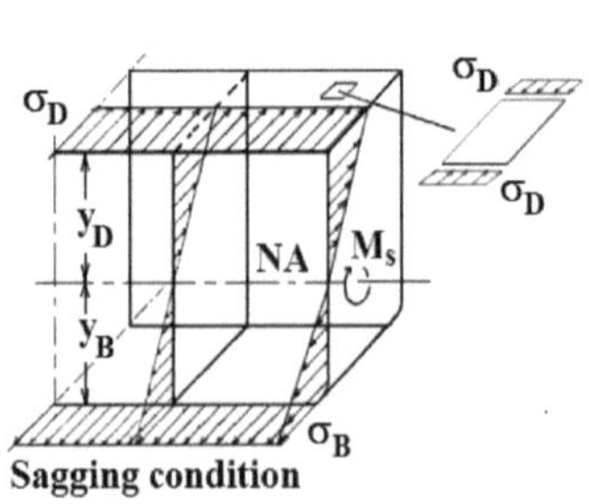

Fig. 6.5 Hull girder primary compressive bending stress over a deck plate panel in a conventional oil tanker due to a sagging bending moment

The compressive hull girder stress over the plate panel is given by:

$$\sigma_D = M_s \cdot \frac{y_D}{I_x}$$

Where: y_D = distance of deck plating from the neutral axis of the ship section

I_x = moment of inertia of ship section about the neutral axis of the ship section

Because of the presence of shear stresses, the primary hull girder stresses should be corrected for the shear lag effect, see Fig. (6.6). The hull girder stress is reduced over the inside deck plating and is increased over the outer deck plating. The increase in stress is about 10%.

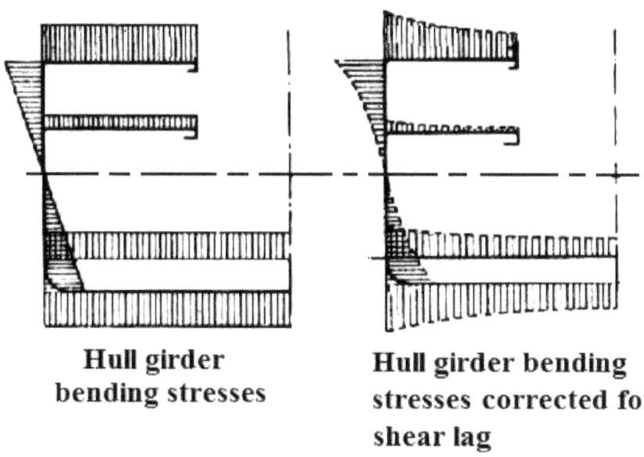

Fig. 6.6 Primary hull girder stresses corrected for the shear lag effect

3 Geometrical and Flexural Properties of Ship Sections

The main ship section geometrical and flexural properties are:

- Shear area
- Total sectional area
- Position of neutral axis
- Moment of inertia of ship section
- Section modulus

The calculation of the geometrical and flexural properties of the ship section is carried out with reference to neutral axis of the ship section shown in fig. (6.7).

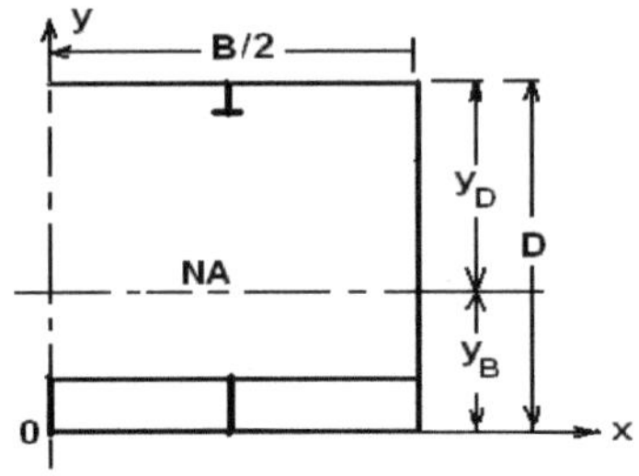

Fig. 6.7 Idealized transversely framed ship section of a single decker ship

Table 6.1 Geometrical and flexural properties of ship section

Item	scantlings	Area m^2	Vertical Distance, y	ay	ay^2	i_y	x	ax^2	i_x
		$\sum a$		$\sum ay$	$\sum ay^2$	$\sum i_y$		$\sum ax^2$	$\sum i_x$

The calculation of the geometrical and flexural properties of the ship section is carried out first with reference to the base line of the ship section. The second moment of area of the ship section about the neutral axis (NA) of the ship section is given by:

$$I_{NA} = I_X - A \cdot y_B^2$$

$$I_X = 2\left[\sum ay^2 + \sum i_y\right]$$

Where: I_X = second moment of area of ship section about X-axis
I_{NA} = second moment of area of ship section about NA
y_B = distance of NA from base line
I_y = second moment of area about center line of ship section
A = sectional area of all effective longitudinal strength members

It should be noted that in the calculation of the geometrical and flexural properties of the ship section, only strength members contributing to the longitudinal strength of ship hull girder are taken into account. For transversely framed ship structures, the deck, bottom and side shell plating as well as the longitudinal girders are the main strength members to be used in the calculation of the geometrical and flexural properties of the ship section. In longitudinally framed ship structures, the strength members to be considered are the shell plating, longitudinals, and longitudinal girders. The calculation of the geometrical and flexural properties of ship sections is normally carried out as given in table 6.1:

The total sectional area of all strength members sustaining vertical hull girder bending is given by:

$$A = 2\sum a$$

The distance of the neutral axis of the ship section from the base line of ship section is given by:

$$y_B = \frac{\Sigma ay}{\Sigma a}$$

The distance of the deck from the neutral axis of the ship section isd given by:

$$y_D = D - y_B$$

The moment of inertia about the base line of the ship section is given by:

$$I_X = 2\left[\sum ay^2 + \sum i_y\right]$$

The moment of inertia about the neutral axis of the ship section is given by:

$$I_{NA} = I_X - A \cdot y_B^2$$

The moment of inertia of ship section about the vertical centerline of the ship (y-axis)

$$I_y = 2\left(\sum a.x^2 + \sum i_x\right)$$

3.1 Flexural Properties of Longitudinally Framed Deck and Bottom Structures

The calculation of the flexural characteristics ship sections of longitudinally framed deck and bottom structures could be simplified by idealizing the deck and bottom structures using the concept of effective thickness. The effective thickness is the sum of the actual plating thickness and the additional thickness representing the contribution of the attached longitudinals and can be calculated as follows, see Fig. (6.8):

$$t_{D_e} = t_D + \delta t$$

Where: δt = virtual increase in plating thickness due to the presence of longitudinals, see Fig. (6.8)
t_D = actual thickness of deck plating
t_{De} = effective thickness of deck plating

$$\delta t = \frac{\sum n_D . a_l}{S}$$

Where: n_D = number of deck longitudinals
a_l = sectional area of each longitudinal

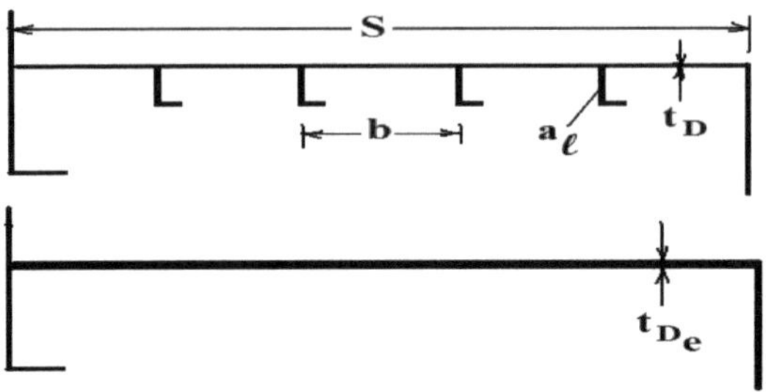

Fig. 6.8 Basic concept of effective thickness

The effective thickness of bottom plating is given by:

$$t_{Be} = t_B + \frac{\sum n_B . a_\ell}{B}$$

The procedure of calculation of the flexural properties of an oil tanker is as follows, see Fig. (6.9):

The moment of inertia of the ship section is given by:

$$I = A_D \cdot D^2 + \frac{t_s D^3}{12} + A_s \left(\frac{D}{2} - y_D \right)^2$$

Where: I = Moment of inertia about the neutral axis of ship section
y_D = Distance of deck plating from neutral axis of ship section
y_B = Distance of bottom plating from N. A. of ship section

$$y_B = \frac{A_D \cdot D + A_S \cdot \frac{D}{2}}{A_t},$$

$$Y_D = D - y_B$$

Where: $A_t = A_B + A_S + A_D$
A_D = Sectional area of deck structure
A_B = Sectional area of bottom structure
A_S = Sectional area of side shell and longitudinal bulkhead structure

$$A_B = B \cdot t_B + \sum a_{\ell B}$$

$$A_D = B \cdot t_D + \sum a_{\ell D}$$

$$A_s = D(t_s + t_{Bl}) + \sum (a_{ls} + a_{lL})$$

Where: t_B = Thickness of bottom plating
t_D = Thickness of deck plating
t_s = Thickness of side shell plating
t_{BL}= Thickness of longitudinal bulkhead plating
$a_{\ell j}$ = Sectional area of one longitudinal, j= B,D,S,BL
a_{lB} = sectional area of each bottom longitudinal
a_{ls} = sectional area of each side longitudinal
a_{lL} = sectional area of each longitudinal of longitudinal bulkheads

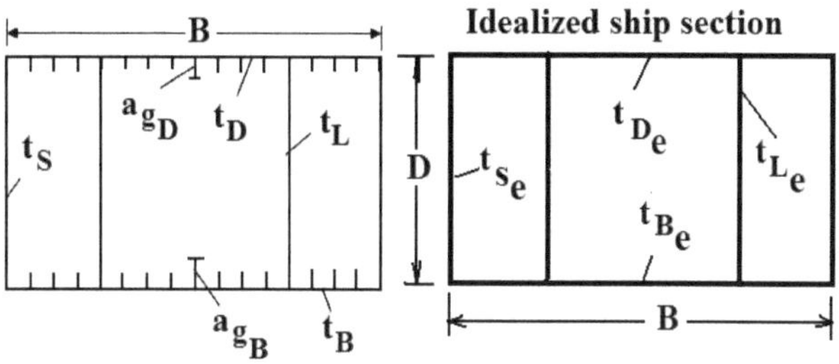

Fig. 6.9 Idealized section of a conventional oil tanker using the concept of effective thickness

4 Hull Girder Stresses When the Ship Is In the Inclined Condition

When a ship is in an inclined condition, one of the upper deck corners and the opposite bottom corner will experience greater hull bending stresses, see Fig. (6.10). In order to calculate the stresses in the top and bottom corners of the ship section furthest from the neutral axis, the vertical B.M is resolved into two components along the centroidal vertical and horizontal axes of the ship section as follows:

$$M_y = M\cos\theta$$

And $$M_x = M\sin\theta$$

Because the x–y axes are principal centroidal axes, the bending stress at an arbitrary point P on the ship section is given by:

$$\sigma_P = \frac{M\cos\theta \cdot y}{I_x} + \frac{M\sin\theta \cdot x}{I_y}$$

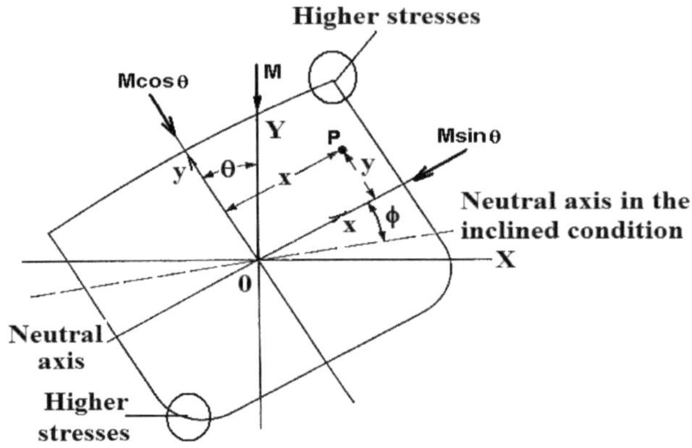

Fig. 6.10 Hull girder vertical bending moment in the inclined condition

The stress is zero at the N.A. corresponding to the inclined position. Thus position of the N.A. is determined from the condition that

$$0 = \frac{M\cos\theta \cdot y}{I_x} + \frac{M\sin\theta \cdot x}{I_y}$$

$$\text{i.e. } y = -\frac{I_x}{I_y}\tan\theta \cdot x$$

$$\text{or } y = \tan\varphi \cdot x$$

$$\tan\phi = -\frac{I_y}{I_y}\tan\theta$$

5 Hull Girder Stresses due to Horizontal Bending Moment

Apart from the vertical hull girder bending moments, ships steaming among waves will also experience horizontal bending and torsional moments. The hull girder horizontal bending moment M_H will induce normal stresses as shown in Fig.(6.11).

The horizontal bending moments induce tensile stresses in either the port side shell plating and compressive stresses in the starboard side shell plating or vice versa.

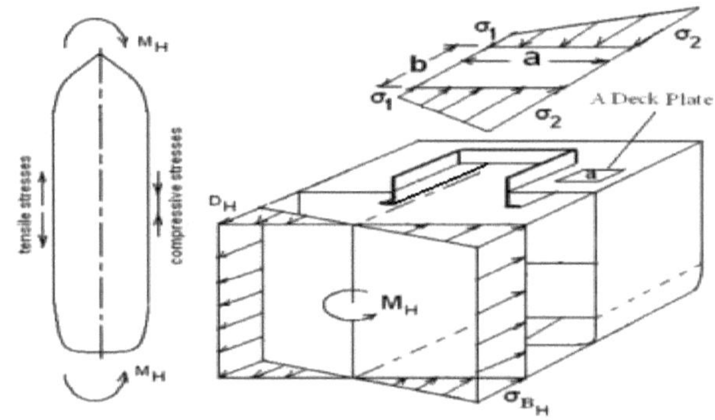

Fig. 6.11 Hull girder stresses induced by horizontal bending moment

The hull girder stresses induced in the side shell plating by horizontal bending moment is given by, see Fig. (6.12):

$$\sigma_S = \frac{M_H \cdot B/2}{I_y}$$

Where: M_H = horizontal bending moment
B = ship breadth
I_y = second moment of area of ship section about the y-axis.

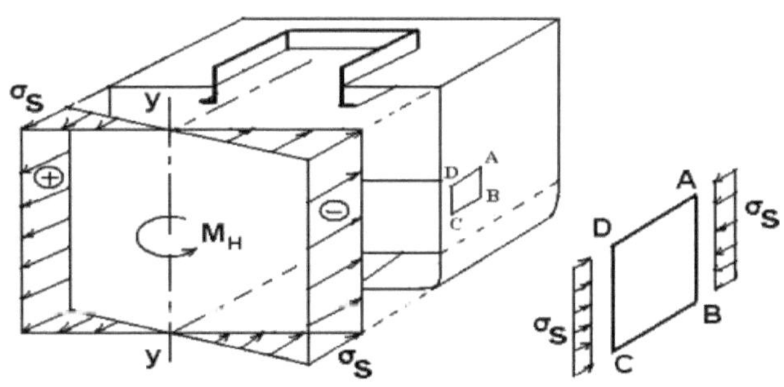

Fig. 6.12 Idealized section of a conventional oil tanker

6 Hull Girder Shear Stresses

The shear flow could be calculated using the general formula given by:

$$q = \frac{F}{I} \sum a.\overline{y}$$

The shear stress is given by:

$$\tau = q/t$$

Where: q = shear flow

Hence, the shear stress is given by:

$$\tau = \frac{F}{I.t}\sum a.\overline{y}$$

Where: F = vertical shear force

τ = shear stress at a specified point on ship section

I = Second moment of area about neutral axis of ship section

t = thickness of plating

$\sum a.\overline{y}$ = Moment of section area about ship section neutral axis of strength members above the specified point on ship section

The calculation of shear flow distribution over a ship section could be greatly simplified by using an idealized ship section based on effective thicknesses of deck, bottom, side and longitudinal bulkheads. The shear stress distribution over a section of a general cargo ship, a bulk carrier and an oil tanker are shown in Figs. (6.13).

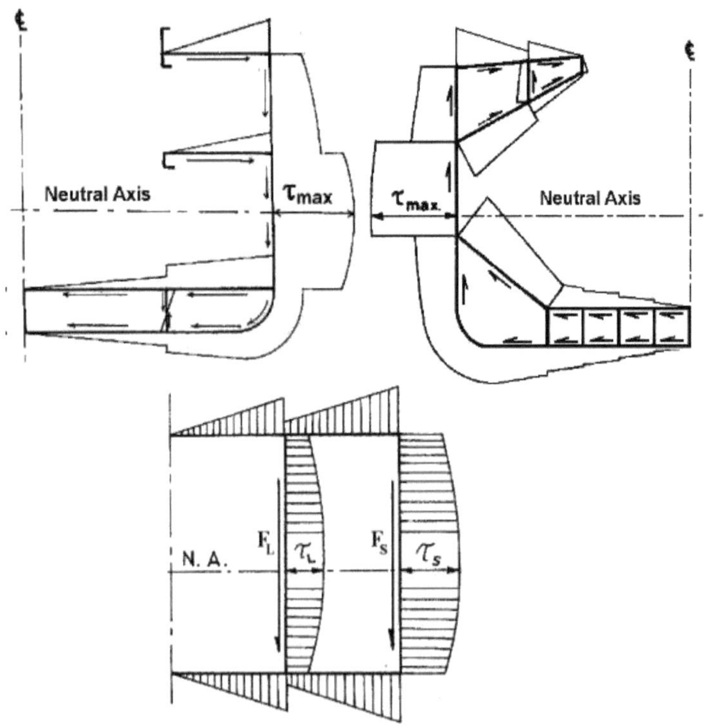

Fig. 6.13 Shear stress distribution over ship sections of a general cargo ship, a bulk carrier and an oil tank

Chapter 7
Secondary Loading and Stresses

1 Introduction

This chapter gives a full analysis of the secondary loadings and stresses induced in the strength members of cargo ships and conventional oil tankers. Strength members sustaining secondary loadings and stresses of transversely and longitudinally stiffened bottom and deck ship structure assemblies are specified. The secondary stresses induced by the bending of double and single bottom structures are presented for hogging and sagging conditions. The loadings and stresses induced in deck and bottom girders, longitudinals and plating are highlighted. Secondary loading and stresses induced in bottom girders, longitudinals and plating as well as tank top longitudinals and plating are defined. Secondary loading and stresses in bottom structure assemblies of oil tankers are identified.

2 Strength Members of Ship Bottom Assemblies Sustaining Secondary Loadings

The strength members of transversely stiffened bottom structure sustaining secondary stresses are bottom girders, bottom and tank top plating. For longitudinally stiffened bottom structures, the strength members sustaining secondary stresses are bottom girders, bottom and tank top longitudinals and plating.

It should be mentioned here that the shear stress is an important factor in determining the scantlings of bottom girders by virtue of their relatively large depths.

2.1 Secondary Loading and Stresses in Bottom Assemblies

The secondary stresses in bottom assemblies are induced by the secondary moments and shear forces. These loadings could be determined by analyzing a specified length of the secondary structure assembly such as a hold length. Alternatively, a structure assembly comprising a full hold length plus half hold length aft and forward of the selected hold. An example is given for the case when the full hold length is fully loaded and the two half hold lengths are empty, see Fig. (7.1).

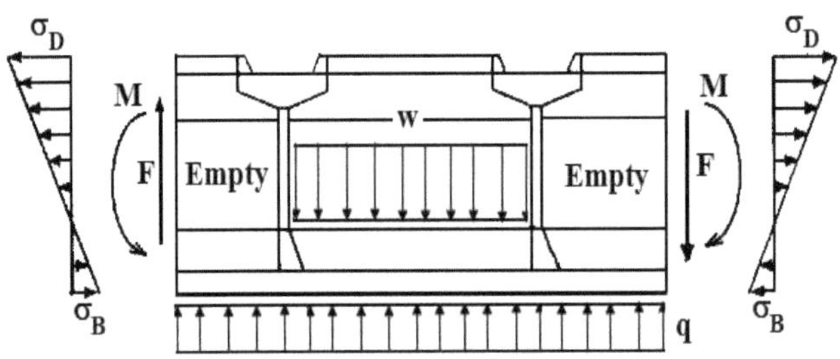

Fig. 7.1 A structural assembly of one fully loaded cargo hold with two half empty holds of a bulk carrier

For bulk carriers loaded in alternate holds, the double bottom structure of the loaded holds could bend outwards and the empty holds bend inwards, see Fig. (7.2).

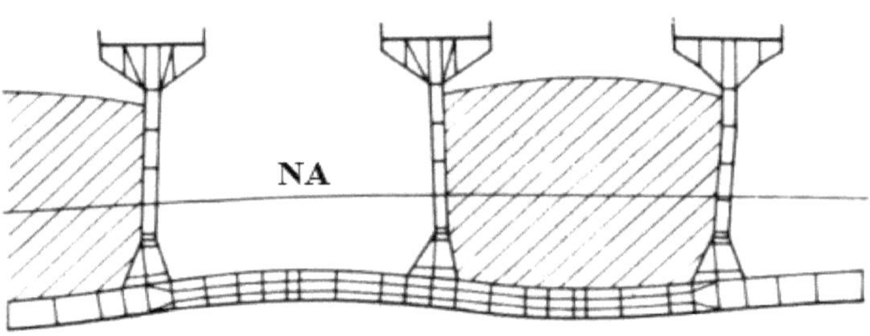

Fig. 7.2 Deformation of the double bottom structure of a bulk carrier loaded in alternate holds

2.2 Secondary Loading and Stresses in Transversely Stiffened Bottom Assemblies

The secondary bending moments and shear forces induced in the strength members of the secondary structure assemblies could be determined by analyzing the idealized secondary structure using one of the following methods:

- 3-D frame analysis
- 2-D grillage analysis
- Continuous beam analysis

The 3-D frame analysis gives more accurate results than the other two methods but is more costly and time consuming. The 2-D grillage analysis gives approximate values of the bending moments, shear forces and shape of deformation of the strength members. The 2-D grillage analysis does not take into account the effect of the supporting strength members of the 3-D structure assembly. Using continuous beam analysis gives less accurate results as it does not take into account the 2-D and 3-D effects of the supporting strength members of the whole structure assembly. It should be used only to know the general behaviour of the strength member under consideration. In all these methods, the main outcome is the maximum bending moments and shear forces at the supporting end of the strength member.

2.2.1 Secondary Loading and Stresses in Bottom Girders

For transversely stiffened double bottom structures, the bottom girders could be bent either inwards or outwards depending on the intensities of cargo loading and the external water pressure, irrespective of the direction of hull girder bending, see Figs. (7.3).

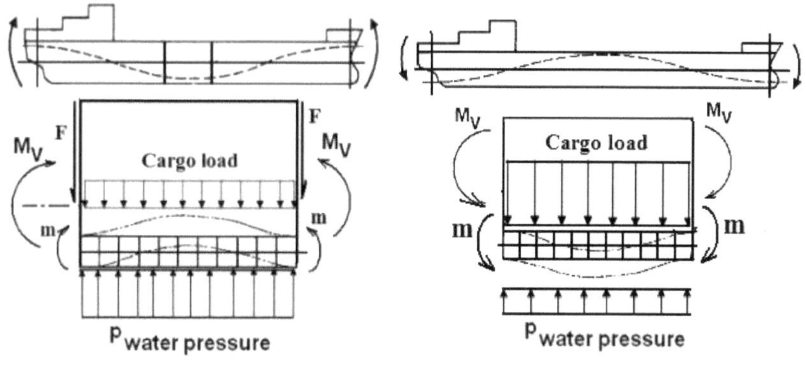

Fig. 7.3 Bottom girder bent inwards/outwards in a ship in sagging condition

When the bottom girder bends downwards as shown in Fig. (7.4), the secondary stress induced in the top edge of the girder web plate at the clamped end is given by:

$$\sigma_f = \frac{m_2 . y_f}{i_g}$$

Where: σ_f = secondary stress induced at the top edge of the web plate of the bottom girder
m_2 = bending moment induced at the clamped end of the girder at the transverse bulkhead

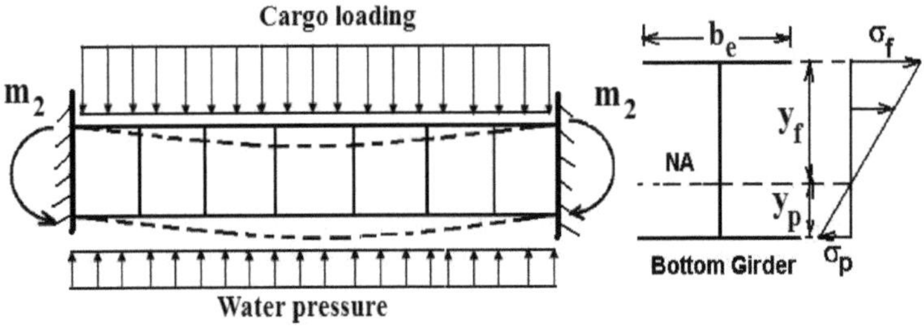

Fig. 7.4 Bottom girder bent outwards due to the net loading q along its length acting downwards

2.2.2 Secondary Stresses in Bottom Plating

The secondary stress induced in the bottom plating of a transversely stiffened bottom structure is given by, see Fig. (7.5):

$$\sigma_p = \frac{m_2 . y_p}{i_g}$$

Where: m_2 = secondary bending moment induced in bottom girder
y_p = distance of bottom plating from the neutral axis of the bottom girder section
i_g = second moment of area of girder cross-section

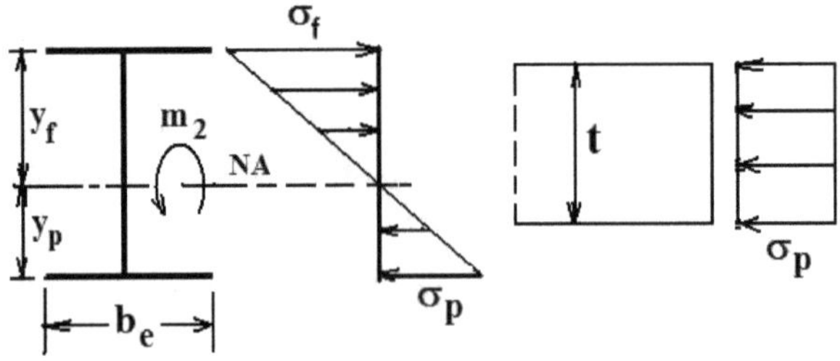

Fig. 7.5 The secondary stress induced in tank top plating of transversely stiffened bottom structure

2.2.3 Secondary Stresses in Tank Top Plating

For transversely stiffened bottom structure, the secondary stress induced in the tank top plating at the clamped end of the girder is given by, see Fig. (7.6):

$$\sigma_f = \frac{m_2 . y_f}{i_g}$$

Where: m_2 = secondary bending moment induced in bottom girder
y_f = distance of tank top plating from the neutral axis of the bottom girder section

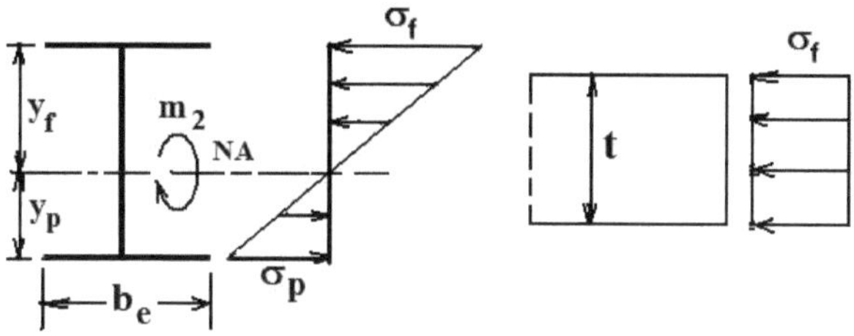

Fig. 7.6 Secondary stress induced in the tank top plating at the clamed end

2.3 Secondary Stresses in Longitudinally Stiffened Double Bottom Structure

Longitudinally stiffened double bottom structure assemblies can be idealized by a grillage structure as shown in Fig. (7.7). Because of the symmetry of the double bottom structure, the solution of the grillage structure could be greatly simplified by dealing with only ¼ of the structure. The girders, floors and longitudinals are idealized as beam elements.

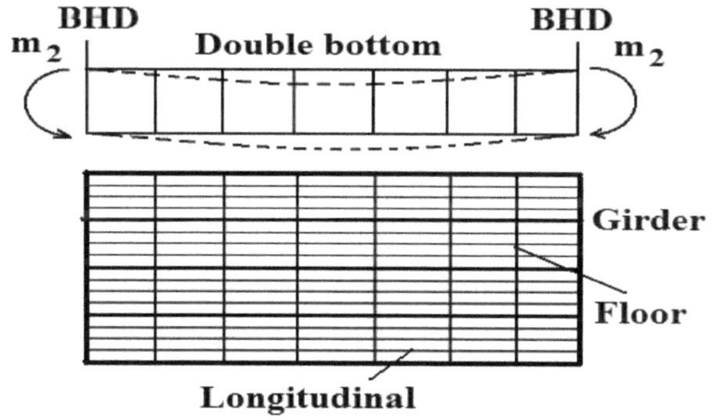

Fig. 7.7 Main elements of longitudinally stiffened double bottom structure assemblies

The geometrical and flexural properties of each idealized member are calculated with the corresponding effective breadth of the attached plating, see Fig. (7.8). The analysis of the grillage structure should give the magnitude and distribution of bending moments and shear forces along the length of each strength member of the bottom structure. The maximum values are expected to occur at the clamped ends of the girders.

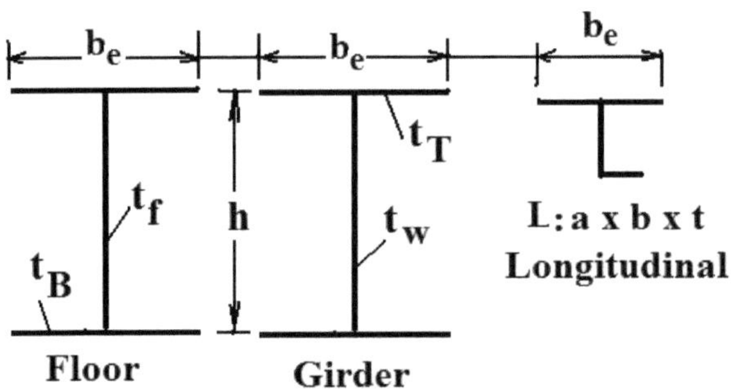

Fig. 7.8 Sections of the idealized strength members of the longitudinally stiffened double bottom structure

2.3.1 Secondary Stresses in Bottom Girders

The bending moments at the ends of the double bottom assembly of a longitudinally stiffened double bottom structure is shown in Fig. (7.9). The bending moments at the ends of the double bottom assembly are determined from a grillage analysis of the double bottom structure.

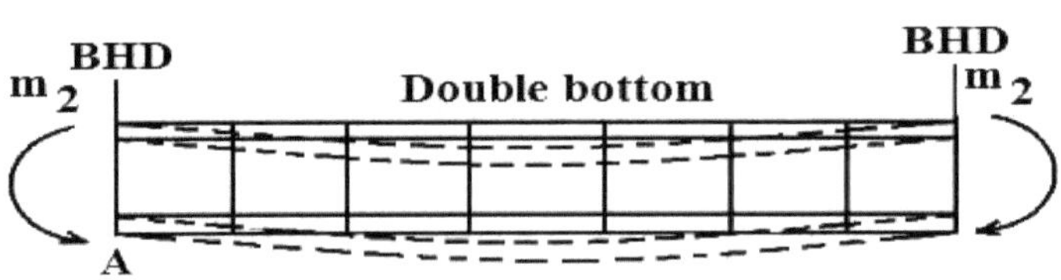

Fig. 7.9 Bending moments at the ends of the double bottom assembly

The secondary stress at the fixed end of the top edge of the web plate of the bottom girder due to bending of the bottom structure assembly is given by, see Fig. (7.10):

$$\sigma_f = \frac{m_2 . y_f}{i_g}$$

Where: i_g = second moment of area of girder cross-section about its neutral axis

The bending stress is either tensile or compressive and is linearly distributed over the depth of the bottom girder.

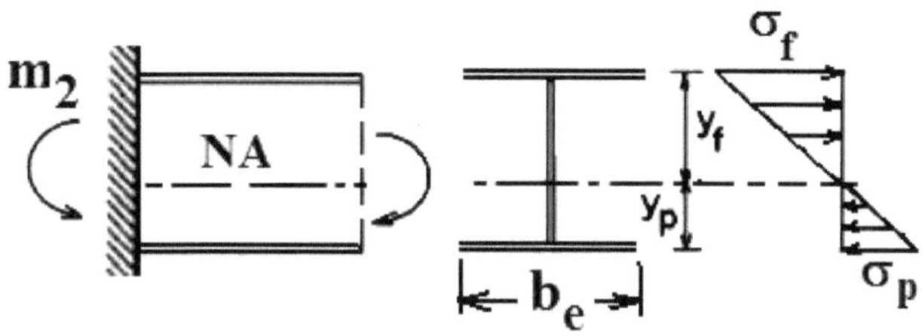

Fig. 7.10 Secondary stress distribution over a section of a bottom girder

2.3.2 Secondary Stresses in Bottom Longitudinals

The secondary stress induced at mid-span in the flange of a bottom longitudinal is given by, see Fig. (7.11):

$$\sigma_{fB} = \frac{m_3 . y_f}{i_g}$$

Where: m_3 = bending moment at mid-span of the longitudinal
i_g = second moment of area of bottom girder section about its neutral axis

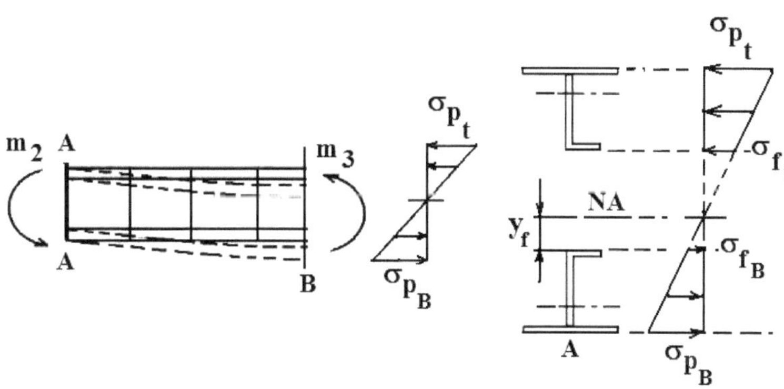

Fig. 7.11 Secondary stress induced in a bottom longitudinal at mid-span

The secondary stress induced at the clamped end in the flange of a bottom longitudinal is given by, see Fig. (7.12):

$$\sigma_{fA} = \frac{m_2 . y_f}{i_g}$$

The bending moments m2 and m3 can be determined by solving the double bottom structure assembly as a grillage structure subjected to the internal cargo loads and the external water pressure.

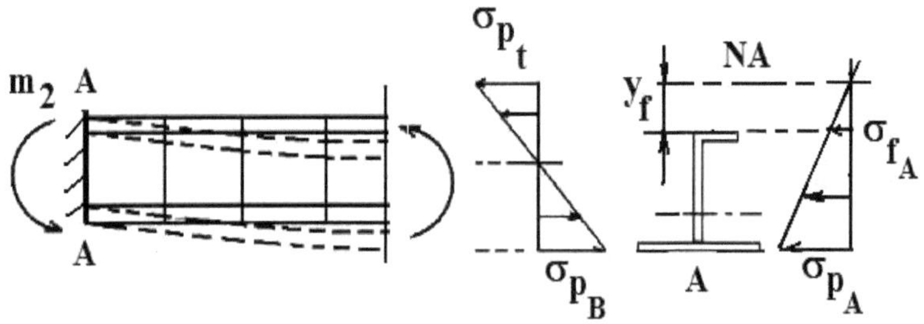

Fig. 7.12 Secondary stress induced in a bottom longitudinal at the clamed end in longitudinally stiffened double bottom assembly

2.3.3 Secondary Stresses in Bottom Plating

The secondary stress induced in bottom plating of longitudinally stiffened bottom assemblies can be calculated in the same way as for transversely stiffened bottom assemblies as follows, see Fig. (7.13):

$$\sigma_p = \frac{m_2 . y_p}{i_g}$$

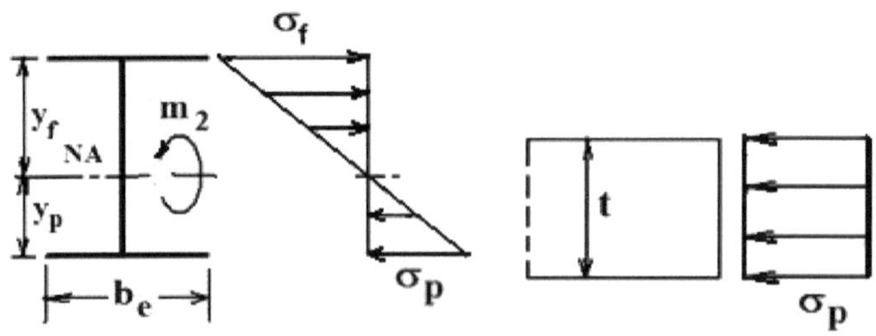

Fig. 7.13 Secondary stress induced in bottom plating of longitudinally stiffened bottom assemblies

2.3.4 Secondary Stresses in Tank Top Longitudinals

The secondary stress induced in the flange of a tank top longitudinal in longitudinally stiffened double bottom assembly is given by, see Fig. (7.14):

$$\sigma_{f1} = \frac{m_2 . y_{ft}}{i_g}$$

Where:

σ_{pt} = secondary stress induced at the tank top plating and is given by, see Fig. (7.14):

σ_{f2} = Secondary stress induced in the flange of a bottom longitudinal in longitudinally stiffened double bottom assembly

y_{ftt} = distance of tank top longitudinals from the neutral axis of the double bottom structure

i_g = second moment of area of bottom girder

$$\sigma_{pt} = \frac{m_2 . y_p}{i_g}$$

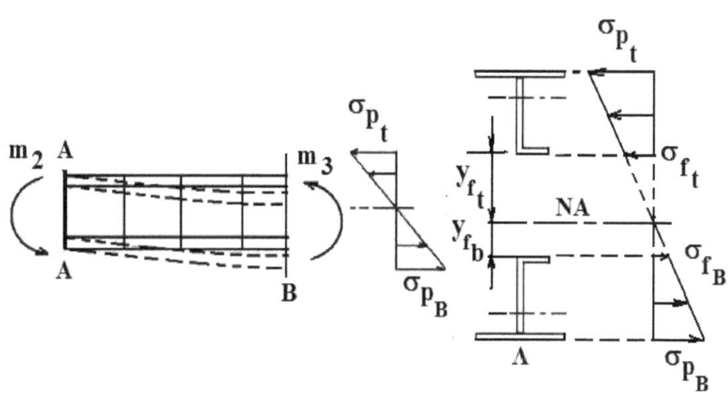

Fig. 7.14 Secondary stresses in tank top longitudinals

2.3.5 Secondary Stresses in Tank Top Plating

The induced secondary stress in the tank top plating is given by, see Fig. (7.15):

$$\sigma_f = \frac{m_2 . y_f}{i_g}$$

Where: m_2 = bending moment at the clamped end of the bottom girder

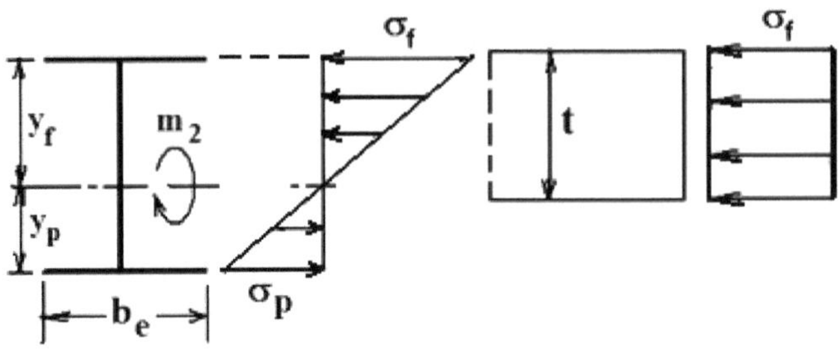

Fig. 7.15 Secondary stress induced in the tank top plating of a longitudinally stiffened bottom structure

3 Secondary Loading and Stresses in Deck Structure Assemblies

Deck girders are essential strength members in the deck structure of cargo ships carrying deck cargo. The deck structure can be modeled as a grillage structure. The intensity of loading of deck cargo on deck girders can be calculated from the magnitude of the cargo load and the loaded area of the deck girder. The deck girder can be assumed clamped at both ends at the connection with transverse bulkheads.

3.1 Secondary Stresses in Transversely Stiffened Deck Structure

Strength members of transversely stiffened deck structure sustaining secondary stresses are deck girders and deck plating. The strength members of longitudinally stiffened deck structure assemblies are the deck girders, deck longitudinals and deck plating.

3.1.1 Secondary Stresses in Deck Girders

The secondary stress induced in the face plate of deck girder is given by, see Fig. (7.16):

$$\sigma_f = \frac{m_2 . y_f}{i_g}$$

Where: m_2 = bending moment at the clamed end of the deck girder

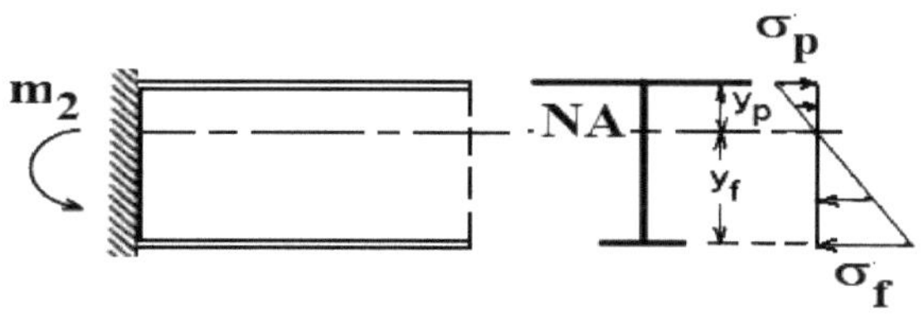

Fig. 7.16 Secondary stress induced at the clamped end of a deck girder

3.1.2 Secondary Stresses in Deck Plating

The secondary stress induced in the deck plating can be calculated from the induced stresses over the section of deck girder, see Fig. (7.17).

The induced stress in the deck plating is given by:

$$\sigma_p = \frac{m_2 . y_p}{i_g}$$

Where: m_2 = bending moment at the clamed end of the deck girder
σ_p = secondary stress induced in deck plating
y_p = Distance of deck plating from the neutral axis of section of the deck girder

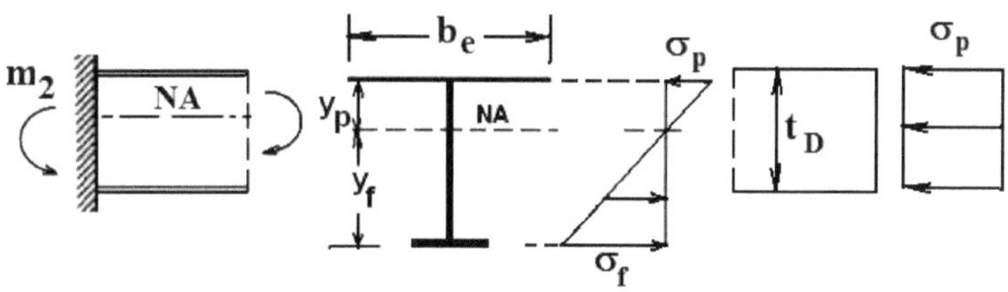

Fig. 7.17 Secondary stress induced in the deck plating

3.2 Secondary Stresses in Longitudinally Stiffened Deck Structure

3.2.1 Secondary Stresses in Deck Girders

The secondary stress induced in the face plate of deck girder at the clamped end is given by, see Fig. (7.18):

$$\sigma_f = \frac{m_2 . y_f}{i_g}$$

Where: σ_f = secondary stress induced in the face plate of the deck girder
m_2 = bending moment at the clamped end of the girder

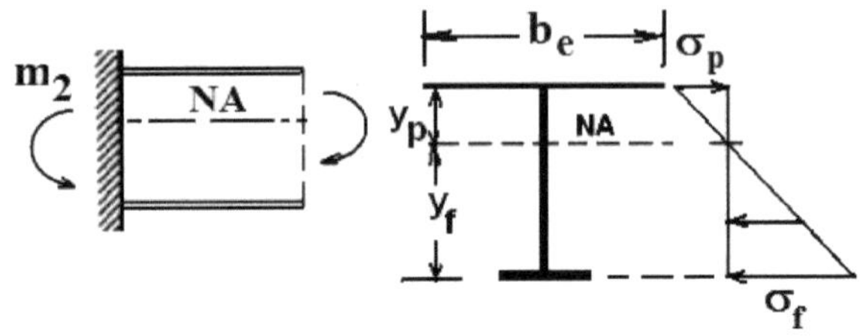

Fig. 7.18 Secondary stress at the face plate of the deck girder

3.2.2 Secondary Stresses in Deck Longitudinals

The secondary stress induced at the flange of a deck longitudinal at the clamped end is given by, see Fig. (7.19):

$$\sigma_f = \frac{m_2 . y_f}{i_l}$$

Where:

σ_f = Tensile secondary stress induced in the flange of deck longitudinal
i_l = second moment of area of the cross-section of deck longitudinal about its neutral axis
m_2 = bending moment at the clamed end of the deck longitudinal

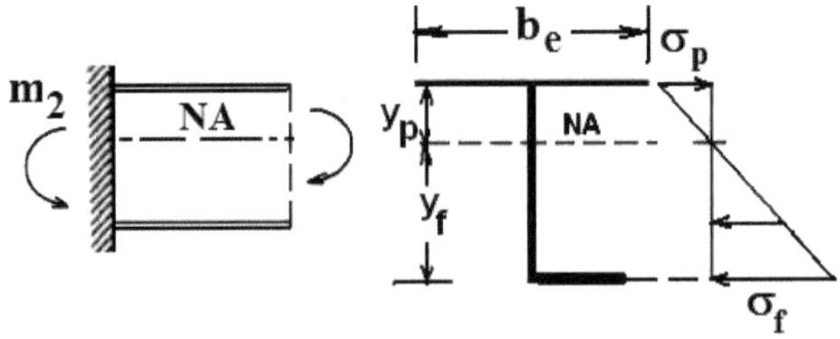

Fig. 7.19 Secondary stress induced over the cross-section of a deck longitudinal

4 Secondary Stresses in Bottom Structure Assemblies of Oil Tankers

The strength members of the bottom structure assemblies of single hull oil tankers sustaining secondary stresses are: girders, longitudinals and bottom plating.

4.1 Secondary Stresses in Bottom Girders

The secondary stresses in bottom girders are induced by the bending moments obtained from the analysis of the modeled bottom structure as a grillage structure, see Fig. (7.20).

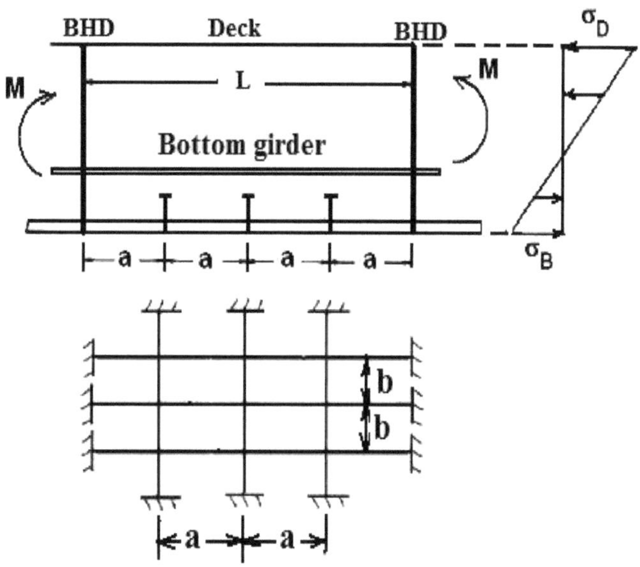

Fig. 7.20 Idealization of bottom structure by a grillage structure

The secondary stress induced in the face plate of bottom girders is calculated at section A, assumed clamed at the connection with the transverse bulkhead, and at mid-span, section B, see Fig. (7.21).

The secondary stress in the face plate of the bottom girder at the clamped end is given by, see Fig. (7.21):

$$\sigma_{fA} = \frac{m_2 . y_f}{i_g}$$

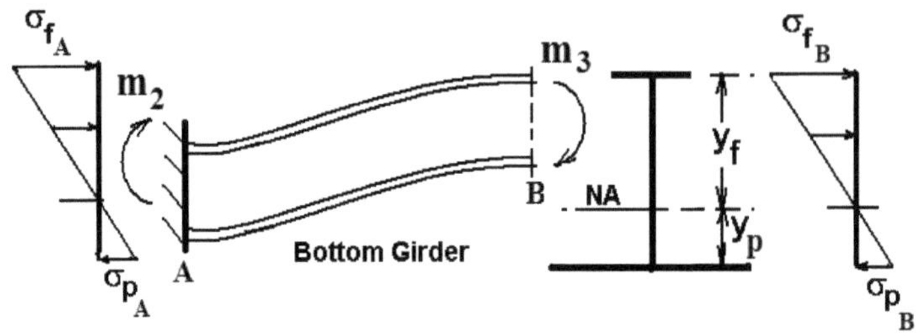

Fig. 7.21 Secondary stresses induced at section A, the clamped end, and at section B, mid-span of the bottom girder

The secondary stress in the face plate of the bottom girder at mid-span is given by, see Fig. (7.21):

$$\sigma_{fB} = \frac{m_3 . y_f}{i_g}$$

Where: σ_{fA} = Compressive secondary stress induced in the face plate of the bottom girder at section A

σ_{fB} = Tensile secondary stress induced in the face plate of the bottom girder at section B

m_2 = bending moment at the clamped end of the girder

m_3 = bending moment at mid-span of the girder

4.2 Secondary Stresses in Bottom Longitudinals

The secondary stresses in bottom longitudinals could be determined by analyzing the bottom structure as a grillage structure, see Fig. (7.22). Alternatively, the secondary stresses induced in bottom longitudinals could be determined assuming a bottom longitudinal as a continuous beam clamped at both ends at the transverse bulkheads and supported by the heavy transverse members, see Fig. (7.23).

The secondary stress induced in the face plate of a bottom longitudinal is given by, see Fig. (7.23):

$$\sigma_f = \frac{m_2 . y_f}{i_l}$$

Where: σ_f = secondary stress induced in the face plate at the clamed end of bottom longitudinal

Alternatively, the secondary stresses induced in bottom longitudinals could be calculated assuming that the bottom longitudinal is a continuous beam clamped at both ends and supported on elastic supports provided by the heavy transverse

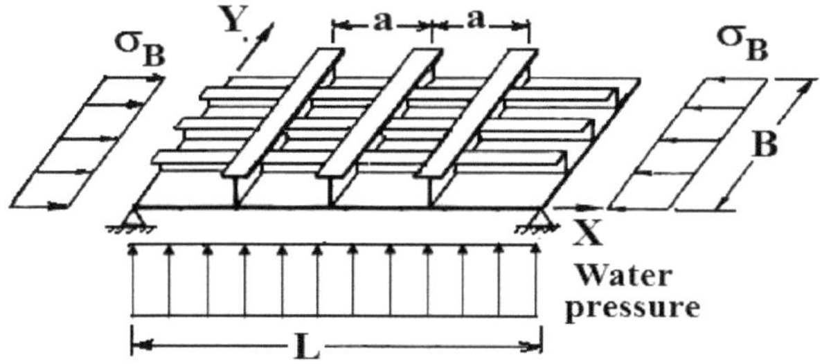

Fig. 7.22 Configuration and loading on a part of bottom structure of an oil tanker

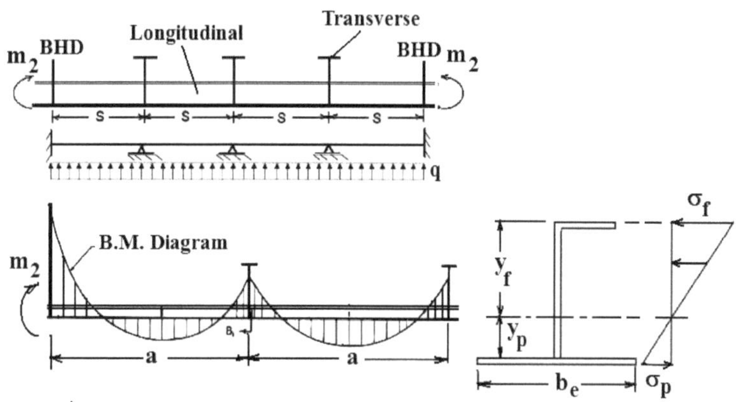

Fig. 7.23 Secondary stresses induced in a bottom longitudinal at the clamped end at the transverse bulkhead

members, see Fig. (7.24). The constants of elastic supports for a beam simply supported at both ends and loaded by a concentrated load at mid-span is given by, see Fig. (7.25):

$$k_1 = \frac{48EI_1}{a^3}$$

Where: k_1 = spring stiffness of the beam

I_1 = second moment of area of beam cross-section about its neutral axis

E = Modulus of elasticity of beam material

a = span od beam

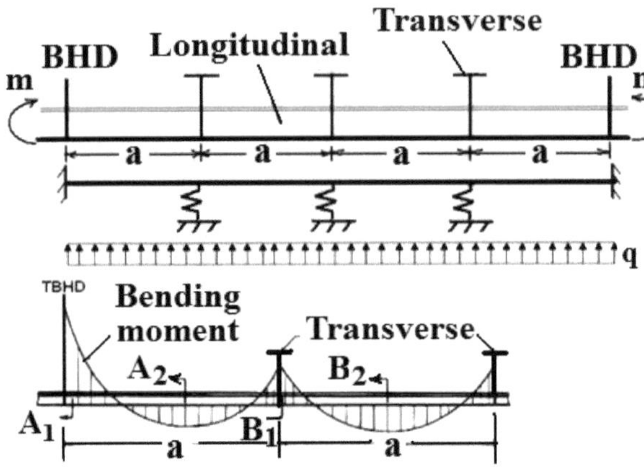

Fig. 7.24 A bottom longitudinal idealized by a continuous beam clamped at both ends and supported by elastic supports

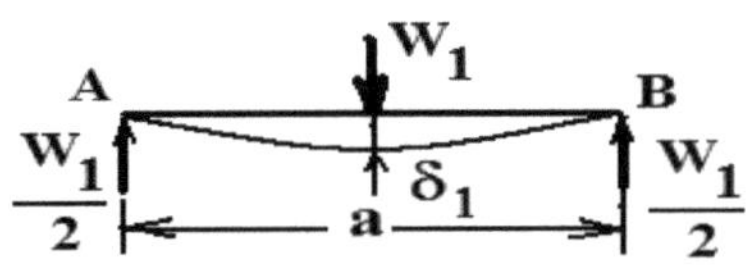

Fig. 7.25 A beam simply supported at both ends and loaded by a concentrated force at mid-span

4.3 Secondary Stresses in Bottom Plating

The secondary stress induced in the bottom plating of conventional oil tanker is the secondary stress induced in the bottom flange of the bottom girder and is given by, see Fig. (7.26):

$$\sigma_p = \frac{m_2 . y_p}{i_g}$$

Where: σ_p = secondary stress induced in the bottom side of the bottom plating

i_g = second moment of area of the bottom girder cross-section about its neutral axis

m_2 = bending moment at the clamped end of the girder

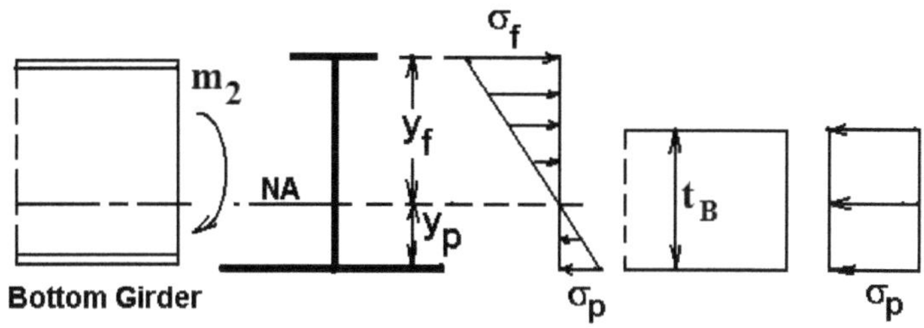

Fig. 7.26 Secondary stress induced in the bottom plating of oil tanker

5 Grillage Structure

The simplest problem of a grillage structure is a grillage composed of two cross connected members loaded by a concentrated force at their point of intersection as shown in Fig. (7.27).

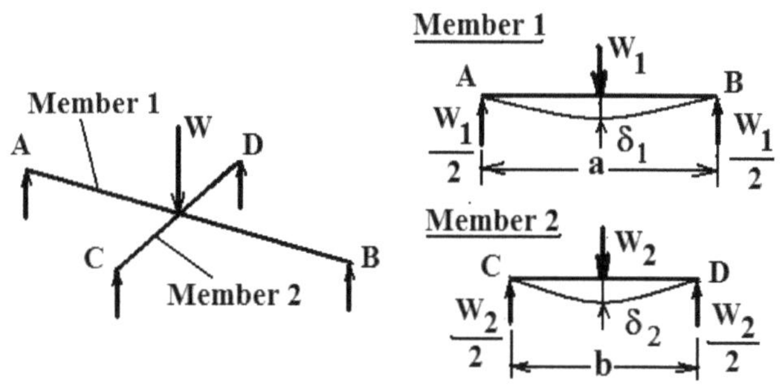

Fig. 7.27 A grillage structure composed of two cross connected members

In order to determine the bending moments at mid-span of each member, the following information must be available for each member: length, second moment of cross-section area of each member and modules of elasticity.

Assume that member 1 is loaded at its mid-span by the load W1 and the second moment of area of its cross-section is I_1 and member 2 is loaded at its mid-span by the load W2 and the second moment of its cross-section is I_2. From Fig. (7.27), it is evident that:

$$W = W_1 + W_2$$

Member 1 is deflected by δ1 and member 2 is deflected by δ2. The deflections at mid-span for members 1 and 2 are given by:

$$\delta_1 = \frac{W_1 a^3}{48 E I_1}$$

$$\delta_2 = \frac{W_2 b^3}{48EI_2}$$

For compatibility conditions, we have:

$$\delta_1 = \delta_2$$

Hence: $\dfrac{W_2 b^3}{I_2} = \dfrac{W_1 a^3}{I_1}$

Substituting we get:

$$\delta_1 = \delta_2 = \frac{Wa^3}{48EI_1} \frac{1}{\left(1 + \dfrac{I_2 a^3}{I_1 b^3}\right)}$$

Hence: $W_1 = \dfrac{48EI_1\delta_1}{a^3}$

The maximum bending moment for members 1 and 2 are given by:

$$M_1 = \frac{W_1 a}{4} = \frac{Wa}{4\left(1 + \dfrac{I_2 a^3}{I_1 b^3}\right)} \quad \textit{and} \quad M_2 = \frac{W_2 b}{4} = \frac{Wb}{4\left(1 + \dfrac{I_1 b^3}{I_2 a^3}\right)}$$

Chapter 8
Tertiary Loading and Stresses in Strength Members of Ships

1 Introduction

This chapter gives a comprehensive analysis of the tertiary loadings and stresses induced in the various strength members of longitudinally and transversely stiffened ship structure. The strength members of longitudinally and transversely stiffened deck and bottom structures are specified. The tertiary loadings and stresses induced in deck, bottom and tank top longitudinals and plating are presented. The local loadings and stresses induced in transversely and longitudinally stiffened bottom plating is explained. A method is given to determine the minimum required thickness of bottom plating of ship structure.

2 Local Loading in Transversely Stiffened Bottom Plating

Bottom plating is the strength member sustaining local loading and stresses in transversely stiffened bottom structures in addition to the primary and secondary hull girder stresses. Local loading on bottom plating comes from the hydrostatic and hydrodynamic pressures of the sea water, see Fig. (8.1).

The external hydrostatic pressure exerted on bottom plating is given by:

$$p = (T + h).\gamma$$

Where: p = Local pressure loading, t/m^2
T = draft of ship, m
h = half wave height, m
γ = sea water density t/m^3

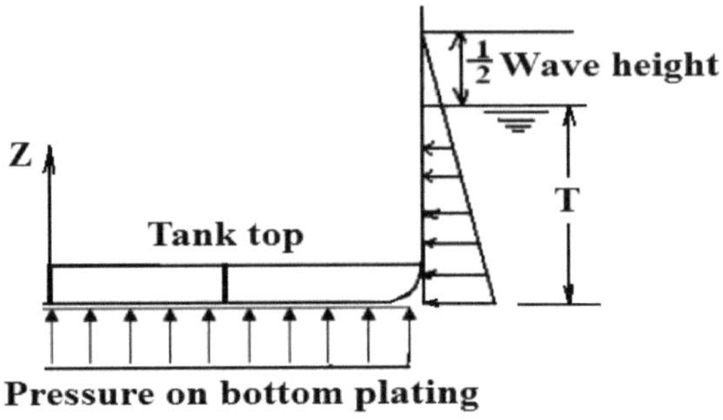

Fig. 8.1 External loading on bottom plating due to water pressure

3 Local Stresses in Transversely Stiffened Bottom Plating

In order to calculate the local stresses induced in bottom plating, it is necessary to determine the bending moments developed in the plate panels of the outer shell plating due to local loadings. Assume a strip in a plate panel of thickness "t", length "b" (m) and width 1.0 (m) in the longitudinal direction of the ship, see Fig. (8.2). The strip can be assumed as a beam simply supported or fixed at both ends. The strip will be bent transversely by the external load of water pressure.

Assuming the strip is simply supported at both ends, the maximum bending moment occurs at mid span of the strip and is given by:

$$m = \frac{q.b^2}{8} \quad t/m$$

Where: m = bending moment at midspan of the strip

b = floor spacing, m

q = uniform load on the strip exerted by the external sea water pressure

$q = p \times 1.0 = p \quad t/m$

p = Local external water pressure, t/m^2

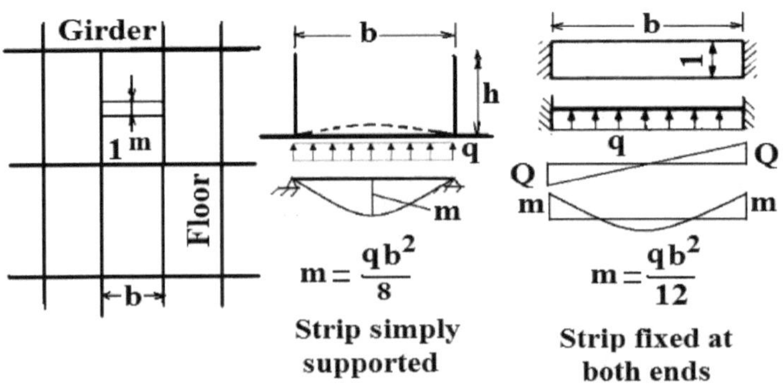

Fig. 8.2 Local bending moment in transversely stiffened bottom plating

The bending stress at midspan is given by:

$$\sigma_{pl} = \frac{m}{z} = \frac{100pb^2}{8} \bigg/ \frac{t^2}{6} = 75p\left(\frac{b}{t}\right)^2 \ t/cm^2$$

Where: σ_{pl} = Local bending stress, t/cm^2

$$z = \frac{i}{y} = \frac{100 \times t^3/12}{1000 \times t/20} = \frac{t^2}{6} \quad cm^3$$

Where: i = moment of inertia of the strip section about its neutral axis.

When the strip is assumed fixed at both ends, the bending moment is given by, see Fig. (8.2):

$$m = \frac{q.b^2}{12} \quad tm$$

The bending moment at midspan of the strip is given by:

$$m = \frac{pb^2}{24}$$

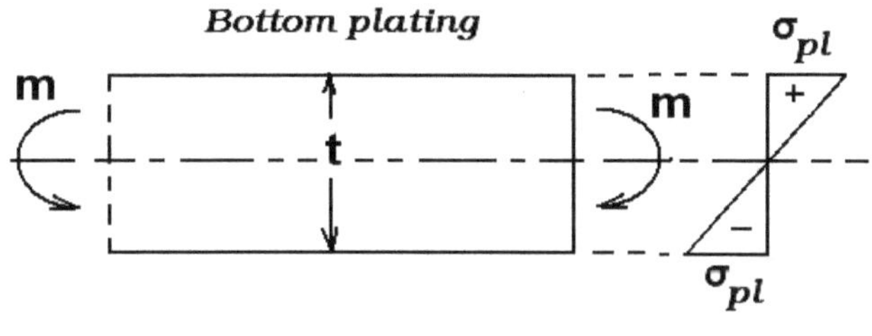

Fig. 8.3 Local bending stress over plate thickness

The local bending stress at the clamped ends is given by, see Fig. (8.3):

$$\sigma_{pl} = \frac{m}{z} = \frac{100pb^2}{12} \bigg/ \frac{t^2}{6} = 50p\left(\frac{b}{t}\right)^2 \ t/cm^2$$

The local bending stress is acting along the direction of ship length, i.e, in the same direction of the primary and secondary stresses.

Maximum deflection at mid-span of the strip assumed simply supported at both ends is given by:

$$y_B = \frac{5b^3}{384Ei}$$

4 Tertiary Loading and Stresses in the Strength Members of Longitudinally Stiffened Bottom Structure

In longitudinally stiffened bottom structures, the strength members sustaining the primary hull girder and secondary stresses as well as the tertiary stresses are the bottom and tank top longitudinals and plating.

4.1 Tertiary Loading on Bottom Longitudinals

Longitudinals are continuous members supported by equidistantly spaced floors of rigidity much greater than that of the longitudinals. Because of the high rigidity of the bottom floors, heavy transverses and transverse bulkheads, compared with the rigidity of the longitudinals, it is acceptable to assume that the longitudinals are fixed at both ends.

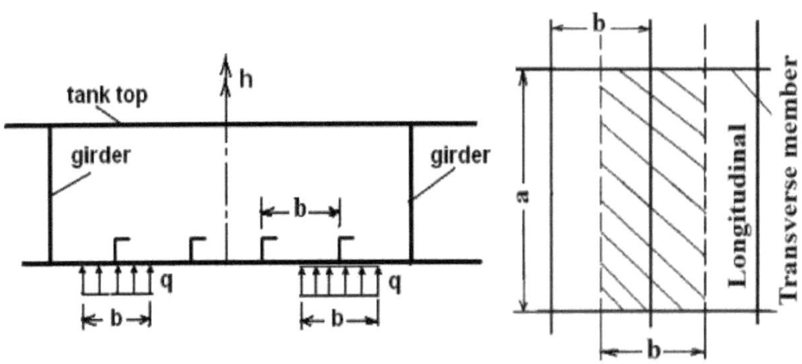

Fig. 8.4 Longitudinal Loading area subjected to the external water pressure

Each bottom longitudinal is subject to a load which is spread over a rectangular area of length “a” equal to the longitudinal length between two floors, and breadth “b” equal to the sum of the half-spaces between two adjacent longitudinals or between a longitudinal and the adjacent girder, see Fig. (8.4).

Bottom longitudinals are also subjected to in-plane normal loadings exerted on the longitudinal cross section. These normal tensile or compressive stresses are induced by hull girder and secondary bending moments.

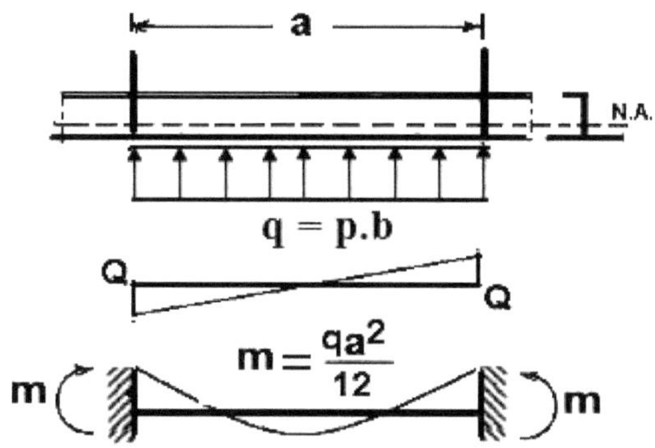

Fig. 8.5 Hydrostatic bottom loadings, shear forces and bending moments exerted on a bottom longitudinal

The hydrostatic pressure exerted on the outer shell plating represents the main local loading on bottom longitudinals. The design loads are expressed in terms of bending moments and shear forces and can be calculated from the well-known formulae given by the simple beam theory. Thus, the bending moment at a point of support, which can be either a transverse bulkhead or a floor, assuming the longitudinal clamped at both ends, is given by, see Fig. (8.5):

$$m = \frac{q.a^2}{12}$$

The bending moment at mid-span is given by:

$$m = \frac{q.a^2}{24}$$

Where: q = uniform load induced by the external hydrostatic pressure on the unsupported length of the longitudinal, tons/m

a = unsupported length of the longitudinal

b = longitudinal spacing, m

The uniform loading of the external hydrostatic pressure is given by:

$$q = p.b \;\; t/m$$

Where: $p = \gamma.h$ = intensity of water pressure acting upwards, t/m^2

h = water head over the longitudinal, see Fig. (8.6)

γ = water density

$$h = T + h_w / 2 \; m$$

h_w = wave height

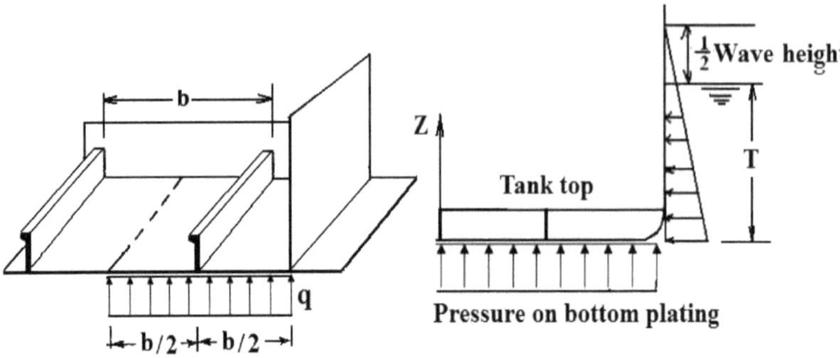

Fig. 8.6 Total width of subjected to the external water pressure inducing hydrostatic loading on bottom longitudinals

4.2 Tertiary Stress in Bottom Longitudinals

In order to calculate the bending stresses due to the straining of longitudinals, it is necessary to determine the section modulus of the longitudinal with the effective breadth "b_e" of the attached plating. The effective breadth is usually assumed equal to the smallest values of longitudinal spacing "b" or a/6, where "a" is the span of longitudinal, see Fig. (8.7).

The bending stresses due to the straining of bottom longitudinals are calculated at the outer free edge of the flange, or face plate, of the longitudinal and also at the

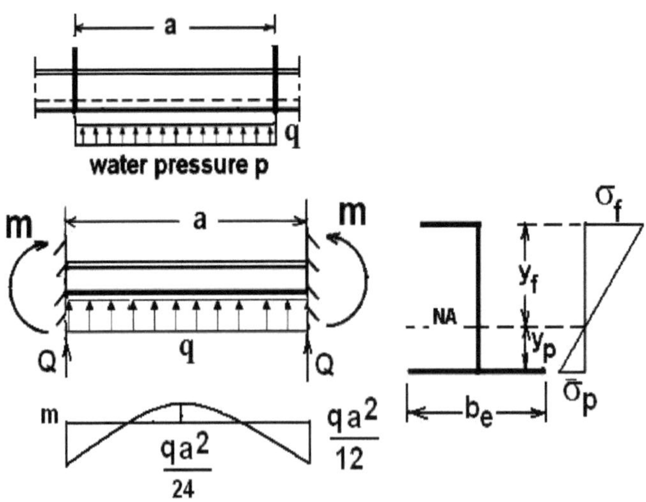

Fig. 8.7 Tertiary stresses over a section of a bottom longitudinal

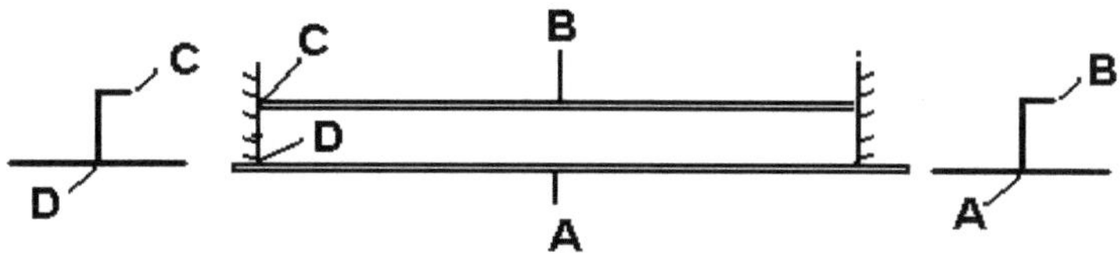

Fig. 8.8 Locations on a bottom longitudinal fort calculating tertiary stresses

bottom plating. These stresses are calculated at the two points of end support which are provided by transverse bulkheads, heavy transverse members or floors, see Fig. (8.8).

The bending stress is assumed linearly distribution over the web depth and is assumed to be uniformly distributed over the effective breadth of the plating and over the width of the face plate, see Fig. (8.9).

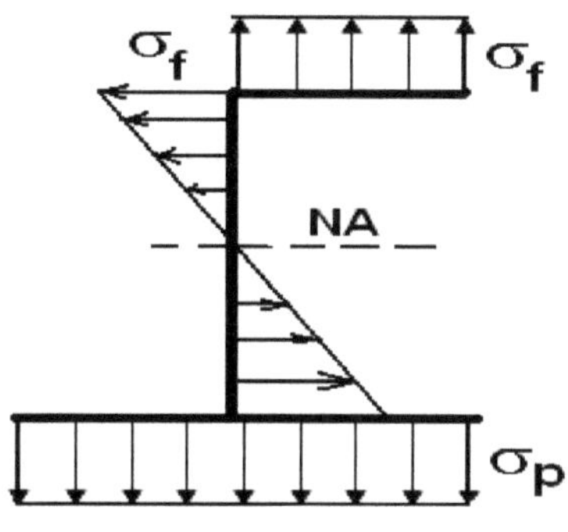

Fig. 8.9 Bending stress distribution over a typical longitudinal angle section

Consider a bottom longitudinal of an angle section with the attached plating, see Fig. (8.10). The bending stress at the attached bottom plating is given by:

$$\sigma_p = \frac{m.y_p}{i_l}$$

The bending stress at the flange of the longitudinal section is given by:

$$\sigma_f = \frac{m.y_f}{i_l}$$

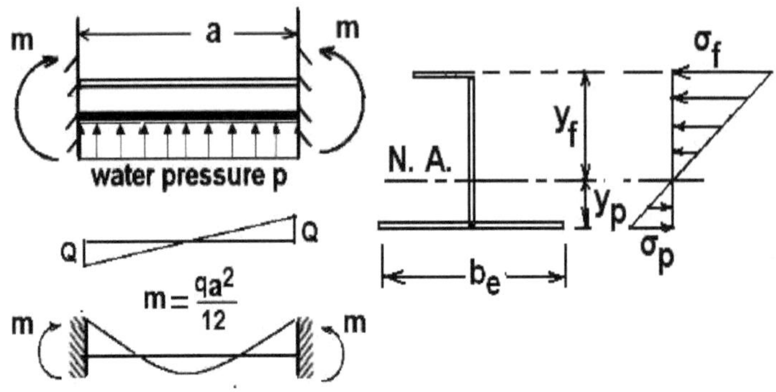

Fig. 8.10 Geometry and stress distribution over a bottom longitudinal of an angle section with the attached plating

Where: i_l = second moment of area of longitudinal section with attached plating

y_f = distance of top flange from the neutral axis of the section

y_p = distance of the attached plating from the neutral axis of the longitudinal section

4.3 Tertiary Loading on Tank Top Longitudinals

The loading on tank top longitudinals is given by:

$$q = w.b \;\; t/m$$

Where: w = intensity of cargo load/water pressure on the tank top plating, t/m^2

b = width of plating carrying the cargo load and transferring it to the tank top longitudinal, see Fig.(8.11).

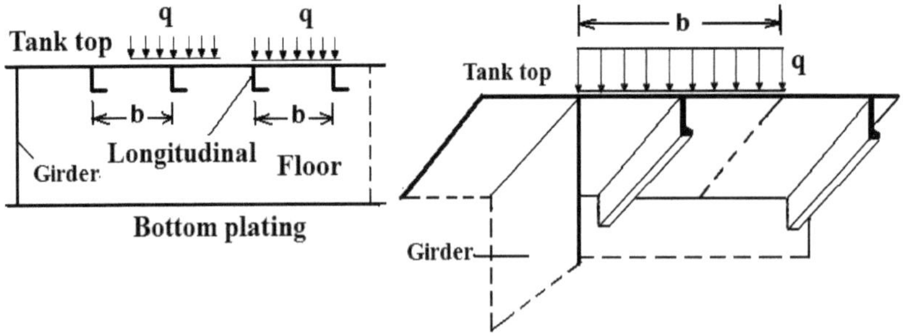

Fig. 8.11 Loading on inner bottom longitudinals

The bending moment and shear force distributions over the longitudinal length are shown in Fig.(8.12).

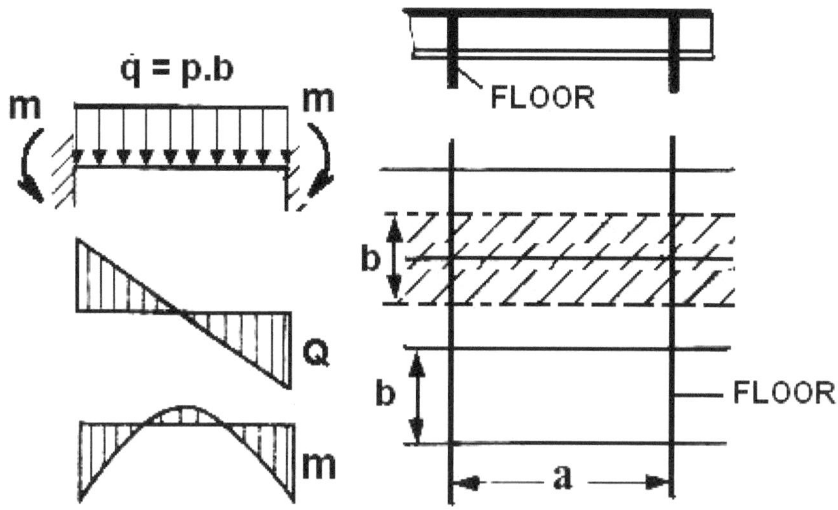

Fig. 8.12 Loading, shear force and bending moment in a strip of width 1 m in a bottom plating panel when the long ends are assumed to be fixed

4.4 Tertiary Loading on Bottom Plating

The outer bottom plating in a longitudinally stiffened bottom structure is subjected to the primary secondary, tertiary and local stresses. The tertiary loading and stresses are induced by the external sea water pressure acting normal to the outer plate surface, see Fig. (8.13). The external hydrostatic water pressure on the bottom plating is assumed to be uniformly distributed over the area of each plate panel and is given by:

$$p = \gamma.h \;\; t/m^2$$

Where: h = head of sea water above the bottom plate panel in m and is given by, see Fig. (8.13):

$$h = T + \frac{1}{2}h_w \;\; m$$

γ = sea water density, t/m^3
h_w = wave height
T = ship draft, m

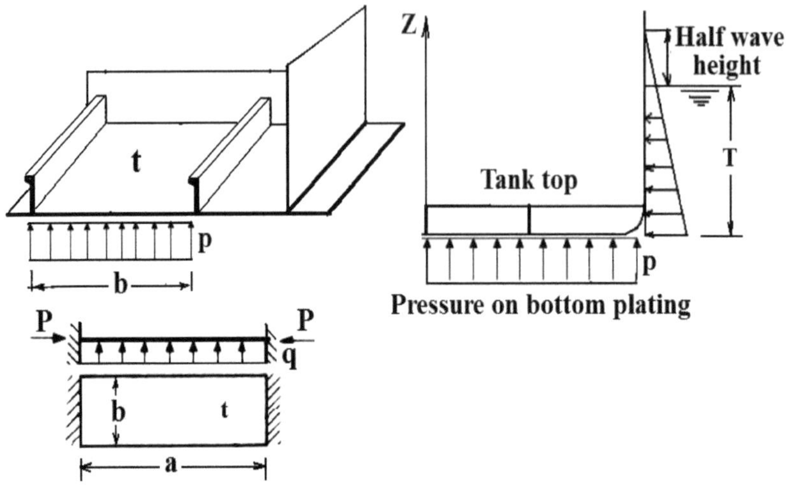

Fig. 8.13 Normal and in-plane loading on a bottom plate panel

The external loading on a bottom plate panel in a longitudinally framed bottom structure is given by, see Fig. (8.13):

$$q = p.b \ t/m$$

Consider a strip of width 1.0 m, length "b" and thickness "t", and assume the strip to be simply supported at both ends see Fig. (8.14). The maximum bending moment at midspan of the plate panel is given by:

$$m \cong \frac{q.b^2}{8} \ t.m$$

Where: m = bending moment at midspan

$$q = p \times 1.0 \ t/m$$

If the long ends of the plate panel are assumed clamped, the bending moment at both ends of the strip are given by:

$$m \cong \frac{q.b^2}{12} \ t.m$$

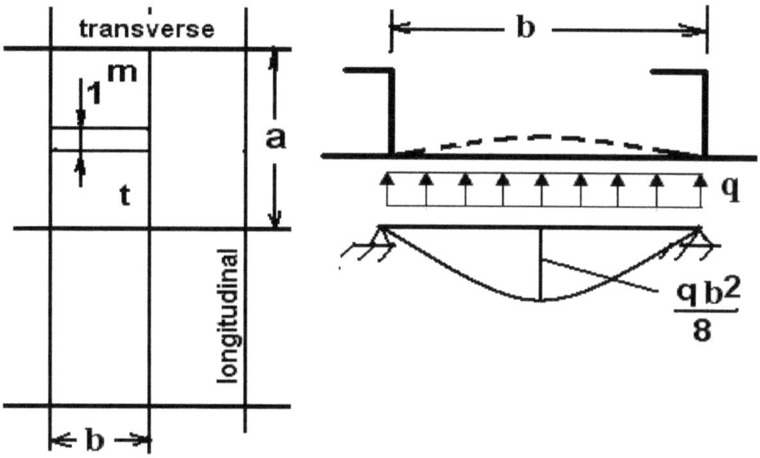

Fig. 8.14 Loading and bending moment in a strip of width 1m in a bottom plating panel when the long ends are assumed simply supported

4.5 Tertiary Stress in Bottom Plating

The tertiary stress induced in the bottom plating is the stress at the attached plating of bottom longitudinals, see Fig. (8.15). The tertiary stress could be either tensile or compressive depending on the loading, end conditions and location on the length of the bottom longitudinal. The tertiary stress in the bottom plating is calculated as follows:

$$\sigma_p = \frac{m.y_p}{i}$$

σ_p = normal stress in the attached plating, see Fig. (8.15)
y_p = distance of the bottom plating from the neutral axis of the longitudinal section

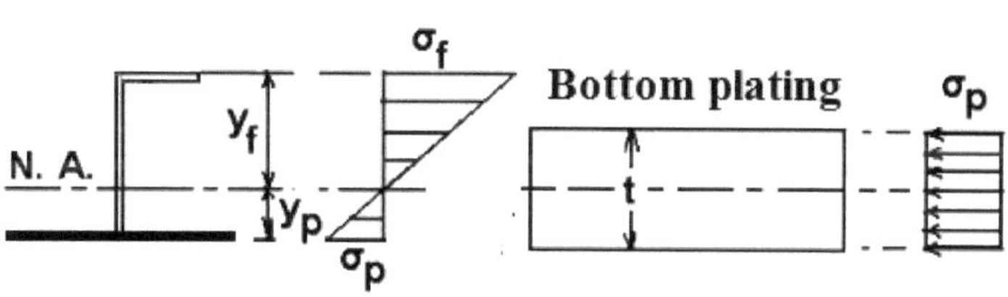

Fig. 8.15 Tertiary stresses in the bottom plating

4.6 Local Stresses in Bottom Plating

The bottom plating is supported by equidistant relatively stiff structural members (longitudinals, girders) and is loaded by the external water pressure. The local bending stresses induced in a panel of bottom plating can be calculated by assuming the panel clamped along its long sides, see Fig. (8.16). The maximum local bending stress occurs in the middle of the long side at point A, see Fig. (8.16).

Longitudinally stiffened plate panel

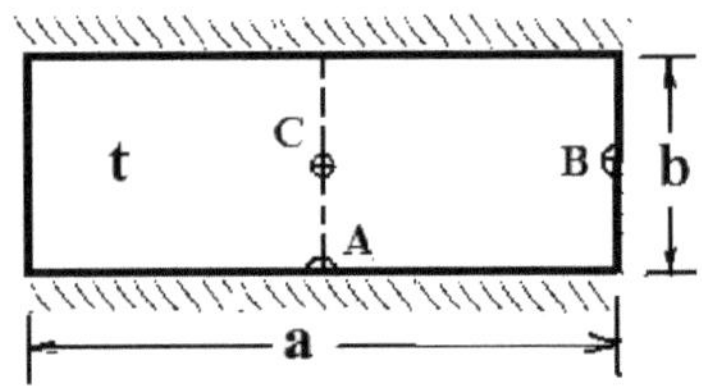

Fig. 8.16 Critical locations in longitudinally stiffened plate panels

The local stress induced in bottom plating is calculated for the condition of maximum straining. When the ratio of long to short side of the plate panel is more than three, the plate panel will be bent to a cylindrical shape. For practical purposes, the plate panel edges are assumed either clamped or simply supported. The choice of end support conditions depends on several factors among them is the torsional rigidity of the longitudinals cross-section. Apart from any rotational restraint provided by the panel supports induced by the longitudinals, there is also the restraint given by the continuity of the plating and its loading.

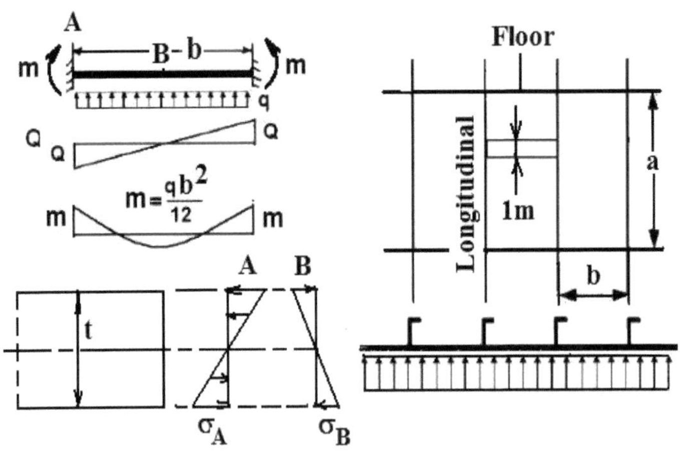

Fig. 8.17 Tertiary stresses induced in a longitudinally stiffened plate panel assumed clamped at the long edges of the plate

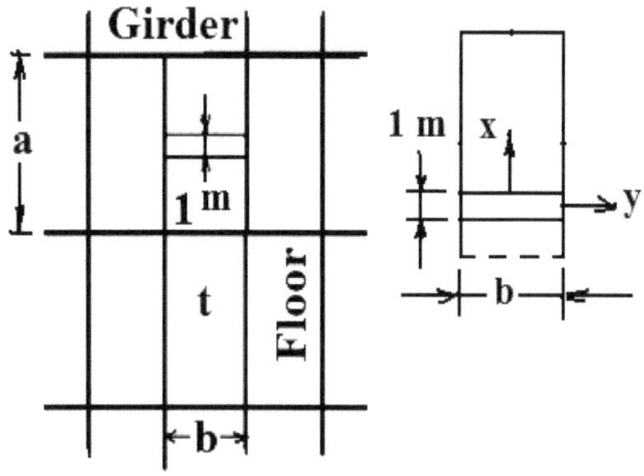

Fig. 8.18 A bottom plate panel in a longitudinally stiffened bottom structure

Consider a plate panel supported along its four edges and loaded by the external sea water pressure acting normal to the outer plate surface. The induced bending stress varies from point to point on the panel, and through its thickness, with maximum stress occurring at the top and bottom extreme fibers of the plate. The local bending stresses are calculated at point mid-way on the long side of the plate (point B), see Fig. (8.17).

For longitudinally stiffened plate panel, the local stresses induced in the bottom plating are due to the bending of the plate panel over its short length, (plate width) i.e. in the transverse direction of the ship. Consider a bottom plate panel in a longitudinally stiffened double bottom structure, see Fig. (8.18). Assume the length of the plate panel is "a" (m) in the longitudinal direction of the ship, breadth "b" (m) in the transverse direction and thickness t (mm). Assume also the plate is fixed over its long edges.

The local bending stress is acting in an orthogonal direction to ship length and is linearly distributed over the plate thickness and is given by, see Fig. (8.19):

$$\sigma = \frac{m.t/2}{i_p}$$

Where: i_p = moment of inertia of the plate strip section

$$i_p = \frac{100 \times t^3}{12 \times 1000} = \frac{t^3}{120} \quad cm^3$$

The local bending stress is given by:

$$\sigma = \frac{m \times t/20}{t^3/120} = \frac{6 \times 100 pb^2}{12t^2} = 50p\left(\frac{b}{t}\right)^2 \quad t/cm^2$$

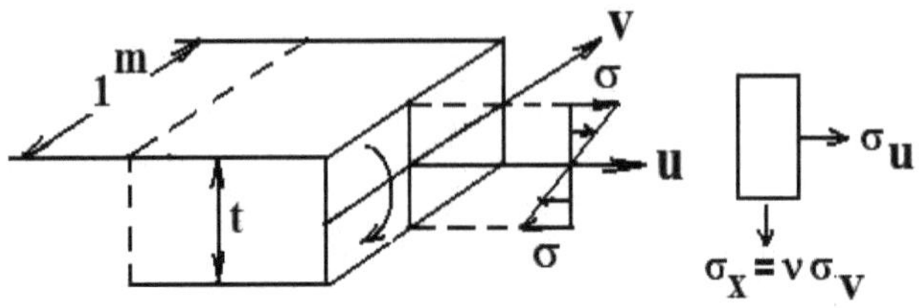

Fig. 8.19 Local bending stresses over plate thickness

For a plate panel fixed at both long edges, the local bending stress in the transverse direction (y-direction) is given by:

$$\sigma = 50\, p \left(\frac{b}{t} \right)^2 \quad t/cm^2$$

Where: p = Local pressure loading, t/m^2

Because of Poisson's effect, the local bending stress in the longitudinal direction of the plate panel (x-direction) is given by, see Fig. (8.19):

$$\sigma_u = \nu \sigma_v$$

4.7 Minimum Required Thickness of Bottom Plating

In the rules of classification societies, the minimum required plate thickness depends on:

- Stiffener spacing
- Design pressure
- Allowable stress

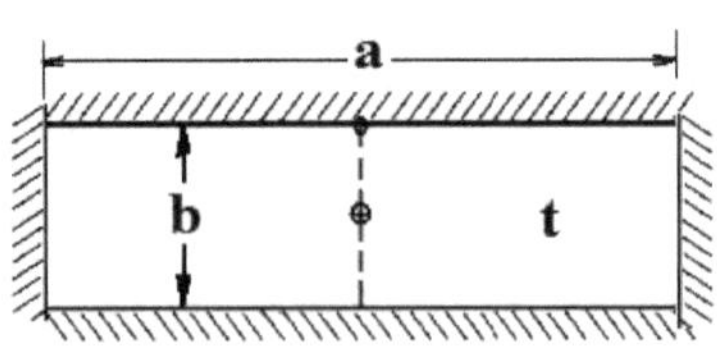

Fig. 8.20 A plate panel fixed at all edges

A plate panel between two stiffeners and two girders can be simplified by a rectangular plate fixed at all edges, see Fig. (8.20). Based on the plate theory, if the plate is subjected to uniform pressure "p", the maximum stresses at the mid-width of the plate, the stress is given by:

$$\sigma = 50p\left(\frac{b}{t}\right)^2 \quad t/cm^2$$

Hence, the required minimum thickness is given by:

$$t = \frac{b\sqrt{50p}}{\sqrt{\sigma}} = b\sqrt{\frac{50p}{\sigma}} \quad mm$$

Where,

σ = the allowable local bending stress at mid-width of plate, t/cm^2

p = the design pressure, t/m^2

b = the width of the plate panel (longitudinal spacing), m

Example: Assume p = q = 2.0 t/m, b = 0.6 m, σ_y = 2.4 t/cm2, factor of safety $\gamma = 2.0$

The allowable stress $\sigma = \sigma_y/\gamma = 2.4/2.0 = 1.2$ t/cm^2

Hence ;

$$t = 0.6\sqrt{\frac{100}{1.2}} = 5.48 \approx 5.5 \quad mm$$

In actual design, a corrosion allowance should be added to the calculated thickness.

4.8 Tertiary Loading and Stresses in Tank Top Longitudinals

The loading, shear force and bending moment distribution over a tank top longitudinal is shown in Fig. (8.21).

The tertiary stress in the face plate of a tank top longitudinal is given by, see Fig. (8.22):

$$\sigma_f = \frac{m_3 \times y_f}{i_l}$$

Where: m_3 = bending moment at the fixed end of the longitudinal

i_l = moment of inertia of the longitudinal section

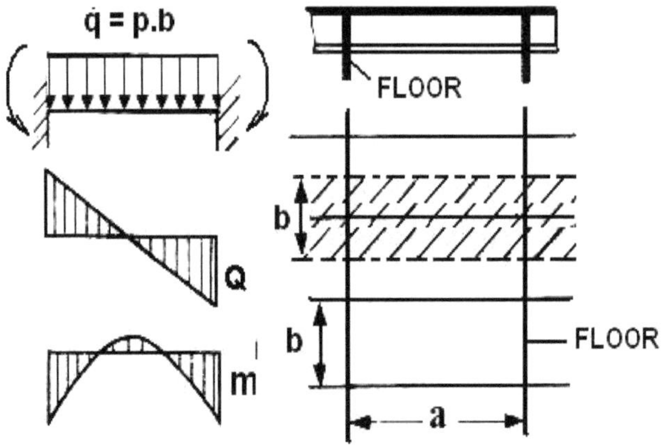

Fig. 8.21 Loading, shear force and bending moment distribution over a tank top longitudinal

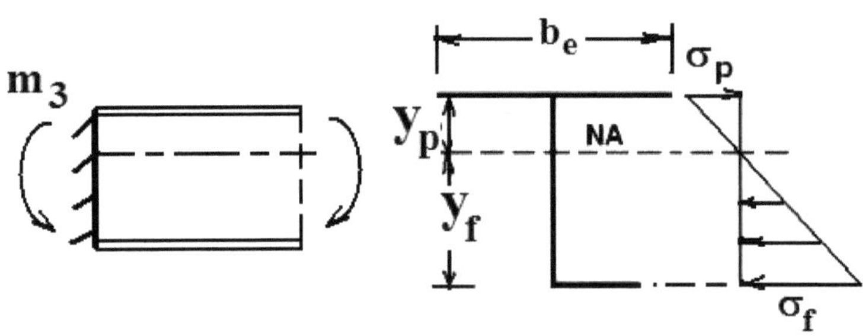

Fig. 8.22 Bending stress distribution over a tank top longitudinal section

5 Tertiary Loading and Stresses in Longitudinally Stiffened Deck Structure

The tertiary strength members of longitudinally stiffened deck structure are the deck longitudinals and deck plating.

5.1 Tertiary Loading on Deck Longitudinals

Deck longitudinals are subjected to the tertiary loading induced by the cargo loading on deck, see Fig. (8.23).

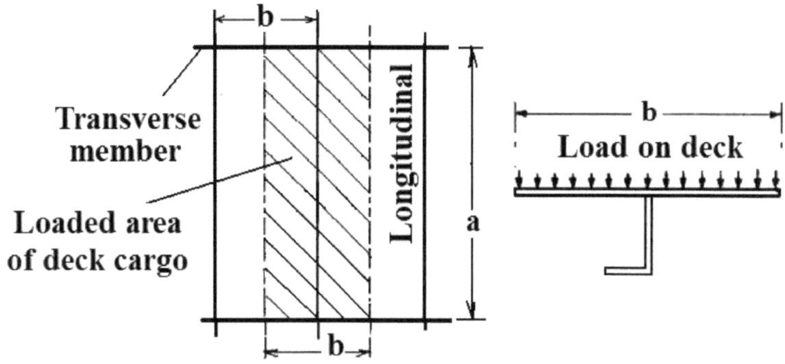

Fig. 8.23 Loaded area on deck longitudinals

The intensity of cargo loading on deck longitudinals could be calculated as follows:

$$q = w.b \ \ t/m$$

Where: $w = W_{cargo}/A_c$ t/m^2

w = Cargo load intensity, t/m^2
W_{cargo} = weight of deck cargo, tons
A_c = cargo loaded area, m^2, see Fig. (8.23)
$A_c = a.b$ m^2
a = unsupported length of deck longitudinal, m
b = longitudinal spacing, m

5.2 *Tertiary Stresses in Deck Longitudinals*

Before calculating the tertiary stresses in deck longitudinals, it is necessary to calculate the local bending moments and shear forces induced by the local loadings carried by the longitudinals. The magnitude and distribution of the bending moments and shear forces induced in deck longitudinals depend on:

- Distribution and intensity of the deck loadings
- Span length of longitudinals
- Degree of constraint at both ends of the longitudinal

Since the deck longitudinals are normally supported by heavy transverse members having sections of much higher rigidity than that of the longitudinals, it is common practice to assume that both support ends of longitudinals are clamped. Hence the bending moment at either support end of the longitudinal is given by, see Fig. (8.24):

$$m = \frac{qa^2}{12}$$

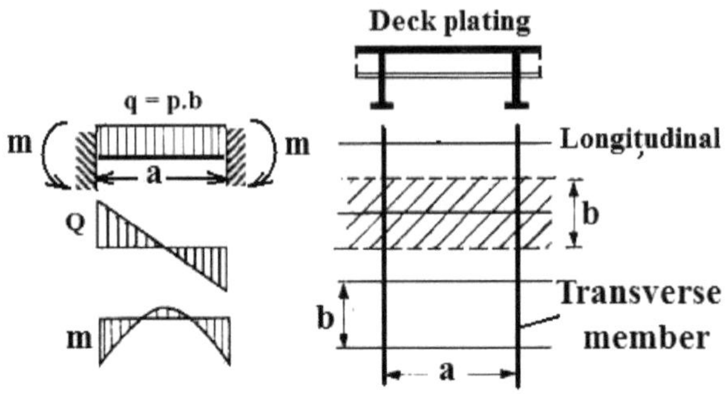

Fig. 8.24 Shear force and bending moment distribution along the unsupported span of deck longitudinals due to cargo deck loadings

The tertiary stresses in deck longitudinals are calculated using the simple beam formula given by:

$$\sigma = \frac{m.y}{i_l}$$

Where:

m = bending moment at the clamped end of the longitudinal
i_l = moment of inertia of the longitudinal section
y = distance of the point at which the stress is calculated from the neutral axis of the longitudinal section

The tertiary stresses induced in deck longitudinals depend on the geometrical configuration and scantlings of the longitudinal section. The longitudinal section is composed of the actual section of the longitudinal and attached plate, see Fig. (8.25). The width of the attached plate is called the effective breadth "b_e" which represents the sector of the deck plate participating in the section modulus of the longitudinal. The moment of inertia of the longitudinal section is therefore based on the actual longitudinal section and the attached effective breadth of deck plating, see Fig. (8.25).

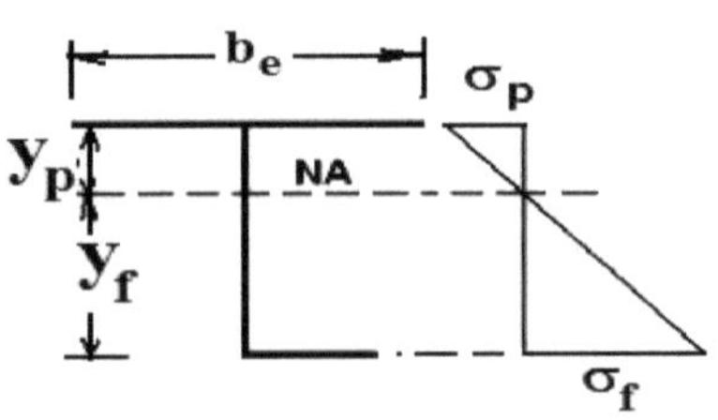

Fig. 8.25 Tertiary stress distribution over a section of deck longitudinal

For a longitudinal having an angle section, the tertiary stress at the flange of the section is given by:

$$\sigma_f = \frac{m.y_f}{i}$$

The tertiary stress at the attached plating of the section is given by:

$$\sigma_p = \frac{m.y_p}{i}$$

Where: y_f = distance of the flange from the neutral axis of the angle section, see Fig. (8.25)
y_p = distance of the attached plate from the neutral axis of the angle section

6 Local Loading and Stresses in Side Longitudinals

The design load for side longitudinals when the ship is on the crest of a wave expressed in terms of the external water pressure is given by, see Fig. (8.26):

$$q = \gamma(T + \frac{h}{2} - z)\ t/m$$

Where: z = distance of the lower edge of the strength member from the base line of the ship section
T = ship draft
h = wave height
γ = water density, t/m^3

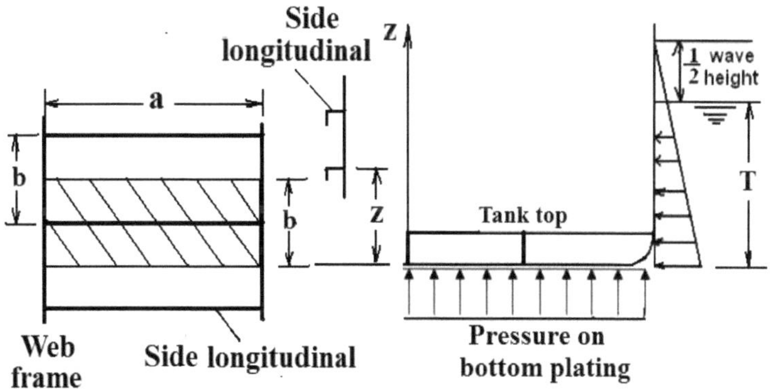

Fig. 8.26 Hydrostatic pressure loading over a ship section

The distribution of the bending moment over the length of longitudinal is as shown in Fig. (8.27) and the magnitude of the bending moment at the supporting ends is given by:

$$m = \frac{qa^2}{12}$$

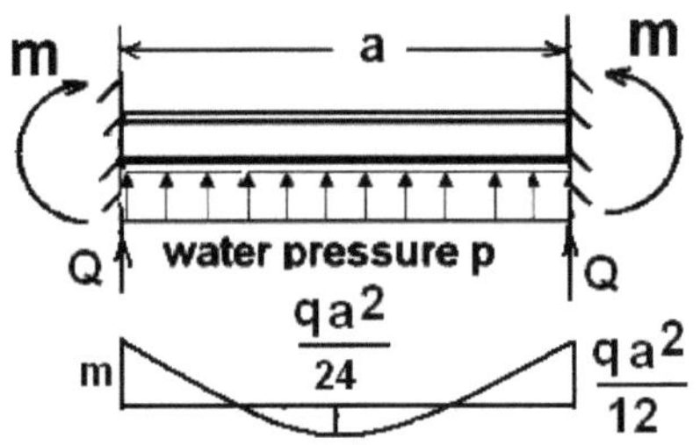

Fig. 8.27 Distribution of bending moment over the length of side longitudinal

The distribution of the bending stress over the longitudinal section is shown in Fig. (8.28). The bending stress at the flange of the longitudinal section is given by:

$$\sigma_f = \frac{m.y_f}{i_l}$$

Where: m = bending moment at the end support
y_f = distance of the flange from the neutral axis of the longitudinal section
i_l = moment of inertia of the longitudinal section about its neutral axis

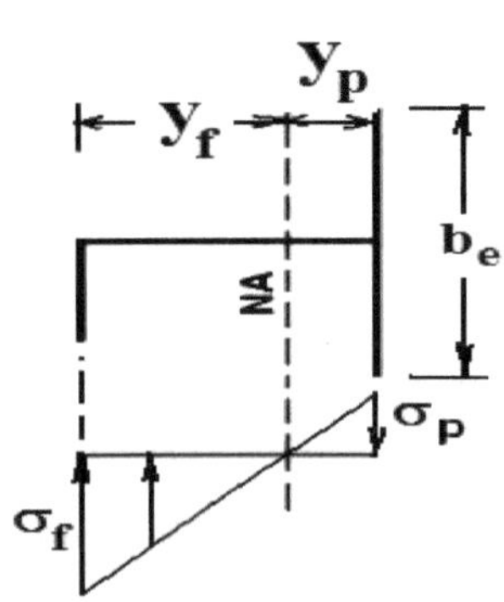

Fig. 8.28 Distribution of bending stress over the section of side longitudinal

Chapter 9
Compounding of Stresses in Ship Strength Members

1 Introduction

This chapter gives a full analysis of the compounding of stresses induced in the various ship strength members of transversely and longitudinally stiffened double bottom and deck structures. The compounding of stresses is carried out for the main ship strength members which includes girders, longitudinals and plating. The compounding of stresses induced in any strength member takes account of the primary, secondary and tertiary stresses. Depending on the location of the strength member in the ship structure, the primary stresses included in the compounding process of stresses are calculated when the ship is in sagging or hogging conditions. The compounding of stresses in tank top longitudinals and plating are also considered. The compounding of stresses of bottom plating takes account of the local stresses induced by the local loading of the external water pressure. The compounding of stresses are carried out at the locations expected to reveal the highest values of compound stresses. The strength members considered for the compounding process includes girders, longitudinals and plating in bottom and deck structures.

2 Various Stresses in Strength Members of Ship Structure

All strength members of ship structure are subjected to several types of stresses induced by external and internal loads, see Fig. (9.1). The fabrication of ship structural members by gas cutting, forming, welding and assembly work induce complicated systems of residual stresses. These residual stresses do not appear in the structural members as they are in equilibrium internally. The hull girder of a ship is subjected to longitudinal vertical and horizontal bending moments, shear forces and torsional moments. These hull girder loadings induce stresses referred to as primary stresses in the primary strength members. Strength members of ship structure assemblies are subjected to cargo and external water pressure loadings which induce stresses called secondary stresses. Tertiary strength members are subjected to tertiary loadings which induce a third type of stresses called tertiary stresses. The bottom plating is subjected to additional local loadings exerted by the external hydrostatic water pressure. These local loadings cause bending of bottom plate panels and thus induce local bending stresses.

The compounding of these stresses for some strength members could reach unacceptable high values of equivalent stresses which may exceed the allowable stresses of the material and cause structural failure.

Under certain conditions, the ship hull girder is subjected to other types of loadings and stresses. The main causes of these stresses are the presence of thermal gradients, launching of ship, occurrence of accidents such as grounding, collision, etc.

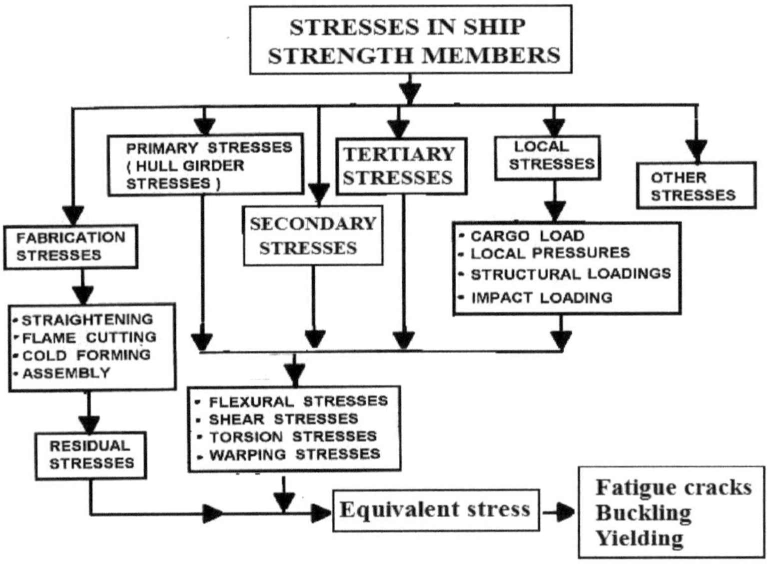

Fig. 9.1 Various types of stresses in ship structure

2.1 Total Stress Induced in Ship Structural Members

The total stress at any point on the ship hull girder is composed of one or more of the following categories of stresses:

- Primary stresses: hull girder stresses
- Secondary stresses: induced in secondary structures
- Local stresses: induced by cargo + local pressures
- Fabrication stresses: residual stresses (Cutting, forming, welding and assembly work)

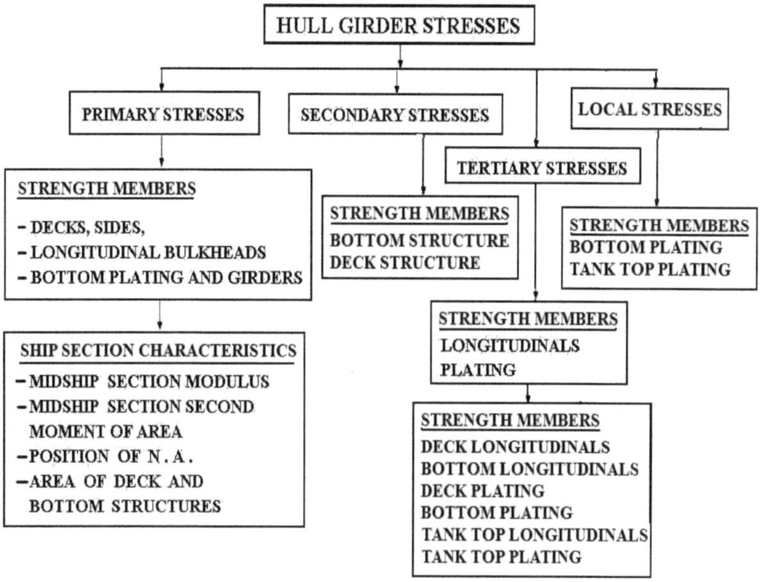

Fig. 9.2 Categories of hull girder and local stresses in strength members

Each category of these stresses is composed of several types of stress components as follows:

- The primary stress includes the flexural stresses, torsion stresses and shear stresses.
- The flexural stress includes the stillwater stress and wave induced stress (vertical, horizontal)
- The torsion stress is composed of the static torsion stress and dynamic torsion stress (shear stress, flexural warping stress)
- The shear stress is composed of the still water stress, wave induced stress and the dynamic stress

Each strength member of ship structure is subjected to one or more types of hull girder and local stresses, see Fig. (9.2).

3 Compounding of Stresses in Ship Strength Members

The most common process of calculating the strength of ship structure is based on the calculation of longitudinal strength and local strength. The stress and the strength of each ship structural member should be estimated on the basis of compounded stresses imposed in them by the longitudinal bending action of the ship hull girder and the induced bending under the local loads, see Fig. (9.3).

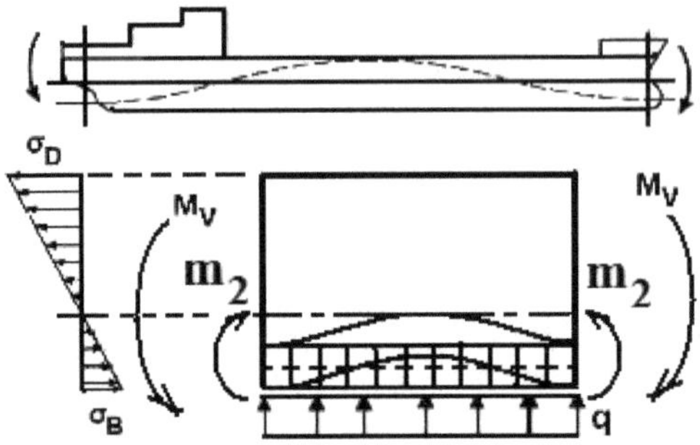

Fig. 9.3 Primary and secondary loadings for a ship in hogging condition

The compounded stresses act along the ship length and should be compared with acceptable allowable values of working stresses.

The stresses that need compounding are:

- Primary hull girder stresses
- Secondary structural assembly stresses
- Tertiary stresses
- Local plating stresses

3.1 Compounding of Stresses in Strength Members of Transversely Stiffened Double Bottom Assembly

The strength members of a transversely stiffened double bottom structure are:

- Bottom girders
- Bottom plating
- Tank top plating

3.1.1 Locations of Compounding of Stresses

For transversely stiffened bottom structures, the compounding of stresses is required at four sections located as follows, see Fig. (9.4):

- points “1” and “2” at a bulkhead
- points “3” and ‘4” in the middle of the extreme frame spacing
- points “5” and “6” at a central floor
- Points “7” and “8” in the middle of the central frame spacing.

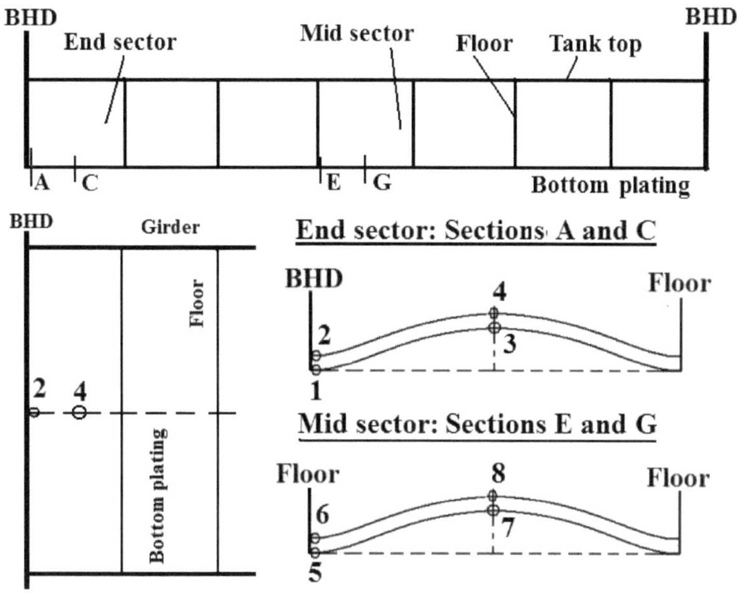

Fig. 9.4 Locations of points requiring compounding of stresses for a transversely stiffened bottom structure

3.1.2 Compounding of Stresses in Bottom Girders of Transversely Stiffened Double Bottom Structure

The compound stress in a bottom girder of transversely stiffened double bottom structure when the cargo hold is empty and the ship is in a sagging condition is composed of the primary and secondary stress components, see Fig. (9.5). The stresses are added/subtracted depending on the type of stress (tensile/compressive). The compound stress is calculated at the connection with the transverse bulkhead, see Fig. (9.5).

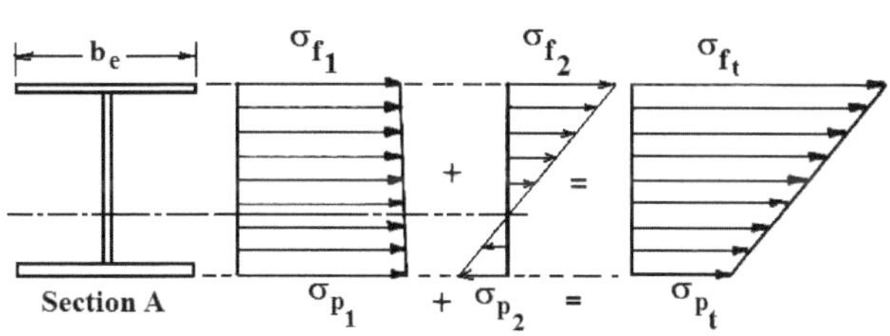

Fig. 9.5 Compound stress of the bottom girder at the section of connection with transverse bulkhead

In addition to calculating the compound bending stresses in the bottom girders, it is also necessary to calculate the shearing stresses in these bottom girders at their sections of support.

a- Primary stresses

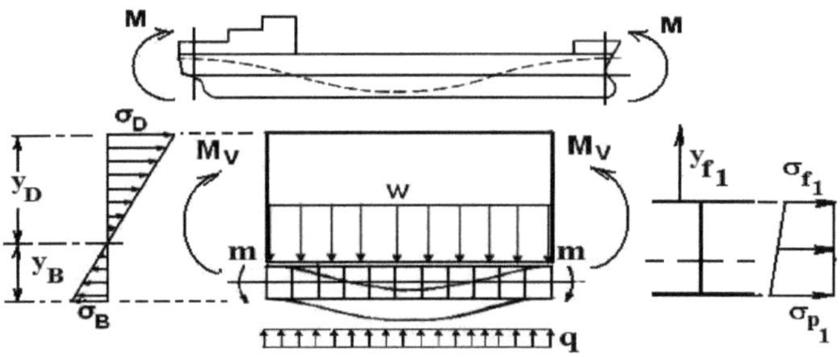

Fig. 9.6 Primary stress in a bottom girder in a sagging condition

The primary hull girder stress induced in the outer flange of the bottom girder of a ship in a sagging condition is given by, see Fig. (9.6):

$$\sigma_{p_1} = \sigma_B = \frac{M_V \cdot y_B}{I}$$

Where: M_V = Vertical hull girder sagging bending moment

y_B = Distance of the bottom plating from the neutral axis of the ship section

I = the second moment of area of the ship section about its neutral axis

$\sigma_B = \sigma_{p1}$ = Hull girder bending stress at the bottom plating

The primary stress induced in the upper flange of the bottom girder is given by, see Fig. (9.6):

$$\sigma_{f_1} = \frac{M_V \cdot y_f}{I}$$

Where: y_f = Distance of the tank top plating from the neutral axis of the ship section

$y_f = y_B - d$

d = depth of bottom girder

σ_{f1} = Hull girder bending stress at the tank top plating

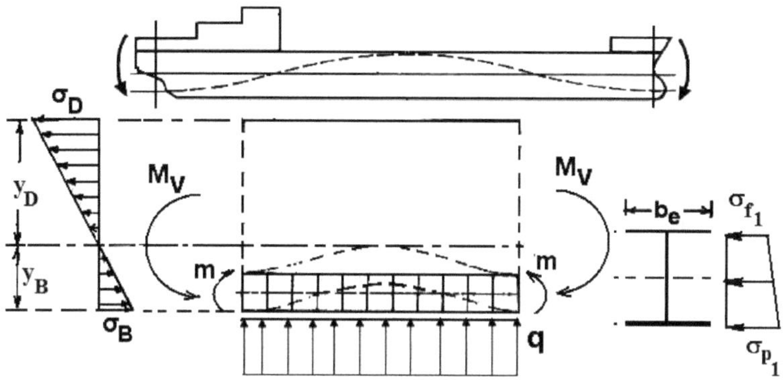

Fig. 9.7 Primary stress in a bottom girder of a ship in a hogging condition

The stress distribution over the section of bottom girder is linear with maximum value at the bottom plating, see Fig. (9.6). The primary stress in the bottom girders is tensile when the ship is in a sagging condition, see Fig. (9.6). On the other hand, when the ship is in a hogging condition, the primary stress in the bottom girders is compressive, see Fig. (9.7).

b- Secondary stresses

The secondary stresses in bottom girders could be calculated by idealizing the bottom structure over a hold length between two transverse bulkheads as a grillage structure loaded by the secondary loading and supported at the transverse bulkheads and the side shell of the ship, see Fig. (9.8).

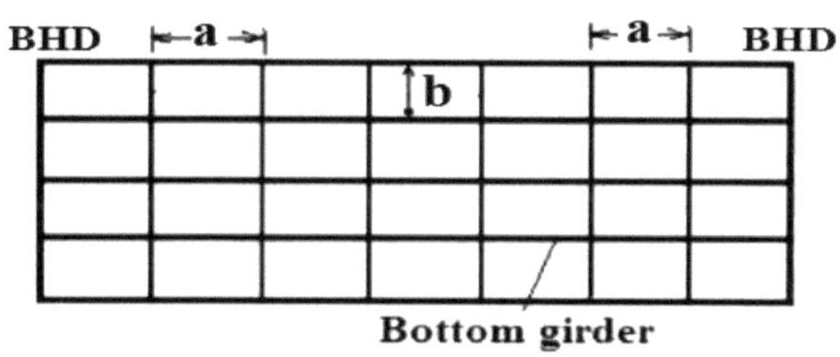

Fig. 9.8 Idealized bottom structure as a grillage structure

The secondary loading is the resultant loading induced by the cargo load and the external water pressure on the outer bottom of the shell plating and is given by:

$$q = q_{cargo} - q_w$$

Where:

q = secondary loading on the double bottom assembly, t/m

q_{cargo} = intensity of cargo loading on the loaded area of the bottom girder, t/m

q_w = intensity of water pressure loading on the loaded area of the bottom girder

When the double bottom structure assembly is subjected only to the loading of the external water pressure i.e. the hold is empty, the double bottom structure will be bent inwards, see Fig. (9.9).

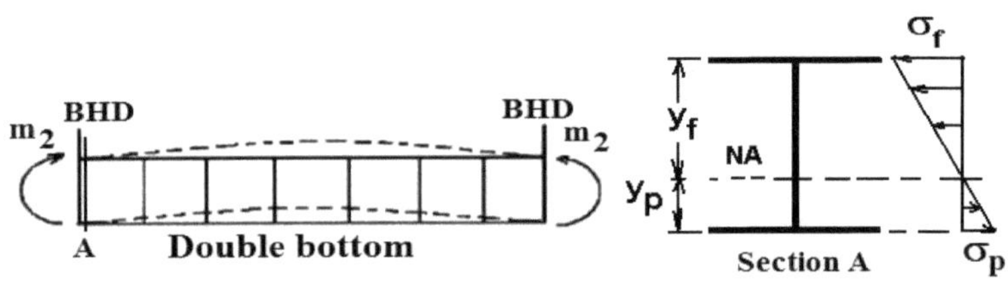

Fig. 9.9 Secondary stress of the bottom girder for an empty cargo hold at section A (at the aft transverse bulkhead)

On the other hand, when the cargo hold is fully loaded, the double bottom is bent outwards, see Fig. (9.10).

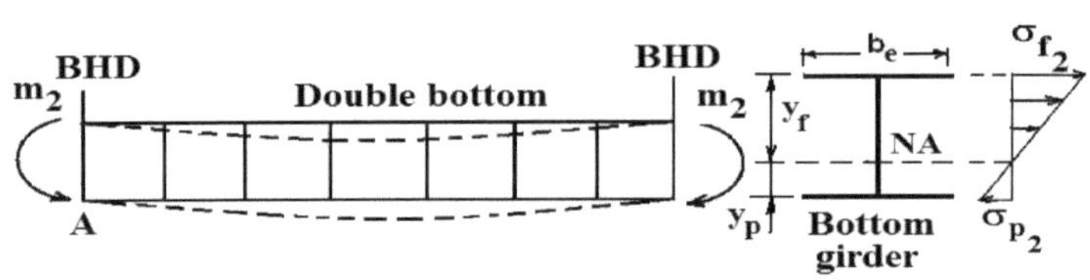

Fig. 9.10 Secondary stresses at section A of the bottom girder for a fully loaded cargo hold

The maximum secondary stress in the bottom girder occurs at section A close to the transverse bulkhead. The secondary stress at the bottom flange of the girder section is given by:

$$\sigma_p = \frac{m_2 . y_p}{i_g}$$

The secondary stress at the top flange of the girder section is given by:

$$\sigma_f = \frac{m_2 . y_f}{i_g}$$

Where: m_2 = secondary bending moment at a section close to transverse bulkhead
y_p = distance of the bottom plating from the neutral axis of the girder section, see Fig. (9.10)
i_g = second moment of area of the girder section about its neutral axis
σ_{f1} = hull girder stress at the top flange of the bottom girder
σ_{p1} = hull girder stress at the bottom flange of the girder

3.1.3 Compounding of Stresses in Bottom Plating

The compound stress in the bottom plating is composed of the primary, secondary and local stresses induced in the bottom plating, see Fig. (9.11).

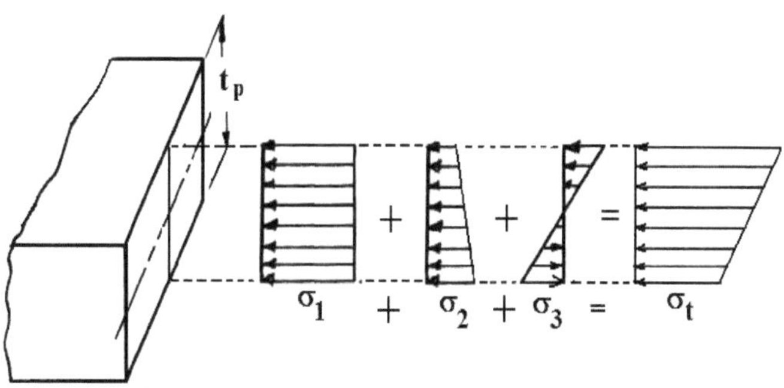

Fig. 9.11 Stress components and compound stresses in the bottom plating of a transversely stiffened double bottom structure

The compounding of stresses of the bottom plating is given here for a cargo ship having a transversely stiffened double bottom structure. The ship is assumed to be subjected to a hogging moment. The cargo hold at the midship region is assumed full and that the intensity of the cargo loading is larger than the intensity of the external water pressure. The compound stress is obtained at the section close to the transverse bulkhead.

a-Primary stress

The primary compressive stress induced in the outer bottom plating for a ship in a hogging condition is given by, see Fig. (9.12):

$$\sigma_B = \frac{M_V . y_B}{I}$$

Where: M_V = Longitudinal vertical hull girder hogging bending moment
y_B = Distance of the bottom plating from the neutral axis of the ship section
σ_B = hull girder stress in the bottom plating

The primary stress is linearly distributed over the plate thickness, see Fig. (9.12). The difference in magnitude between the primary stresses at the top and bottom surfaces is very small and could be ignored.

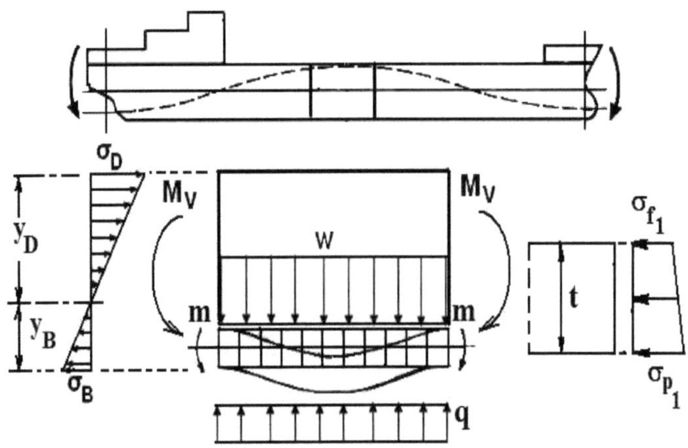

Fig. 9.12 Primary stresses in the bottom plating of a ship in sagging condition

b-Secondary stress

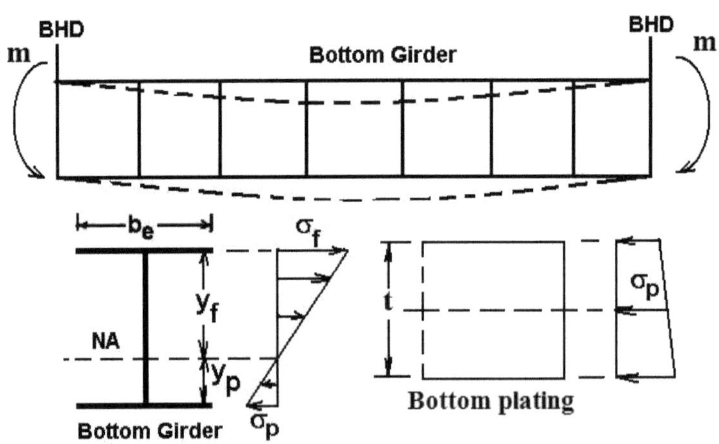

Fig. 9.13 Secondary stress in the bottom plating of a transversely stiffened double bottom structure

The secondary stress induced in the bottom plating is linearly distributed and is given by, see Fig. (9.13):

$$\sigma_p = \frac{m.y_p}{i_g}$$

Where: m = Secondary bending moment at the end bulkhead
y_p = distance of the bottom plating from the neutral axis of the girder section
i_g = second moment of area of the girder section about its neutral axis
σ_p = secondary compressive stresses at the bottom plating

The double bottom structurs is assumed heavily loaded and the stresses are determined close to the transverse bulkhead, see Fig. (9.13).

c-Local stress

The maximum local stress induced in the outer bottom plating for a transversely stiffened bottom structure occurs at a section (1-2) close to the transverse bulkhead. The calculation of the local stress could be performed using a strip of the bottom plating having a width 1.0 m and length "b" m. Because of the high rigidity of the BHD and the floor structures compared with that of the bottom plating, the strip is assumed clamped at both ends, see Fig. (9.14).

The local stress is linearly distributed over the plate thickness, see Fig. (9.14), and could be calculated as follows:

$$\sigma_{p_3} = \frac{m_3.t/2}{i}$$

Where: m_3 = bending moment for a strip of thickness "t, width 1.0 m, length "b" m clamped at both ends

$$m_3 = \frac{q.b^2}{12}$$

$$i = \frac{1.0 \times t^3}{12}$$

Hence: $\sigma_{p_3} = \frac{q.b^2.t/2}{12t^3/12} = 50q\left(\frac{b}{t}\right)^2 t/cm^2$

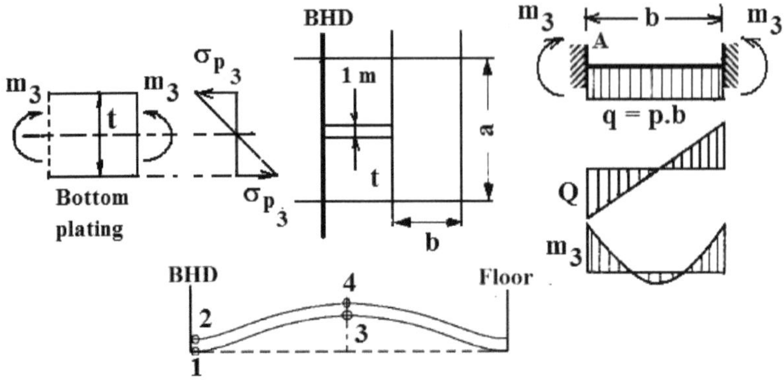

Fig. 9.14 Local stresses in the bottom plating of a transversely stiffened double bottom structure

4 Compounding of Stresses in Tank Top Plating

The compound stress in the tank top plating of transversely stiffened bottom structure is composed of the primary and secondary stresses, see Fig. (9.15).

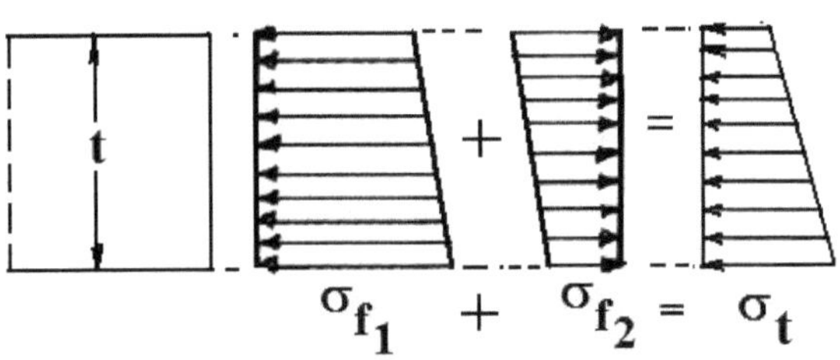

Fig. 9.15 Compound stress of tank top plating of transversely stiffened bottom structure

<u>**a-Primary stress**</u>

The primary stress induced in the tank top plating is given by, see Fig. (9.16):

$$\sigma_{f_1} = \frac{M_V . y_{TT}}{I}$$

Where: y_{TT} = distance of tank top plating from the neutral axis of ship section

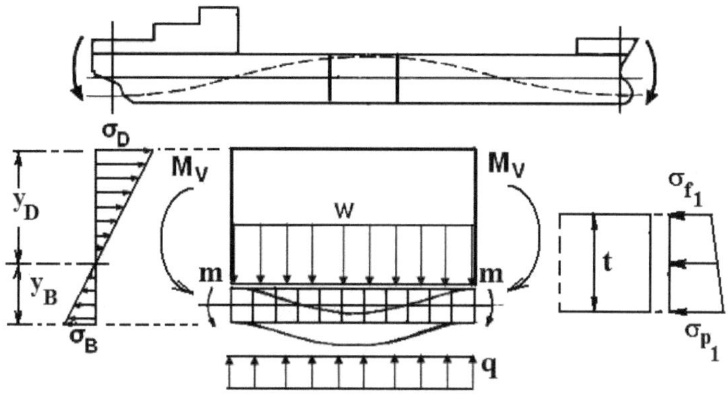

Fig. 9.16 Primary stress induced in the tank top plating

b-Secondary stress

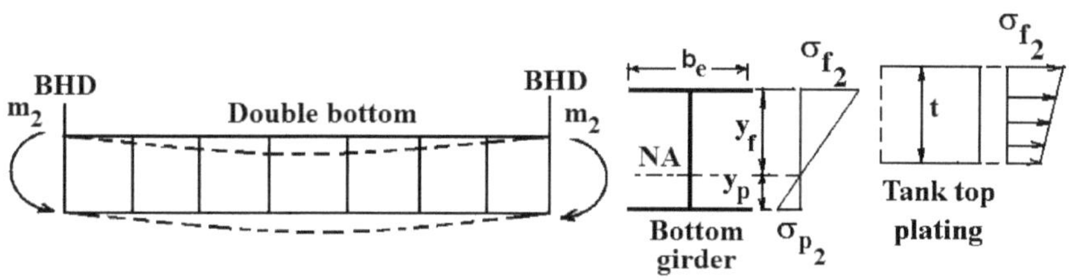

Fig. 9.17 Secondary stress induced in the tank top plating

The secondary stress induced in the tank top plating at the transverse bulkhead is given by:

$$m_{f_2} = \frac{m_2 . y_f}{i_g}$$

Where: m_2 = bending moment at the clamped end of the bottom girder
y_f = distance of the tank top plating from the neutral axis of the bottom girder section
i_g = second moment of area of the bottom girder section
σ_{f2} = secondary stress induced in the tank top plating

The secondary stress induced in the tank top plating results from the bending of the double bottom assembly, see Fig. (9.17). It is accepted to assume that the bottom girder is clamped at the transverse bulkhead.

5 Compounding of Stresses in the Strength Members of Longitudinally Stiffened Double Bottom Structure

The strength members of longitudinally stiffened double bottom structures are:

- Bottom girders
- Bottom longitudinals
- Bottom shell plating
- Tank top longitudinals
- Tank top plating

The strength members of longitudinally stiffened bottom structure are subject to the following stresses:

- Primary stress σ_1 due to the bending of ship hull girder
- Secondary stress σ_2 due to the straining of the bottom structure assembly
- Tertiary stress σ_3 due to the straining of the bottom longitudinals
- Local stress σ_4 due to the straining of the outer bottom plating

These stresses acting at certain sections and points over the section of the strength member are either added to or subtracted from one another depending on the condition of hull girder loading.

Depending on the types of stresses coexisting in the structural strength member of a hull girder, each one of these strength members can be assumed to fall under one of the following groups:

Group 1: Strength members subject to the stress σ_1 only
Group 2: Strength members subject to the stresses σ_1 and σ_2
Group 3: Strength members subject to the stresses σ_1, σ_2 and σ_3
Group 4: Strength members subject to the stresses σ_1, σ_2, σ_3 and σ_4

The bottom plating stiffened by a transverse system of framing is regarded as a strength member of the third group. On the other hand, longitudinally stiffened bottom plating is strength member of the fourth group.

The compounding of stresses is made for the following two conditions of maximum straining of ship hull girder:

- The ship is on the crest of a wave (hogging condition)
- The ship is on the trough of a wave (sagging condition)

The stresses to be taken into account are those which are directed along the ship and act over the transverse sections of plate panels.

In ship structure, the ratio of long to short side of the rectangular plate panel formed by intersecting stiffening girders is normally more than three. Plate panels, therefore, may be regarded as bent to a cylindrical shape.

5.1 Locations of Compounding of Stresses for Longitudinally Stiffened Bottom Structure

The compounding of stresses for bottom longitudinals in strength members of a longitudinally stiffened bottom structure is done at four sections which are located as shown in Fig. (9.18).

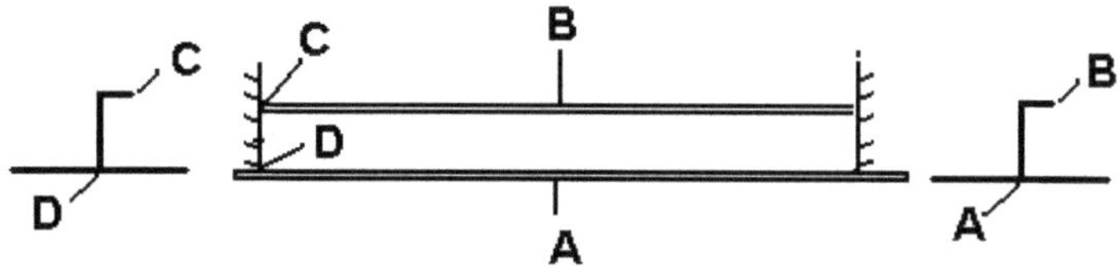

Fig. 9.18 Locations of points on bottom longitudinals requiring compounding for a longitudinally stiffened bottom structure

The details of the locations on the four sections of the bottom longitudinals requiring compounding of stresses are as follows, see Fig. (9.19):

- points "9" and "10" at a transverse bulkhead
- points "11" and "12" in the middle of the extreme frame spacing
- points "13" and "14" at a central floor
- Points "15" and "16" in the middle of the central frame spacing

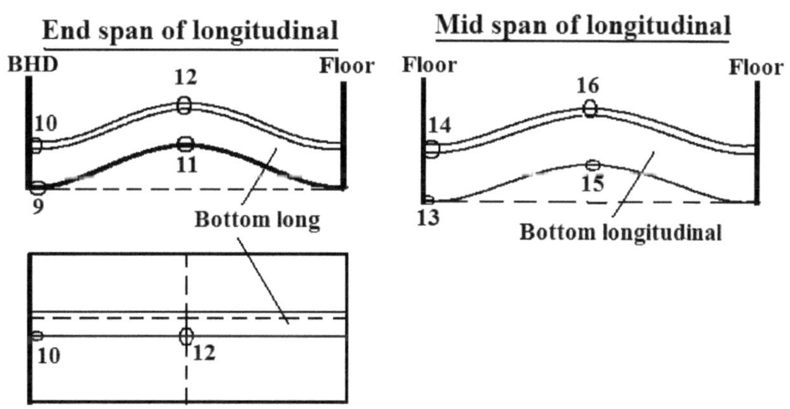

Fig. 9.19 Points of compounding stresses on a bottom longitudinal at a section close to a transverse bulkhead and at the midsection

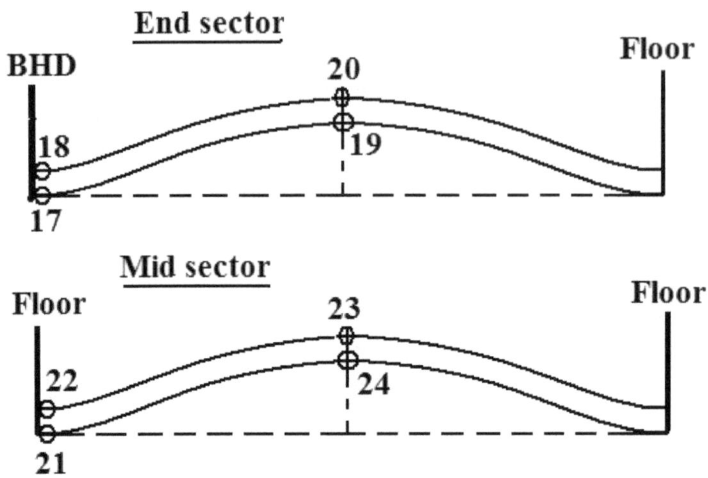

Fig. 9.20 Locations of compounding of stresses in longitudinally stiffened bottom plating

For bottom plating stiffened by longitudinal system of framing, the stresses are calculated at points mid-way on the short sides of plate panels, see Fig. (9.20). The compounding of stresses in the outside shell plating is done at four sections, see Fig. (9.20):

- points "17" and "18" at a bulkhead
- points "19" and "20" in the middle of the extreme frame spacing
- points "21" and "22" at a central floor
- Points "23" and "24" in the middle of the central frame spacing

5.2 Compounding of Stresses in a Bottom Girder

The compound stress in bottom girder of longitudinally stiffened bottom structure is composed of the primary and secondary stress components, see Fig. (9.21).

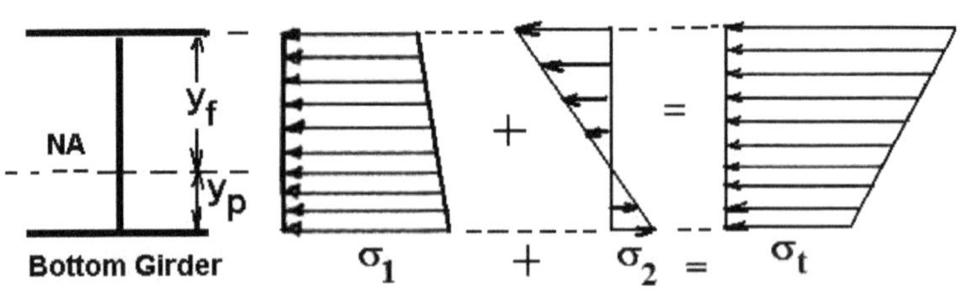

Fig. 9.21 Compounding of stresses in a bottom girder of a longitudinally stiffened bottom structure

a-Primary stress

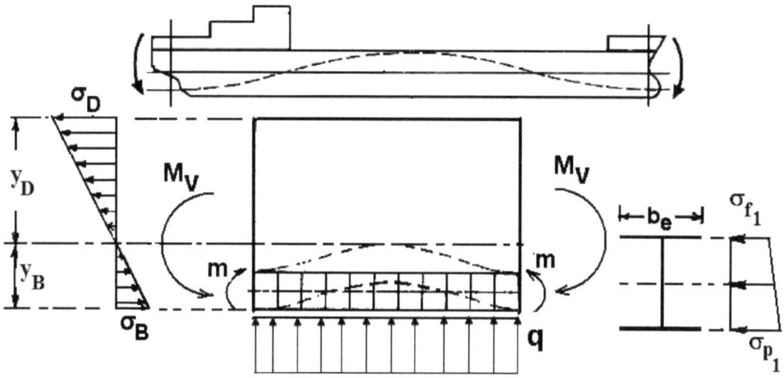

Fig. 9.22 Primary stresses in a bottom girder of a longitudinally stiffened bottom structure

The primary stress induced in bottom girder of longitudinally stiffened bottom structure of a ship in a hogging condition is given by, see Fig. (9.22):

$$\sigma_{p_1} = \sigma_B = \frac{M_V \cdot y_p}{I}$$

Where:

M_v = Longitudinal vertical hull girder bending moment

I = second moment of area of ship section about its neutral axis (NA)

y_p = distance of bottom plating from the NA of the ship section

$\sigma_{p1} = \sigma_B$ = Hull girder bending stress at the ship bottom plating

b- Secondary stress

The secondary stress in the bottom flange of the bottom girder is given by, see Fig. (9.23):

$$\sigma_{p_2} = \frac{m_2 \cdot y_p}{i_g}$$

The secondary bending can be calculated by solving the bottom structure as a grillage structure, see Fig. (9.24).

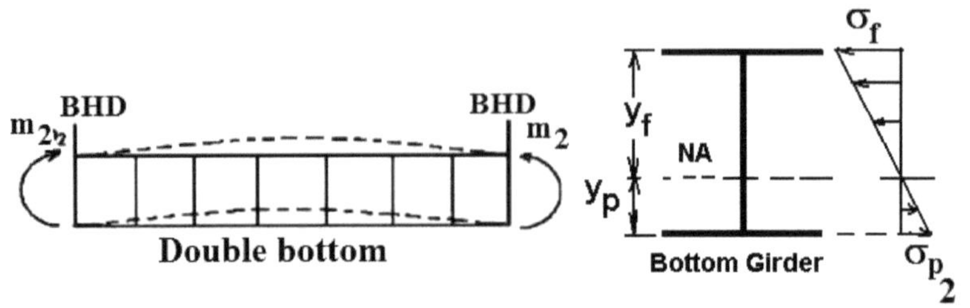

Fig. 9.23 Secondary stresses in the bottom girder of a longitudinally stiffened bottom structure

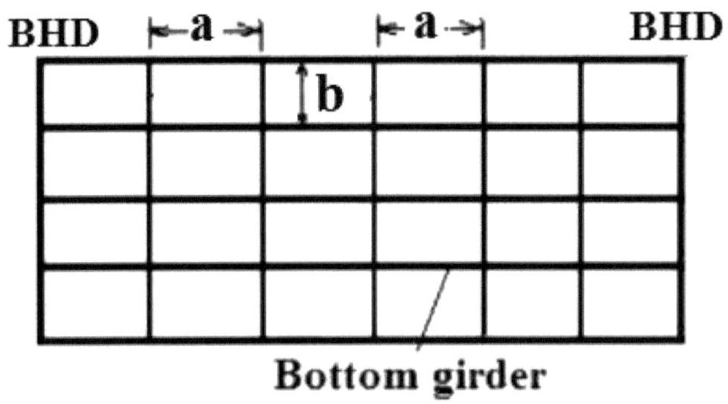

Fig. 9.24 An idealized grillage of longitudinally stiffened bottom structure

5.3 Compounding Stresses in Bottom Longitudinals

The compound stress in a bottom longitudinal of a longitudinally stiffened bottom structure is composed of the primary, secondary and tertiary stress components, see Fig. (9.25).

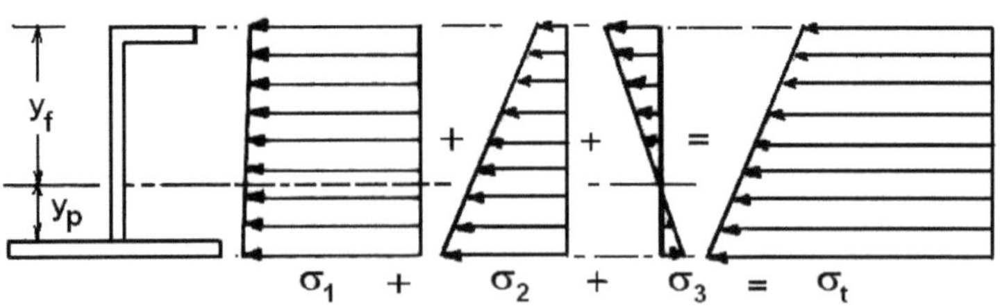

Fig. 9.25 Compound stresses in a bottom longitudinal of a longitudinally stiffened bottom structure

a- Primary stresses

The primary stress induced in a bottom longitudinal of a ship in hogging condition is linearly distributed over the longitudinal section, see Fig. (9.26). The primary stress induced in the attached plating in is given by:

$$\sigma_{p_1} = \sigma_1 = \frac{M_V \cdot y_p}{I}$$

The primary stress at the top flange of bottom longitudinal is given by:

$$\sigma_{f_1} = \frac{M_V \cdot y_f}{I}$$

Where:

$\sigma_{p1} = \sigma_B$ = Hull girder stresses at the bottom plating
y_p = distance of the attached plating from the ship section neutral axis
y_f = distance of the flange of from the ship section neutral axis

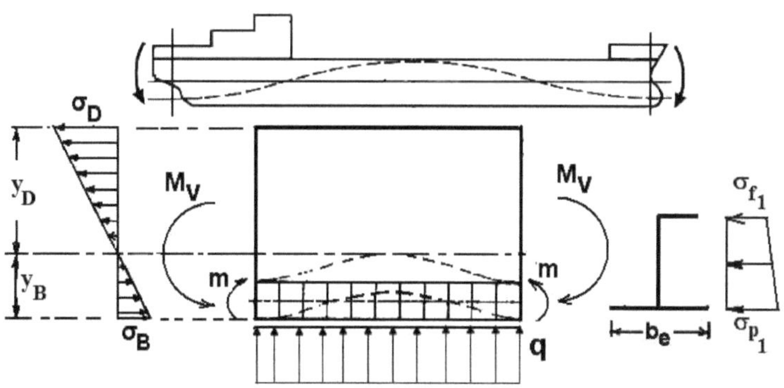

Fig. 9.26 Primary stresses in a bottom longitudinal of a longitudinally stiffened bottom structure of a ship in hogging condition

b- Secondary stress

The secondary stress induced by the longitudinal vertical hull girder bending moment in the attached plating of a bottom longitudinal of a longitudinally stiffened bottom structure is given by, see Fig. (9.27):

$$\sigma_{p_2} = \frac{m_2 \cdot y_p}{i}$$

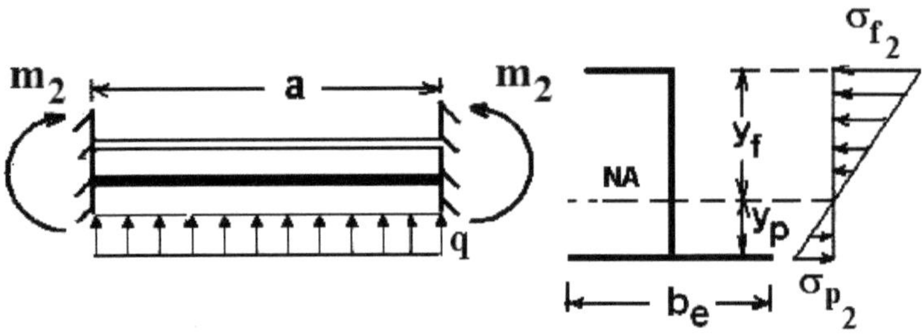

Fig. 9.27 Secondary stress in bottom longitudinal

The secondary stress induced in the flange at the clamed end of the bottom longitudinal is given by, see Fig. (9.27):

$$\sigma_{f_2} = \frac{m_2 . y_f}{i_l}$$

Where: $\sigma_{pl} = \sigma_B$ = Secondary stress at the bottom plating

y_p = distance of the attached plating from the neutral axis of the longitudinal section

y_f = distance of the flange of from the neutral axis of the longitudinal section

m_2 = bending moment at the fixed end of longitudinal and could be determined by solving the bottom structure as a grillage structure

i_l = moment of inertia of longitudinal section

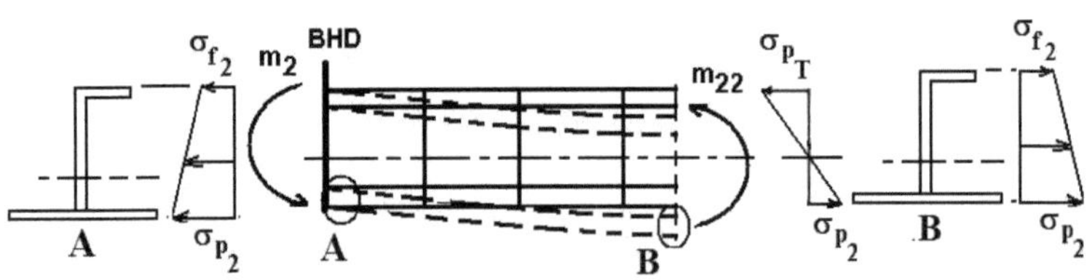

Fig. 9.28 Secondary stress in a bottom longitudinal of a longitudinally stiffened bottom structure

The secondary stress induced in the flange at mid span of the bottom longitudinal (Section B) is given by, see Fig. (9.28):

$$\sigma_{f_{2A}} = \frac{m_{2_2} . y_f}{i_l}$$

Where: m_{22} = secondary bending moment at mid-span
σ_{f2A} = secondary stress in the flange of the longitudinal at section A

c- Tertiary stress

The tertiary stress induced in the attached bottom plating for a longitudinally stiffened bottom structure is given by, see Fig. (9.29):

$$\sigma_{p_3} = \frac{m_3 . y_p}{i}$$

Where:

q = Local loading on the outer bottom of the longitudinal
i_l = Second moment of area of the longitudinal section about its own neutral axis
y_p = distance of the attached plating from the neutral axis of the longitudinal section
m_3 = Bending moment at the assumed clamped end of the longitudinal and is given by:

$$m_3 = \frac{q.a^2}{12}$$

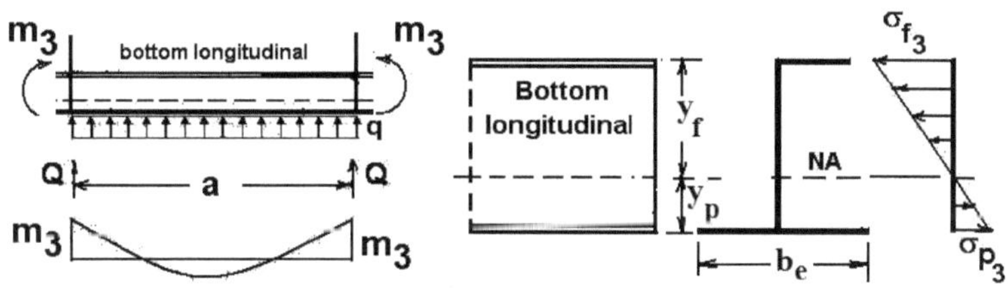

Fig. 9.29 Tertiary stresses in a bottom longitudinal of a longitudinally stiffened bottom structure

The tertiary stress is linearly distributed over the section of the longitudinal, see Fig. (9.29). The tertiary stress at the flange of the longitudinal is given by:

$$\sigma_{f_3} = \frac{m_3 . y_f}{i_l}$$

5.4 Compounding of Stresses in the Bottom Plating

In a longitudinally stiffened bottom structure, the bottom plating is subjected to the four components of stresses, see Fig. (9.30).

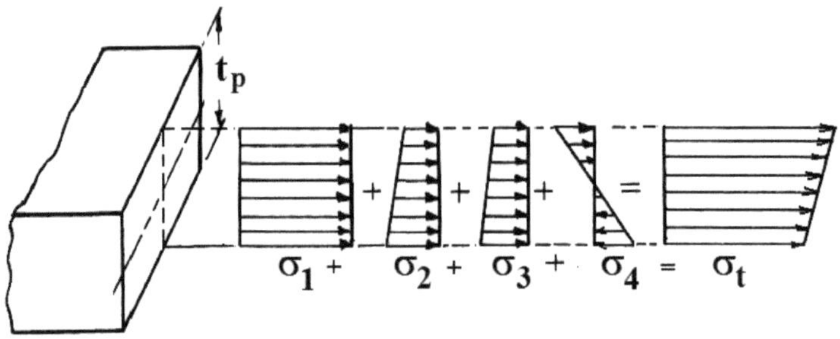

Fig. 9.30 Compound stresses in the bottom plating of a longitudinally stiffened bottom structure

a- Primary stress

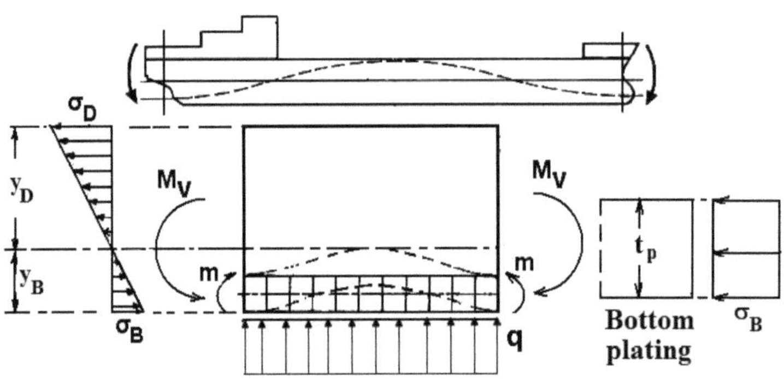

Fig. 9.31 Primary stress in the bottom plating for a ship in a hogging condition

The primary stress induced in the bottom plating for a ship subjected to hogging bending moment is given by, see Fig. (9.31):

$$\sigma_{p_1} = \frac{M_V \cdot y_p}{I}$$

Where: M_V= Longitudinal vertical hull girder bending moment

y_p = Distance of bottom plating from neutral axis of the ship section

b- Secondary stress

The secondary stress in the bottom plating of a longitudinally stiffened bottom structure is given by, see Fig. (9.32):

$$\sigma_{p_2} = \frac{m_2 . y_p}{i}$$

Where: m_2 = secondary bending moment

The secondary stress induced in the bottom plating results from the bending of the double bottom structure outwards or inwards depending on the net loading on the double bottom structure.

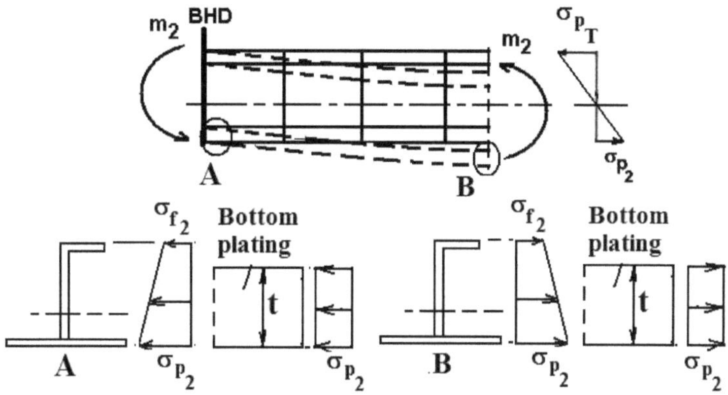

Fig. 9.32 Secondary stresses in the bottom plating of a longitudinally stiffened bottom structure

Bottom longitudinals are subjected to either tensile or compressive stresses depending on the net loading on the double bottom structure and the position along the length of the longitudinal. The secondary stresses induced at the fixed end and at midspan of the longitudinal are shown in Fig. (9.32).

c- Tertiary stresses

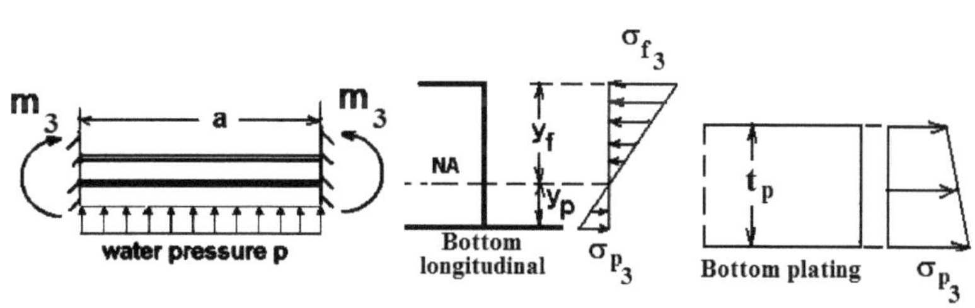

Fig. 9.33 Tertiary stresses in the bottom plating of a longitudinally stiffened bottom structure

The tertiary stress in the attached plating of the bottom longitudinal is calculated by assuming both ends of the longitudinal are clamped as follows:

$$\sigma_3 = \frac{m_3 . y_p}{i_l}$$

Where: m_3 = local bending moment at the assumed clamped end of the longitudinal, see Fig. (9.33)

i_l = second moment of area of the longitudinal section with the attached plating about its own neutral axis

y_p = distance of the attached plating from the neutral axis of the section

d-Local stress

The local stress induced in longitudinally stiffened bottom plating is calculated at mid-width of plate panel, see Fig. (9.34).

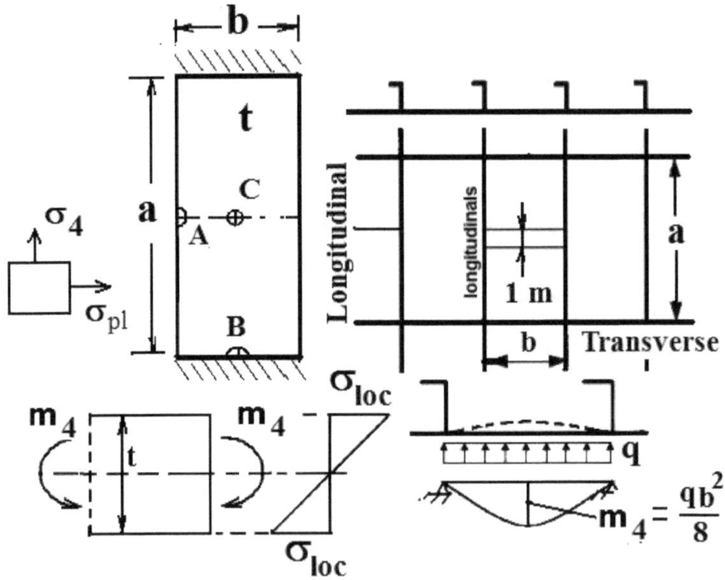

Fig. 9.34 Local bending stress of bottom plating of a longitudinally stiffened bottom structure

The local stress in the bottom plating is acting in the transverse direction of the ship. The local stress component acting in the longitudinal direction of the ship is given by, see Fig. (9.34):

$$\sigma_4 = \nu\sigma_l$$

Where: ν = Poisson's ratio

σ_4 = Local stress induced in the longitudinal direction of the bottom plate panel

σ_l = Local bending stress induced in the transverse direction of the bottom plate panel

5.5 Compounding of Stresses in the Tank Top Longitudinals

The compound stress in tank top longitudinals is composed of the primary, secondary and tertiary stresses, see Fig. (9.35).

The calculation of the primary, secondary and tertiary stress components could be carried out as was given for bottom longitudinals.

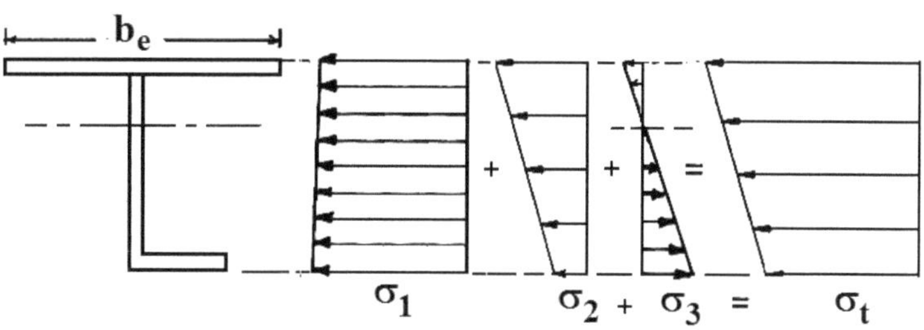

Fig. 9.35 Compound stress in tank top longitudinals

5.6 Compounding of Stresses in the Tank Top Plating

The compound stress in the tank top plating of a longitudinally stiffened bottom structure is composed of primary, secondary and tertiary stress components, see Fig. (9.36).

a-Primary stress

The primary stress induced in the tank top plating of a ship subjected to sagging bending moment is given by, see Fig. (9.37):

$$\sigma_{p_2} = \frac{M_V \cdot y_T}{I}$$

Where: y_T = Distance of tank top from ship section neutral axis, see Fig. (9.38)

$$y_T = y_B - d$$

σ_{p2} = primary stress induced in the tank top plating

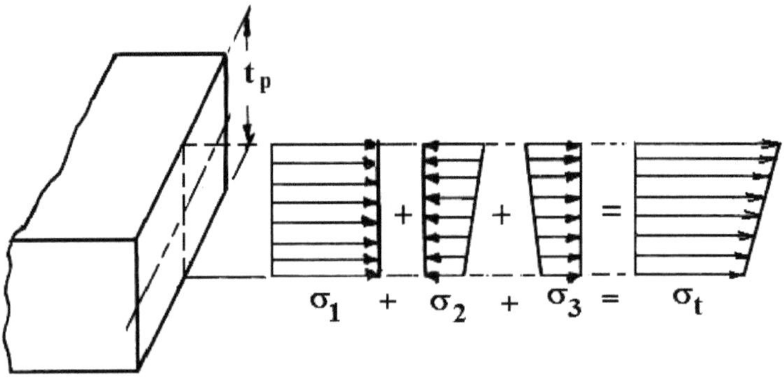

Fig. 9.36 Compound stress of longitudinally stiffened tank top plating

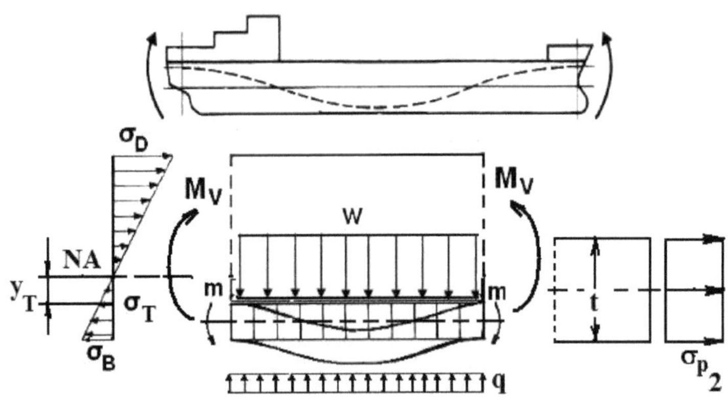

Fig. 9.37 Primary stresses in the tank top plating of a longitudinally stiffened bottom structure of a ship in a sagging condition

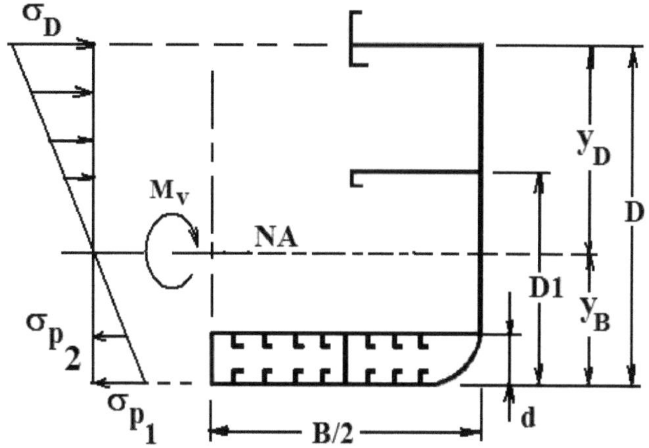

Fig. 9.38 Primary stresses over a ship section of a ship having longitudinally stiffened bottom structure

6 Compounding of Stresses in Longitudinally Stiffened Deck Structure

The strength members of a longitudinally stiffened deck structure are:

- Deck girders
- Deck longitudinals
- Deck plating

6.1 Compounding of Stresses in Deck Girders

The compound stress in deck girders of longitudinally stiffened deck structure is composed of the primary and secondary stress components, see Fig. (9.39).

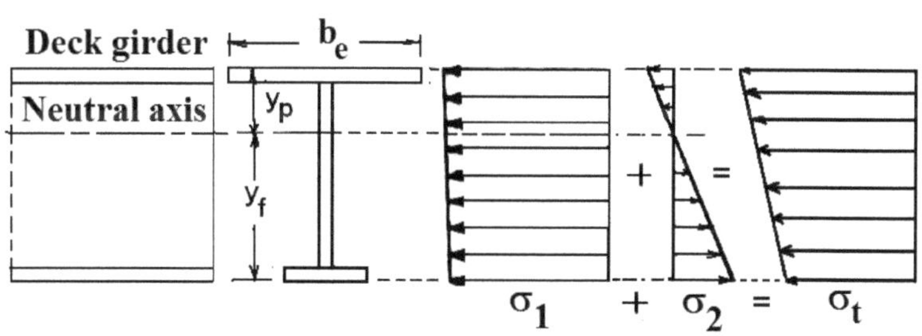

Fig. 9.39 Compound stresses in the deck girder of a longitudinally stiffened deck structure

The flange and web plate of the deck girder are subjected to uniform compressive stresses induced by hull girder sagging bending moments and linear stress distribution induced by the secondary and local loadings, see Fig.(9.39).

a-Primary stress

The primary stress induced in the attached plating of a deck girder for a ship in a hogging condition is given by, see Fig. (9.40):

$$\sigma_D = \frac{M_V . y_D}{I}$$

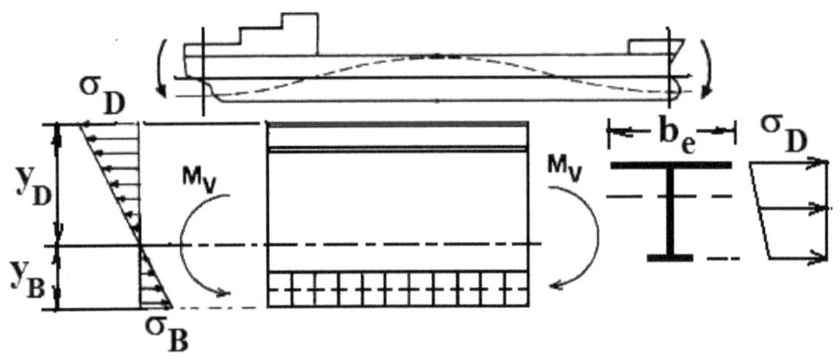

Fig. 9.40 Primary stresses in the deck girder of longitudinally stiffened deck structure of a ship in a hogging condition

b-Secondary stress

The secondary stress induced in deck girders are due to deck loading. The deck loading could be cargo loads, accumulation of ice, shipping green water, etc. In this case, the secondary stress induced in the flange of the deck girder is given by, see Fig. (9.41):

$$\sigma_f = \frac{m_2 . y_f}{i}$$

Where: m = bending moment induced in the deck girder

The secondary bending moment can be determined by solving the deck structure as an open grillage or as a continuous beam supported by the deck transverses and loaded by the deck loading, see Fig. (9.41).

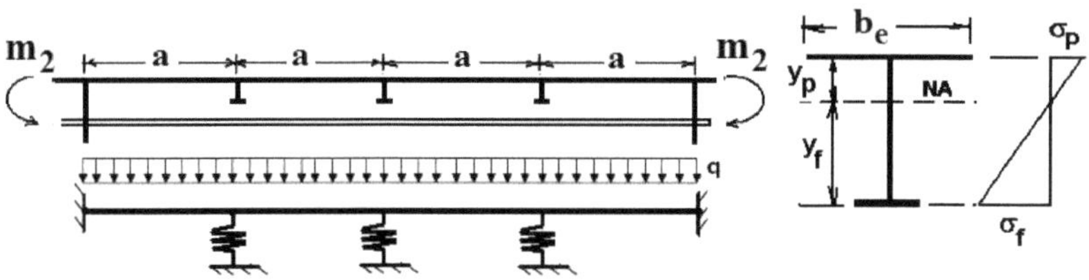

Fig. 9.41 Secondary stresses in the deck girder of a longitudinally stiffened deck structure

6.2 Compounding of Stresses in Deck Longitudinals

The compound stress induced in deck longitudinals at the clamped end is composed of the primary, secondary and tertiary stresses, see Fig. (9.42).

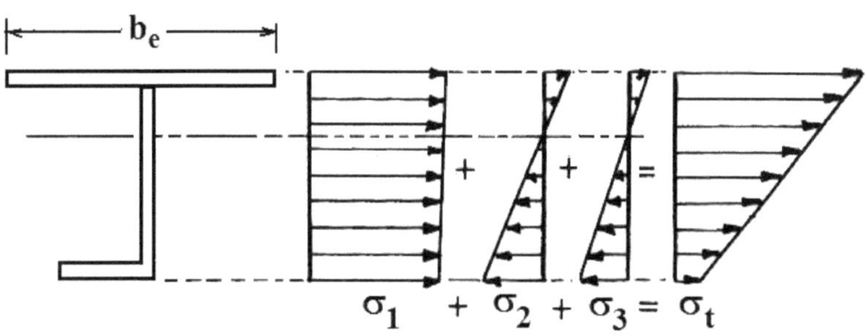

Fig. 9.42 Compound stresses induced in the deck longitudinal section at the clamped end

a-Primary stress

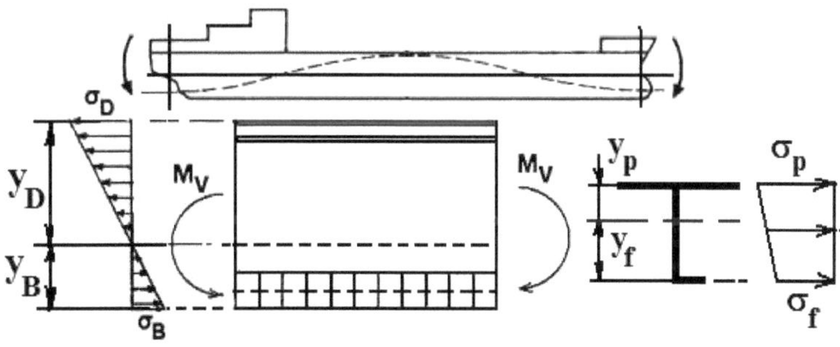

Fig. 9.43 Primary stresses in deck longitudinal of longitudinally stiffened deck structure of a ship in a hogging condition

The primary stress induced in the attached plating of deck longitudinals for a ship in a hogging condition is given by, see Fig. (9.43):

$$\sigma_D = \frac{M_V . y_D}{I}$$

Where:

σ_D = primary stress induced in deck longitudinal
y_D = distance of the deck plating from the neutral axis of ship section

b-Secondary stress

The secondary stress induced in deck longitudinals is given by:

$$\sigma_f = \frac{m_2 . y_f}{i}$$

Where: m_3 = Bending moment obtained from a grillage analysis of deck structure
i = Second moment of area of longitudinal section with attached plating about its neutral axis
y_f = distance of the flange of the longitudinal from the neutral axis of the longitudinal section

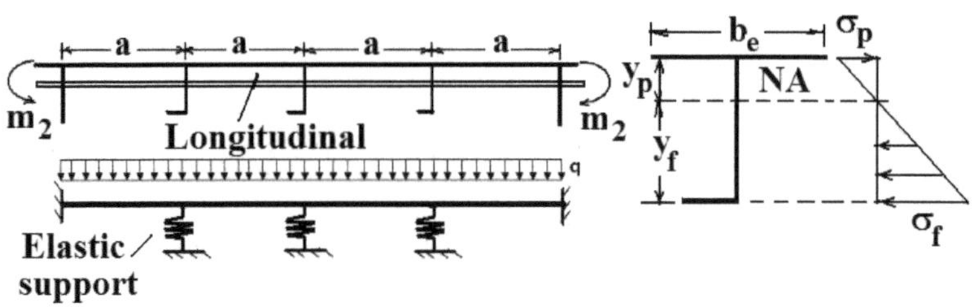

Fig. 9.44 Secondary stresses in a deck longitudinal of a longitudinally stiffened deck structure

The secondary stress could be calculated by idealizing the longitudinal as a continuous beam supported by the deck transverses, see Fig. (9.44). In this case, the secondary stress induced in the flange of the deck longitudinal is given by, see Fig. (9.44):

$$\sigma_f = \frac{m_2 \cdot y_f}{i_l}$$

Where

i_l= moment of inertia of longitudinal section about its neutral axis

The definitions of m_2, y_f, σ_f are as shown in Fig. (9.44)

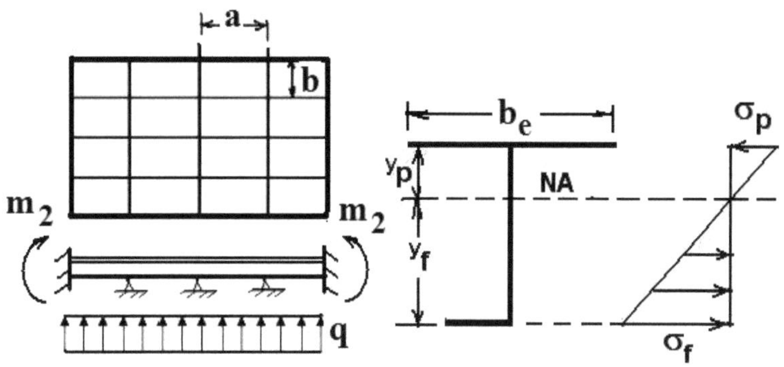

Fig. 9.45 Secondary stress in the deck longitudinals

Alternatively, the secondary stress could be calculated by idealizing the deck structure as an open grillage, see Fig. (9.45).

c-Tertiary stresses

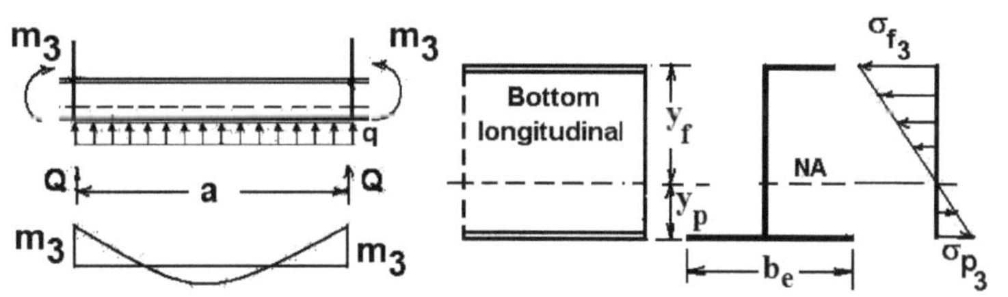

Fig. 9.46 Tertiary stresses in a deck longitudinal of a longitudinally stiffened deck structure

The tertiary stress induced in the attached plating of deck longitudinals at the clamped end is given by, see Fig. (9.46):

$$\sigma_p = \frac{m_3 \cdot y_P}{i_l}$$

Where: σ_p = tertiary stress induced in the attached plating at the clamped end

m_3 = bending moment at the clamped end of the longitudinal

i_l = second moment of area of the section of member

y_p = distance of the attached plating from the neutral axis of the longitudinal section

6.3 Compounding of Stresses in Deck Plating

The compound stress in the deck plating of longitudinally stiffened deck structure is composed of the primary, secondary and tertiary stress components, see Fig. (9.47).

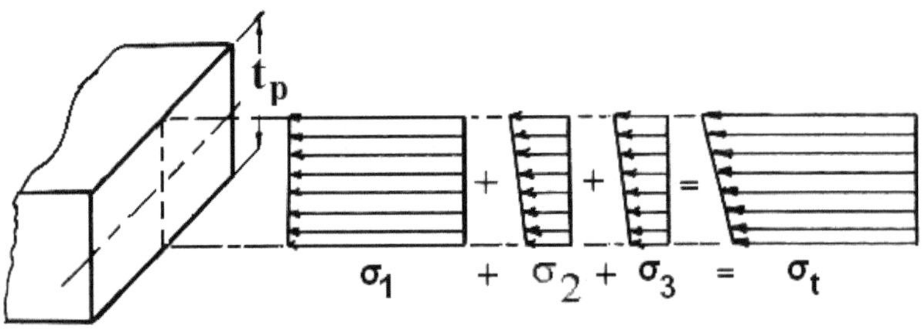

Fig. 9.47 Compound stresses in the deck plating of a longitudinally stiffened deck structure

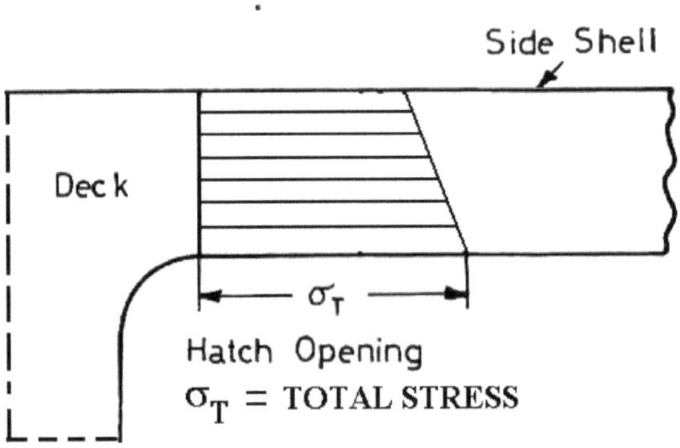

Fig. 9.48 Compound stresses in the deck plating port of hatch opening

For a more rational approach, the compound stress for the port side deck plating of a longitudinally stiffened deck structure should include the following stress components, see Fig. (9.48):

- Primary stress due to longitudinal vertical bending moment, tensile or compressive (vessel in hogging or sagging condition).
- Tensile or compressive hull girder bending stress due to horizontal bending moment
- Flexural warping stress due to torsional hull girder loading
- Secondary stresses due to cargo deck loading
- Tertiary stress due to the local loading of deck longitudinals

a- Primary stress

The primary stress induced in the deck plating of longitudinally stiffened deck structure of a ship in a sagging condition is given by, see Fig. (9.49):

$$\sigma_1 = \sigma_D = \frac{M_V \cdot y_D}{I}$$

Where: M_V = longitudinal vertical sagging bending moment
y_D = distance of deck plating from ship section neutral axis
σ_D = Primary compressive stress induced in the deck plating

b- Secondary stress

The secondary stress induced in the attached plating of deck girders is given by, see Fig. (9.50):

The secondary stress induced in the deck plating

$$\sigma_2 = \sigma_p = \frac{m_2 \cdot y_p}{i_g}$$

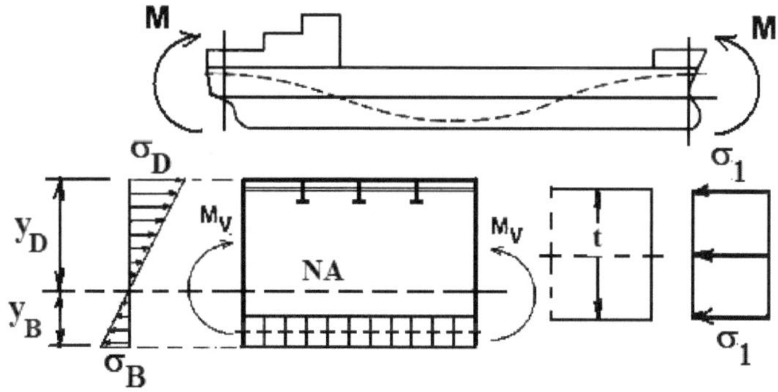

Fig. 9.49 Primary stress in the deck plating of a longitudinally stiffened deck structure of a ship in a hogging condition

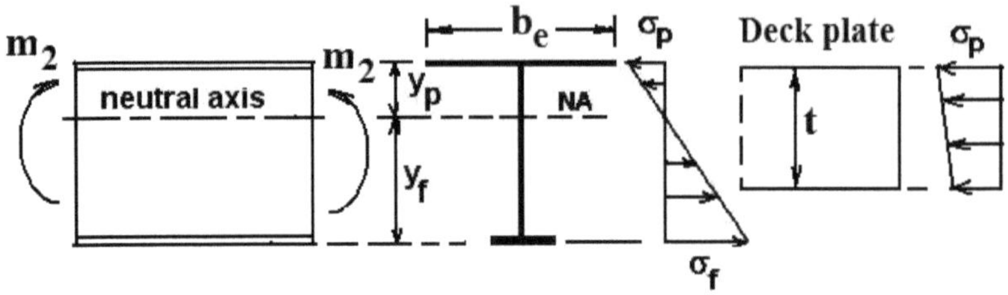

Fig. 9.50 Secondary stresses induced in the deck plating

Where: m_2 = bending moment over the deck girder section
y_p = distance of the deck plating from the neutral axis of the girder section
i_g = second moment of area of the girder section about its own neutral axis

c-Tertiary stress

The tertiary stress in the deck plating of longitudinally stiffened deck structure is induced by the bending of deck longitudinals. The deck longitudinals are supported by heavy deck transverses and carry the local loadings over the loaded area of each deck longitudinal.

The tertiary stress in the deck plating at the clamped end support is given by, see Fig. (9.51):

$$\sigma_{p_A} = \frac{m_{3A} \cdot y_p}{i_l}$$

Where: m_{3A} = local bending moment on bottom longitudinal section at the clamped end support, and is given by:

$$m_{3A} = \frac{q.a^2}{12}$$

The bending moment at midspan of the deck longitudinal is given by:

$$m_{3B} = \frac{q.a^2}{24}$$

Where: q = transverse loading on deck longitudinal in t/m and is given by:

$$q = w \times b, \, t/m$$

w = deck load intensity, t/m^2

b = longitudinal spacing

l = span of longitudinal, m

i = second moment of area of section of the deck longitudinal

y_p = distance of deck plating from neutral axis of the longitudinal section

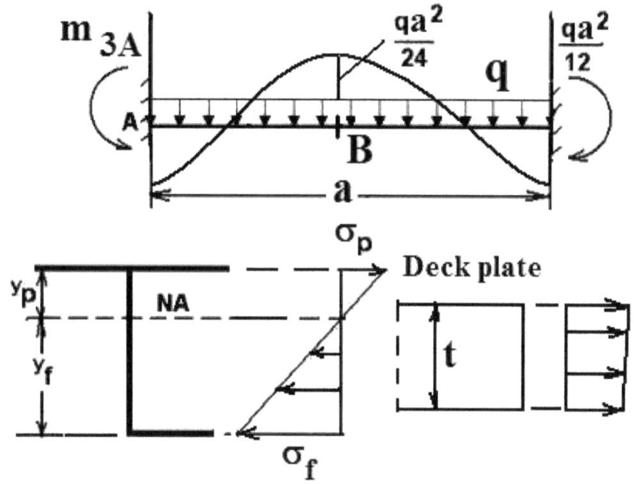

Fig. 9.51 Tertiary stresses in the deck plating of a longitudinally stiffened deck structure

The tertiary stress at section B, midspan, is given by, see Fig. (9.52):

$$\sigma_{p_B} = \frac{m_{3B} \cdot y_p}{i_l}$$

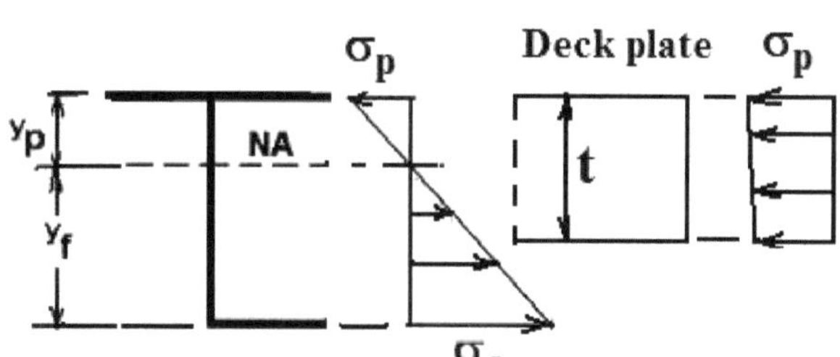

Fig. 9.52 Tertiary stress distribution over the deck plating thickness at mid-span of longitudinal

7 Compounding of Stresses for Oil Tankers

The strength members of longitudinally stiffened bottom structure are:

- Bottom girders
- Bottom longitudinals
- Bottom plating

For oil tankers, the compounding of stresses are calculated for the ship in hogging and sagging conditions.

7.1 Compounding of Stresses in the Bottom Girder

The compound stress in the bottom girder of longitudinally stiffened single bottom structure is composed of the primary and secondary stresses, see Fig. (9.53).

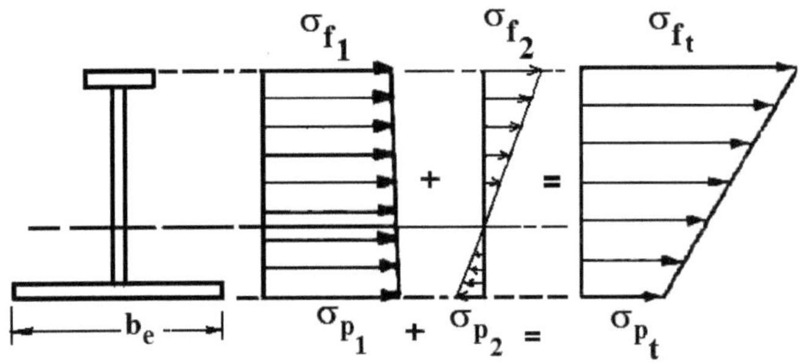

Fig. 9.53 Compound stresses in the bottom girder of a longitudinally stiffened single bottom structure

a- Primary stresses

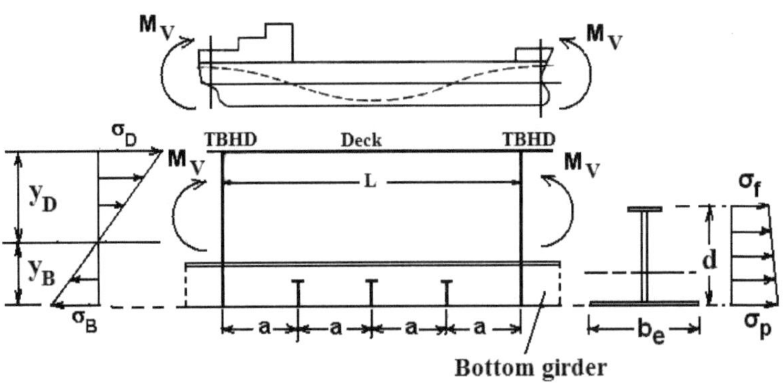

Fig. 9.54 Primary stresses in the bottom girder of a longitudinally stiffened single bottom structure

The primary hull girder stress induced by a sagging bending moment in the face plate of a bottom girder in a longitudinally stiffened single bottom oil tanker is given by, see Fig. (9.54):

$$\sigma_f = \frac{M_V (y_B - d)}{I}$$

Where: σ_f = primary stress induced in the flange of the bottom girder

d = depth of longitudinal

b-Secondary stress

The secondary stress induced in the attached plating of bottom girders is given by, see Fig. (9.55):

$$\sigma_2 = \sigma_p = \frac{m_2 . y_p}{i_g}$$

Where:

m_2 = secondary bending moment

y_p = distance of bottom plating from the neutral axis of the girder

i_g = moment of inertia of the girder section about its neutral axis

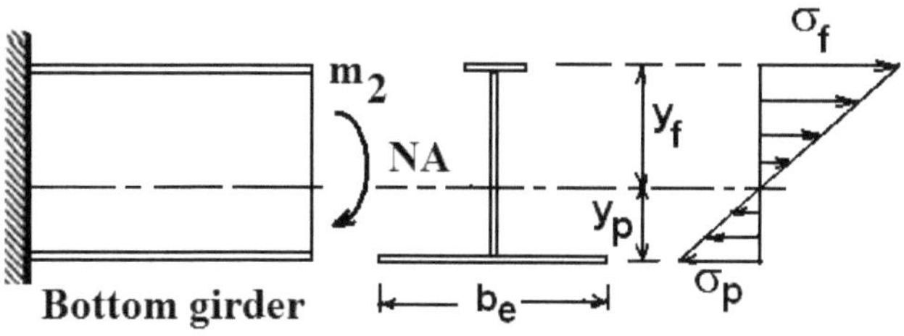

Fig. 9.55 Secondary stresses in a bottom girder of a longitudinally stiffened bottom structure

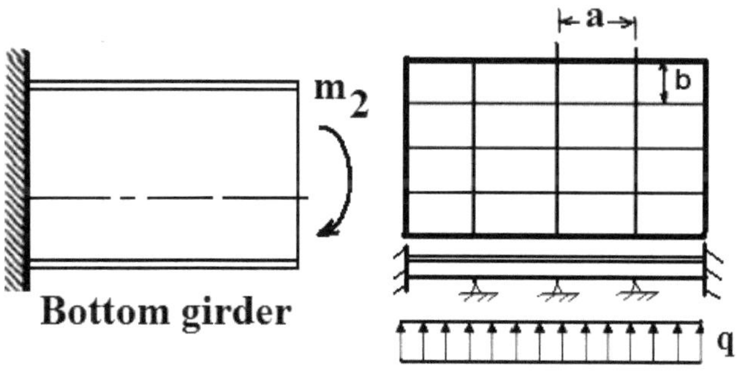

Fig. 9.56 Grillage idealization of the longitudinally stiffened bottom structure

The secondary bending moment could be calculated by solving the bottom structure as a grillage structure, see Fig. (9.56).

7.2 Compounding of Stresses in Bottom Longitudinals

The compound stress in bottom longitudinals in oil tankers is composed of the primary, secondary and tertiary stresses, see Fig. (9.57).

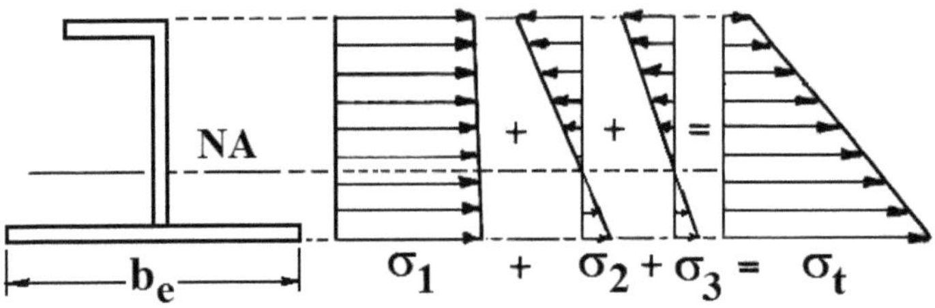

Fig. 9.57 Compound stresses in a bottom longitudinal of a longitudinally stiffened single bottom structure

a-Primary stress

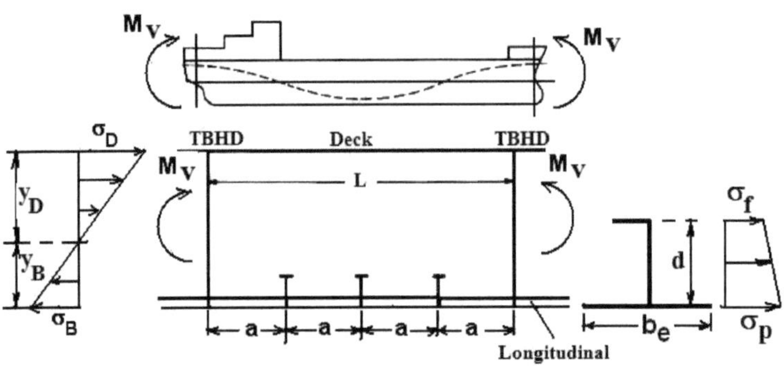

Fig. 9.58 Primary stresses in a bottom longitudinal of a longitudinally stiffened single bottom structure

The primary stress induced in the attached bottom plating in a ship subjected to sagging bending moment is given by, see Fig. (9.58):

$$\sigma_p = \sigma_B = \frac{M_V \cdot y_B}{I}$$

The primary stress induced in the flange of the bottom longitudinal is given by

$$\sigma_f = \frac{M_V (y_B - d)}{I}$$

Where: d = depth of bottom longitudinals
I = moment of inertia of ship section about its neutral axis

b- Secondary stress

The secondary stress induced in bottom longitudinals can be calculated by idealizing the bottom structure as an open grillage subjected to the external hydrostatic pressure. The secondary stress induced in the flange of the bottom longitudinal is given by, see Fig. (9.59):

$$\sigma_{f_2} = \frac{m_2 . y_f}{i}$$

Where: m_2 = secondary bending moment

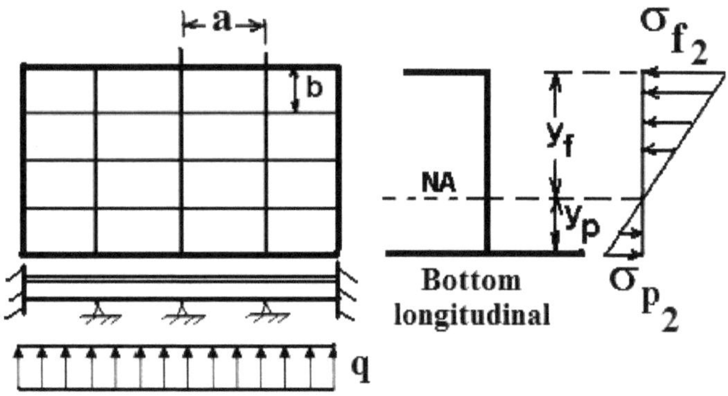

Fig. 9.59 Secondary stresses in a bottom longitudinal of a longitudinally stiffened single bottom structure

c- Tertiary stress

The tertiary stress induced at the flange of a bottom longitudinal is given by, see Fig. (9.60):

$$\sigma_{f_3} = \frac{m_3 . y_f}{i_l}$$

Where: σ_{f3} = Tertiary stress at the flange of the longitudinal

Assuming the bottom longitudinal clamped at both ends, the tertiary stress at the attached bottom plating is given by:

$$\sigma_{p_3} = \frac{m_3 . y_p}{i_l}$$

Where: σ_{p3} = Tertiary stress at the attached plating of the longitudinal
i_l = moment of inertia of longitudinal section about its neutral axis

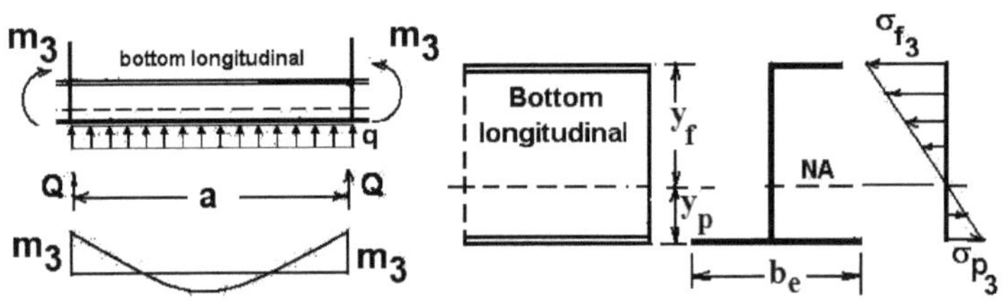

Fig. 9.60 Tertiary stresses in a bottom longitudinal of a longitudinally stiffened single bottom structure

7.3 Compound Stress in Bottom Plating of an Oil Tanker Subjected to Sagging Moment

The compound stress is composed of the primary, secondary, tertiary and local stress, see Fig. (9.61). All these stresses are acting along the direction of ship length.

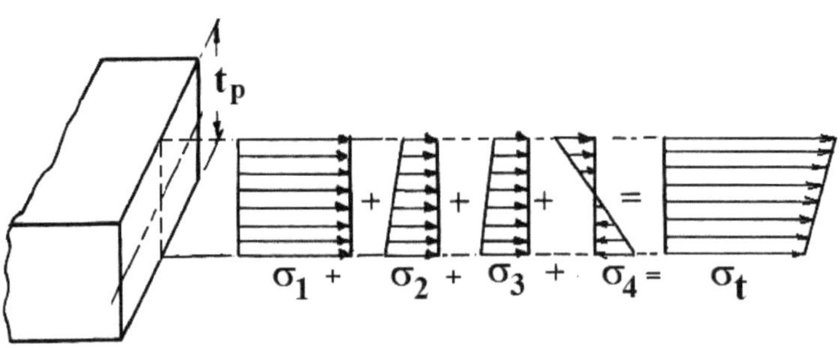

Fig. 9.61 Compound stresses induced in longitudinally stiffened bottom plating

a- Primary stress

The primary stress induced in the bottom plating of an oil tanker subjected to sagging moment is given by, see Fig. (9.62):

$$\sigma_{p_1} = \sigma_B = \frac{M_V \cdot y_B}{I}$$

Where: M_V = Longitudinal vertical sagging bending moment
y_B = Distance of bottom plating from the neutral axis of ship section

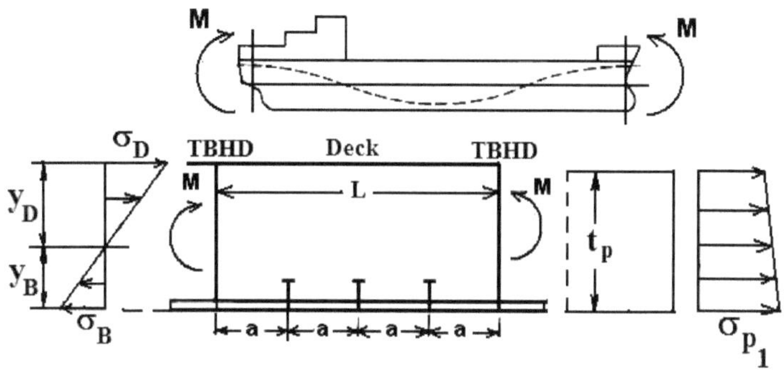

Fig. 9.62 Primary stress in the bottom plating of a longitudinally stiffened single bottom structure in an oil tanker under a sagging moment

b- Secondary stress

The secondary stress in the bottom plating is given by, see Fig. (9.63):

$$\sigma_{p_2} = \frac{m_2 \cdot y_p}{i_l}$$

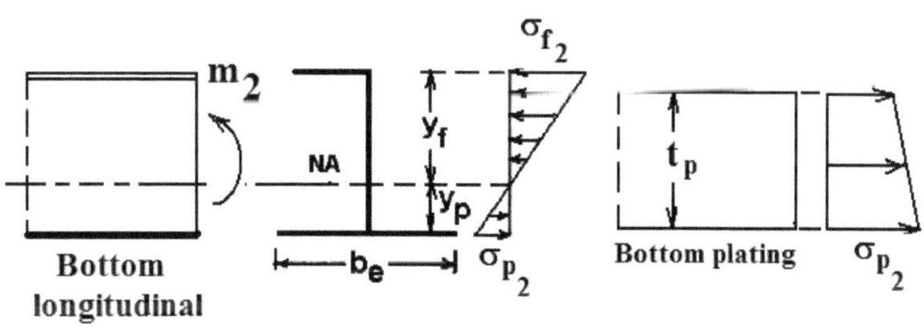

Fig. 9.63 Secondary stress in the bottom plating of a longitudinally stiffened bottom structure assuming the transverses as elastic supports

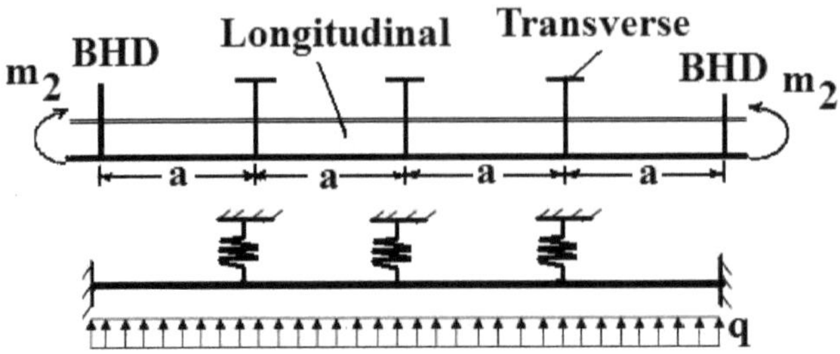

Fig. 9.64 Bottom longitudinal between two transverse bulkhead supported by springs at the bottom transverses

The secondary stress in the bottom plating can be calculated by assuming the bottom longitudinal between two transverse bulkheads as a continuous beam supported by the heavy bottom transverses and loaded by the external water pressure, see Fig. (9.64). The support of the transverses can be assumed as spring supports. The stiffness of these spring supports can be calculated from the flexural rigidities of the transverse members. For a beam of length "S" and flexural rigidity EI loaded with a concentrated force at midspan, the spring constant is given by:

$$k = \frac{48\,EI}{S^3}\ t/m$$

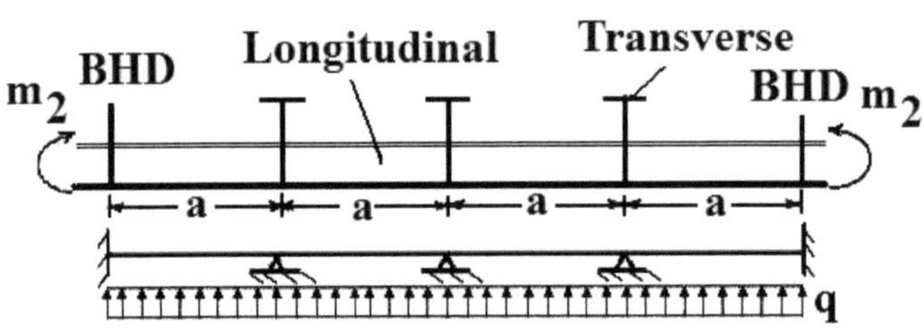

Fig. 9.65 Bottom longitudinal between two transverse bulkhead supported by the bottom transverses

For a beam of length "S" and flexural rigidity EI and loaded with uniformly distributed load, the spring constant is given by:

$$k = \frac{384\,EI}{5S^3}\ t/m$$

Alternatively, the bottom longitudinal can be assumed as a continuous beam supported at the transverse members. This set up can be used to calculate the bending moments and shear forces at the points of support see Fig. (9.65).

c- Tertiary stress

The tertiary stress is the stress induced in the attached plating of the bottom longitudinal assumed as a beam fixed at both ends, see Fig. (9.66).

The tertiary stress is given by:

$$\sigma_3 = \sigma_{p_3} = \frac{m_3 . y_p}{i_l}$$

Where: σ_{p3} = local stress induced in the attached plating

m_3 = local bending moment induced at the fixed end of the longitudinal

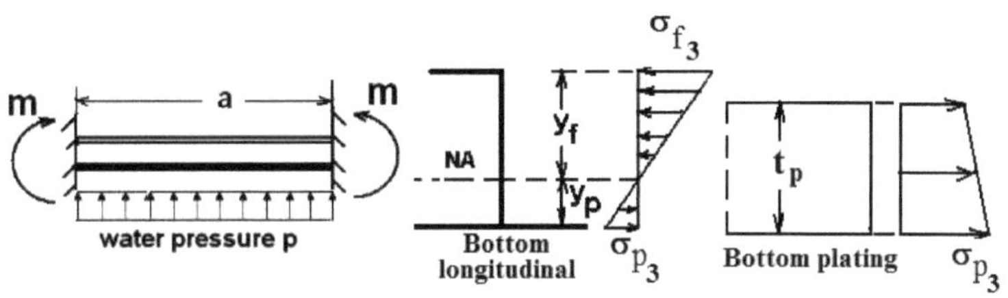

Fig. 9.66 Tertiary stress of bottom plating of a longitudinally stiffened bottom structure

d- Local stress

The local stress induced in the bottom plating of a longitudinally stiffened bottom structure and acting in the longitudinal direction of the ship is given by, see Fig. (9.67):

$$\sigma_4 = \nu . \sigma_{pl}$$

Where: ν = Poisson's effect = 0.3

σ_{pl} = local stress induced in the bottom plating acting transversely and could be calculated as follows:

$$\sigma_{pl} = \frac{m_4 . t/2}{i_p}$$

Where: m_4 = bending moment acting transversely for a strip of width 1.0 m, length "b" m and thickness "t", assumed simply supported at both ends, see Fig. (9.68).

$$m_4 = \frac{q.b^2}{8}$$

$$i_p = \frac{1.0 \times t^3}{12}$$

Where: q = external loading on the bottom plating, t/m

i_p = moment of inertia of the section of the plate strip

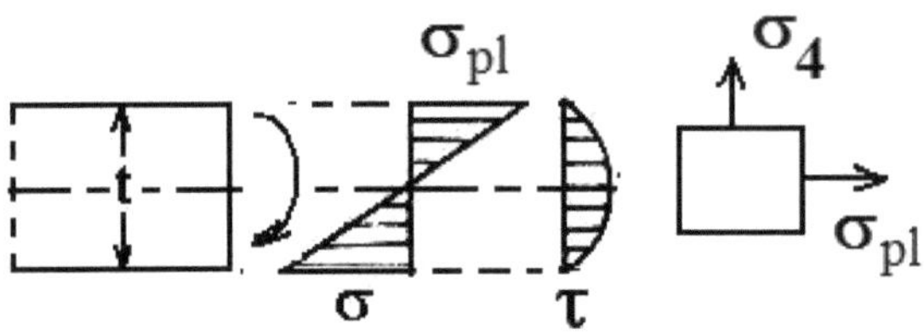

Fig. 9.67 Local stress induced in a longitudinally stiffened bottom plating

Hence: $\sigma_{pl} = \dfrac{12.q.b^2.t/2}{8t^3} \times 100 = 75q\left(\dfrac{b}{t}\right)^2 \quad t/cm^2$

This local stress is induced in a direction orthogonal to ship length. The local stress induced in the direction of ship length is given by:

$$\sigma_x = \nu \sigma_{pl}$$

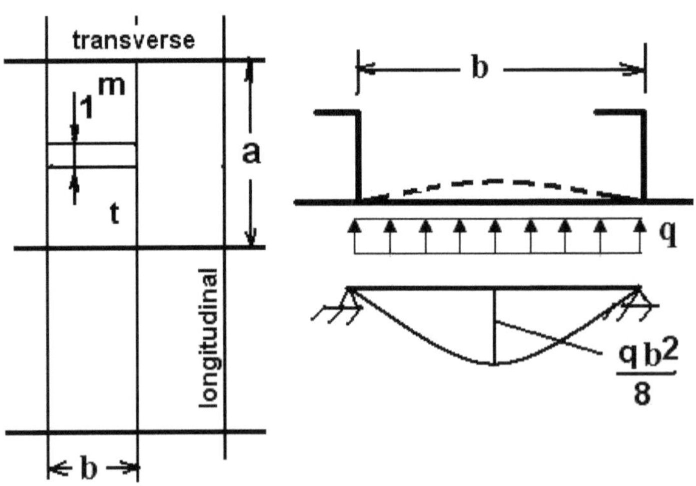

Fig. 9.68 Local stress induced in a longitudinally stiffened bottom plating

Part III
Chapter 10 – Chapter 14

Chapter 10
Columns and Beam Columns

1 Introduction

The stability phenomenon of a structure is defined by its state of equilibrium. The equilibrium of a designed structure is stable if small imperfections or defects will cause correspondingly small deviations from the idealized operating condition. Buckling or structural instability is considered one of the main modes of failure of ship structure members subjected to compressive forces. The design of an idealized column differs from the actual structure of the column. This difference is due to the presence of numerous small imperfections, defects or deviations. The buckling mode of a column corresponds to the lowest buckling load which is referred to as the critical buckling load. Structural stability should be provided for the structure as a whole and also for each of its structural members. Because of the inevitable presence of geometric imperfections in fabricated structures, actual instabilities may be expected to occur at a load rather less than the load of an ideal column.

This Chapter presents the buckling of columns and beam columns. The most common classes of perturbations experienced by beam columns are identified and presented. The physical problem of stability is explained and the concept of critical force and critical stress are explained. The effect of eccentric loading on the critical buckling load and critical stress are given. The load-deflection relationship of beam columns is introduced. The behavior of beam columns under various loading conditions and different types of end supports are investigated.

2 Structural Members Subjected to Compressive Loadings

Depending on the loading condition, a structural member subjected to compressive loadings may be referred to as follows:

- Column: when the structural member is subjected to pure compression
- Beam: when the structural member is subjected to bending
- Beam-column: when the structural member is subjected to bending and compression.

An ideal column is the column that does not experience the presence of any deviations or imperfections. The designer must be confident that despite of the presence of these deviations, the actual column will perform in a manner similar to its corresponding idealized structure for all possible loading combinations.
Buckling modes of columns are categorized as follows:

- Flexural buckling: when the column is subjected to bending about the axis of least resistance
- Torsional buckling: when the column is subjected to twisting without bending
- Flexural-torsional buckling: when the column is subjected to simultaneous twisting and bending
- Lateral-torsional buckling: when the beam is subjected to simultaneous twisting and bending
- Local buckling: when buckling occurs in a thin-walled part of the cross-section (plate-buckling, shell-buckling).

3 Classes of Perturbations

Actual structure differs from the idealized structure because of the inevitable presence of small deviations, defects and imperfections. The most common deformed shapes and the corresponding assumed deflection curves of actual columns are shown in Fig. (10.1).

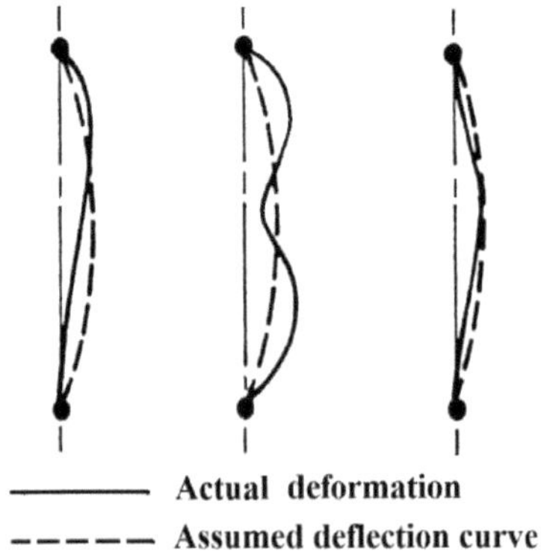

Fig. 10.1 Common deviations of column shapes

Despite of the presence of these deviations, the actual structure should perform in a manner similar to its corresponding idealized structure. Therefore, the size of structure (geometry and scantlings) should be selected to ensure that stable equilibrium would occur even in the presence of any kind of perturbations. For columns subjected to compressive loadings and stresses, the most important classes of perturbations are:

- Initial lateral deflection, see Fig. (10.2)
- Eccentricity of the compressive force

In general, the limiting initial lateral deflection is given by:

$$\delta_o \leq L/750$$

The limiting eccentricity is given by:

$$e \leq 0.1r$$

Where:

L = length of the structural member
δ_o = initial lateral deflection
e = eccentricity of the applied compressive force, see Fig. (10.2)
r = core radius $= Z/A$
Z = section modulus of the column
A = sectional area of the column

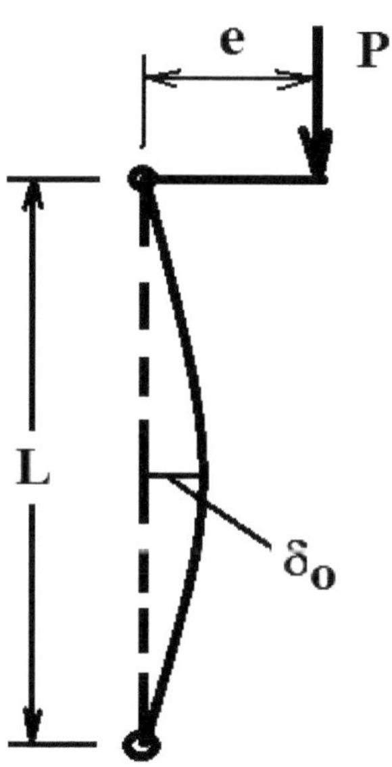

Fig. 10.2 Initial deflection of a column and eccentricity of loading

It should be realized that the common perturbations in practice are not completely deterministic but are subject to statistical variations. Their magnitudes should be defined by their expected mean values and standard deviations.

4 The Problem of Stability

A slender member may lose its load-carrying capacity, i.e., the ability to withstand external forces, not as a result of failure of the material, but due to the loss of stability (buckling). The problem of stability of steel structural members is of great significance as its disregarding may lead to disastrous results.

If a straight rod is compressed by an axially applied force P, see Fig. (10.3), then the rod will initially remain straight, and this is the state of stable equilibrium. The stable state of equilibrium of an elastic rod is characterized by the rod returning to its original position after removal of the external cause. Upon a further increase in the compressive force until it reaches such a value that its work will be equal to the work of deformation in bending induced by any small disturbing factor. In this instance the compressive force reaches its critical value.

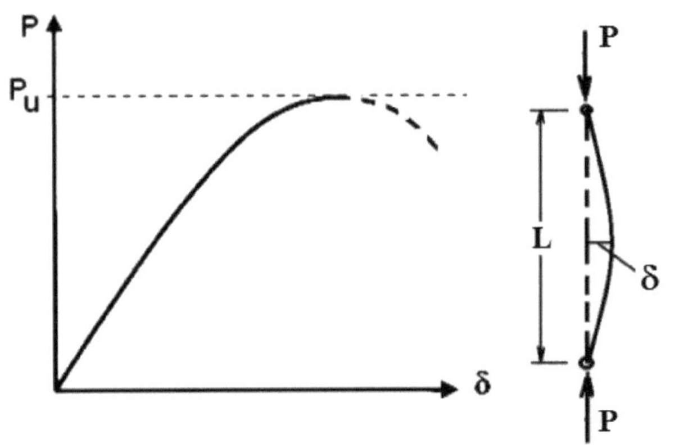

Fig. 10.3 Load deflection curve of an axially loaded column

Thus a perfect straight rod when loaded with a force up to the critical state has a linear shape in the stable state of equilibrium. When the force reaches its critical value, the linear shape of equilibrium stops being stable, the rod may buckle in the plane of least stiffness and it will now have a new curvilinear shape of stable equilibrium. The magnitude of the force which causes the original stable equilibrium of the rod to become unstable is known as the critical force. If there is a slight initial curvature of the rod (or a slight eccentricity of the point of application of the compressive force) then upon an increase in the load the rod will immediately deflect from a straight position.

Structural stability should be provided for the structure as a whole and also for each of its structural members. Because of the inevitable presence of geometric imperfections in fabricated structures, actual instabilities may be expected to occur at a load rather less than the load of an ideal column.

The ultimate load of the column "P_u" is defined as the maximum load that a column can carry. The ultimate load depends on:

- Initial eccentricity of the column
- Eccentricity of the load
- Presence of transverse loads
- Type of end conditions
- Presence of local or lateral buckling
- Inelastic action
- Presence of residual stresses.

4.1 Critical Force and Critical Stress

The critical force P_{cr} for an I-section column in axial compression within the proportional limit of the material, see Fig. (10.4), is determined by Euler's formula:

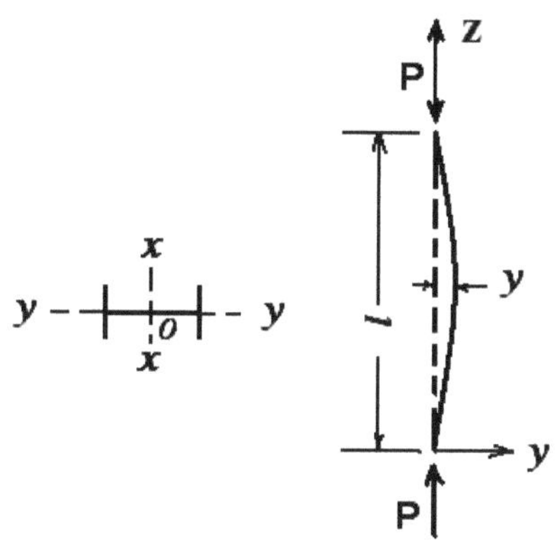

Fig. 10.4 Idealized pin-ended column

$$P_E = \frac{\pi^2 EI}{L_e^2}$$

Where:

E= Young's modulus (modulus of elasticity) of the bar material
I = minimum moment of inertia of the bar cross-section
A = sectional area of the bar
L_e = kL = reduced or effective length of the bar
L = true length of the bar
k = length reduction factor depending on the column end conditions

The effective length is that length at which the length with hinged ends is equivalent in stability to the bar with the given end conditions. The effective length of column for different end conditions is given in Fig. (10.5).

The generalized Euler formula for a column subjected to end compressive force P is given by, see Fig. (10.5):

$$P_E = \frac{\pi^2 EI}{L_e^2}$$

L_e = effective length of column

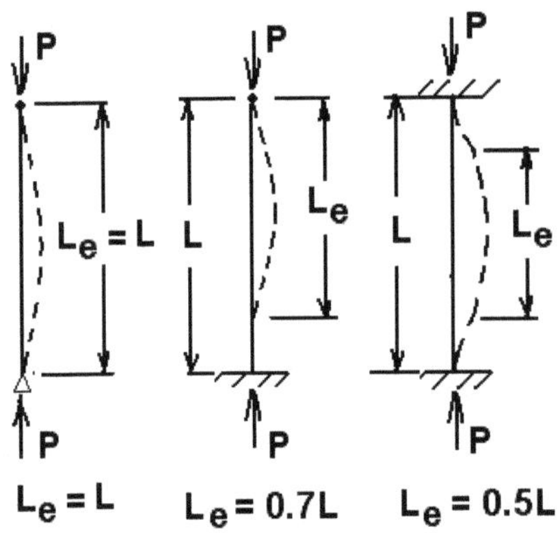

k = 1.0 for hinged supports
k = 0.5 for fixed end supports
k = 0.7 hinged-fixed supports

Fig. 10.5 Effective length of column for different end conditions

The Euler buckling stress σ_E is given by:

$$\sigma_E = \frac{P_E}{A}$$

Where: σ_E = Euler buckling stress and is given by:

$$\sigma_E = \frac{\pi^2 EI}{AL_e^2} = \frac{\pi^2 E}{\lambda^2}$$

Where: $\lambda = \dfrac{L_E}{\sqrt{I / A}} = \dfrac{L_E}{r}$

λ = the slenderness ratio of the bar (an abstract value, characterizing the tendency of the bar to buckle).

I = the minimum moment of inertia of the bar cross section.
r = radius of inertia of the bar cross-section

For elastic buckling of columns, the limiting value of λ is given by:

$$\lambda = \sqrt{\frac{\pi^2 E}{\sigma_{pr}}}$$

Where: σ_{pr} = proportional limit of the material
For shipbuilding steel, we have:
$E = 2.1 \text{x} 10^6$ kg/cm^2
And σ_{pr} = 2000 kg/cm^2
Hence: $\lambda = 3.14\sqrt{\frac{2100000}{2000}} = 105$

If we assume that the limit of elastic buckling occurs at $\sigma_y/2$, the limiting value of λ is given by:

$$\lambda = 131.35$$

The limiting value of λ for elastic buckling is shown in Fig. (10.6).

For inelastic buckling, the effect of plasticity may be accounted for by correcting the Euler buckling stress using the Johnson-Ostenfeld formula. This formula assumes that elastic buckling occurs when the Euler buckling stress is less than half the yield stress of the material i.e. when λ =131.35.

For less values of λ, inelastic buckling occurs.
Hence, the Johnson-Ostenfeld formula is given by:

$For\ \sigma_E \leq 0.5\sigma_y$

$$\sigma_{cr} = \sigma_E$$

For $\sigma_E \geq 0.5\sigma_y$

$$\sigma_{cr} = \sigma_y\left[1 - \frac{\sigma_y}{4\sigma_E}\right]$$

Where: σ_y = lower yield stress

σ_{cr} = critical buckling stress

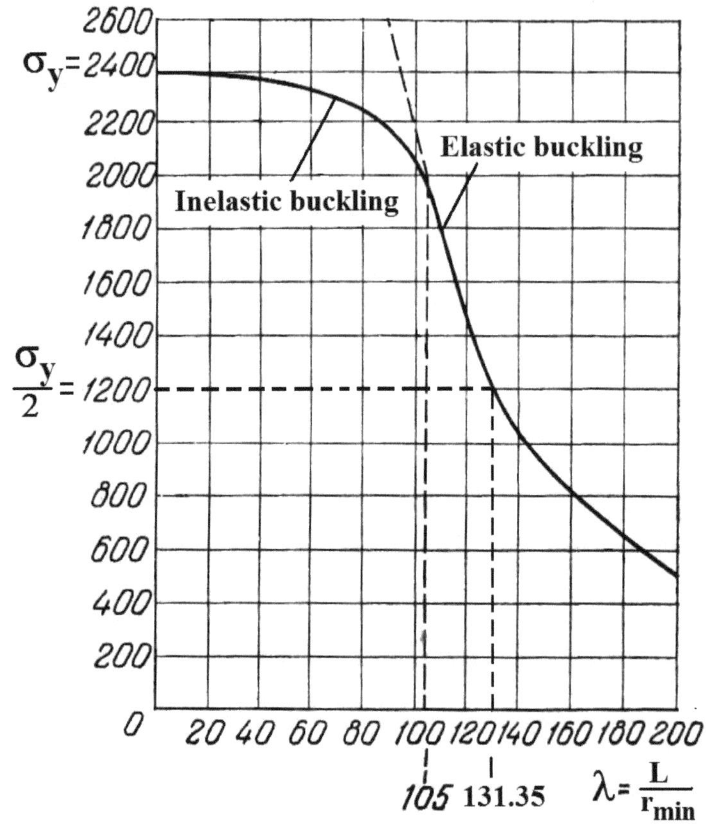

Fig.10.6 Limiting value of λ for elastic buckling

The Euler buckling load for columns is derived on the assumption that failure occurs by bending and could be called flexural buckling. For some types of sections, particularly thin-walled sections, the buckling mode may be torsional buckling or may involve torsion and bending, i.e. torsional flexural buckling. Ordinary angle and T sections are particularly susceptible to torsional buckling and therefore should be avoided as pillars or struts in ship structure. Closed sections such as tubular or box section pillars are strong in torsion and torsional buckling need not be considered. The load-carrying capacity of a compressed member is also affected by the local stability of its elements, which depends on the slenderness ratio of the flanges, webs or other elements forming the cross section of the member. The slenderness ratios of these elements are determined by the ratio between their main dimensions (depth of web of section or width of flanges) and their thickness, see Fig. (10.7).

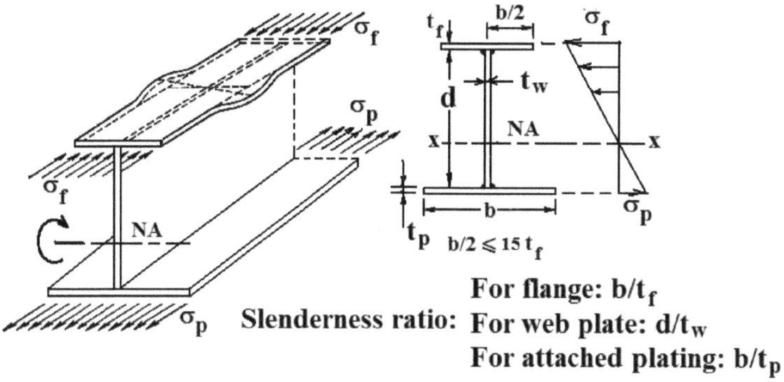

Fig. 10.7 Local buckling of the face plate of a fabricated section

***E*xample 10.1:** Determine the critical load P_{cr} and the critical stress σ_{cr} for an AL rod shown in Fig. (10.8), given: $E = 0.71 \times 10^5 MN/m^2, \sigma_y = 180\ MN/m^2$, L = 1.2 m, D = 4 cm and d = 3cm.

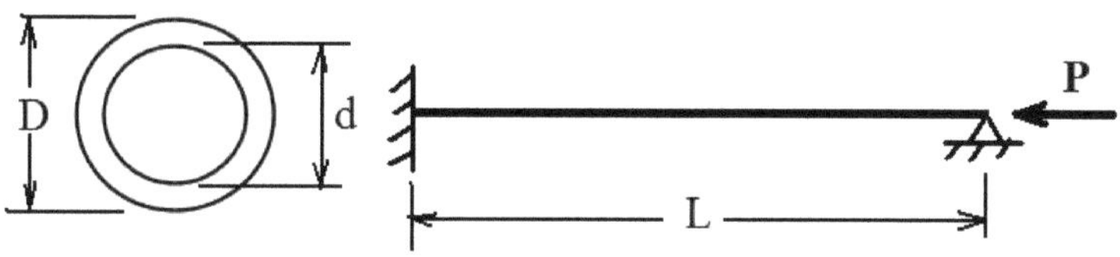

Fig. 10.8 A tubular column hinged at one end and fixed at the other end

Solution:

The cross-sectional area of the bar is given by:

$$A = \frac{\pi}{4}\left(D^2 - d^2\right) = \frac{\pi}{4}\left(4^2 - 3^2\right) = 3.925\ \text{cm}^2$$

The moment of inertia of the bar cross section is given by:

$$I = \frac{\pi}{64}\left(D^4 - d^4\right) = \frac{\pi}{64}\left(4^4 - 3^4\right) = 8.586\ \text{cm}^4$$

The radius of inertia of the bar cross-section is:

$$r = \sqrt{\frac{I}{A}} = 1.479\ cm$$

For the end conditions of the bar, the reduction factor of the length is: k = 0.7

The slenderness ratio of the bar is:

$$\lambda = \frac{kL}{r} = \frac{0.7 \times 120}{1.479} \times = 56.79$$

$$\sigma_E = \frac{\pi^2 E}{\lambda^2} = \frac{\pi^2 \times 0.7 \times 10^5}{62.36^2} = 177.48\, MN/m^2$$

The critical slenderness ratio is given by:

$$\lambda_{cr} = \pi \sqrt{\frac{0.71 \times 10^5}{180}} \approx 62.36$$

Since $\lambda > \lambda_{cr}$

Buckling will occur in the inelastic range. The critical buckling stress can be found by Johnson-Ostenfeld formula:

$$\sigma_{cr} = \sigma_y \left[1 - \frac{\sigma_y}{4\sigma_E}\right]$$

$$\sigma_{cr} = 180\left[1 - \frac{180}{4 \times 177.48}\right] = 134.36\, MN/m^2$$

Example 2: Calculate the critical buckling stress for a column having the following data for the two cases:

1- The column is fixed at both ends
2- The column is hinged at both ends

Column data: Length of column L = 3.0 m, E = 2.1 $\times 10^7$ t/m^2. $\sigma_y = 2.4\ t/cm^2$

Scantlings of column section as shown in Fig. (10.9).

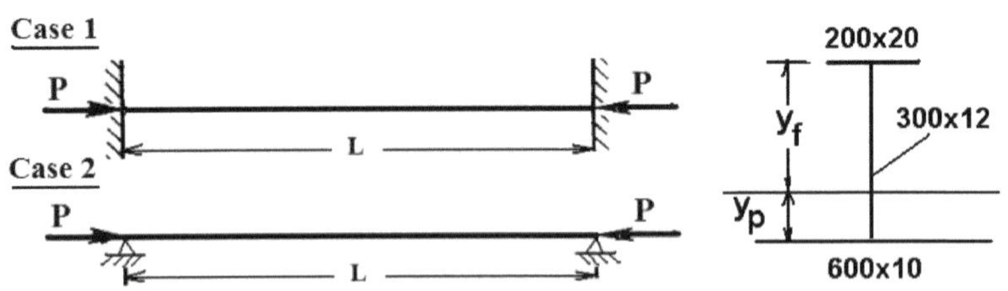

Fig. 10.9 Beam column and scantlings of the column section

Solution:

The geometrical characteristics of the column section are given by:

A = 136 cm^2, y_f = 17.206 cm, y_P = 12.794 , I = 24538.2 cm^4

Case 1: The column is fixed at both ends:

The effective length is given by:

$$L_e = L/2 = 1.5\ m$$

The Euler buckling load P_e is given by:

$$P_e = \pi^2 EI/L_e^2$$

$$= \pi^2 \times 2.1 \times 24538.2 \times 10^{-1} / 1.5^2 = 22580.77\ \ tons$$

The Euler buckling stress σ_e is given by:

$$\sigma_e = \frac{22580.77}{136} = 166.035\ \ t/cm^2$$

$$\frac{\sigma_y}{2} = 1.2\ t/cm^2$$

Since: $\sigma_e > \dfrac{\sigma_y}{2}$

Then the critical buckling stress could be calculated using Johnson Ostenfeld formula:

$$\sigma_{cr} = \sigma_y \left[1 - \frac{\sigma_y}{4\sigma_E} \right]$$

$$= 2.4 \left[1 - \frac{2.4}{4 \times 166.035} \right] = 2.391\ \ t/cm^2$$

Case 2: The column is hinged at both ends

The effective length L_e is given by:

$$L_e = L = 3.0\ m$$

The Euler buckling load is given by:

$$P_e = \frac{22580.77}{9} = 5645.2\ \ t$$

The Euler buckling stress is given by:

$$\sigma_e = \frac{5645.2}{136} = 41.51 \ t/cm^2,$$

Since: $\sigma_e > \frac{\sigma_y}{2}$

Then the critical buckling stress is given by

$$\sigma_{cr} = 2.4\left[1 - \frac{2.4}{4 \times 41.51}\right] = 2.365 \ t/cm^2$$

4.2 Effect of Eccentric Loading

The bending moment at a distance "x" from the top support is given by:

$$M = -P(e + y) \tag{1}$$

$$Hence: EI\frac{d^2y}{dx^2} = -P(y + e) \tag{2}$$

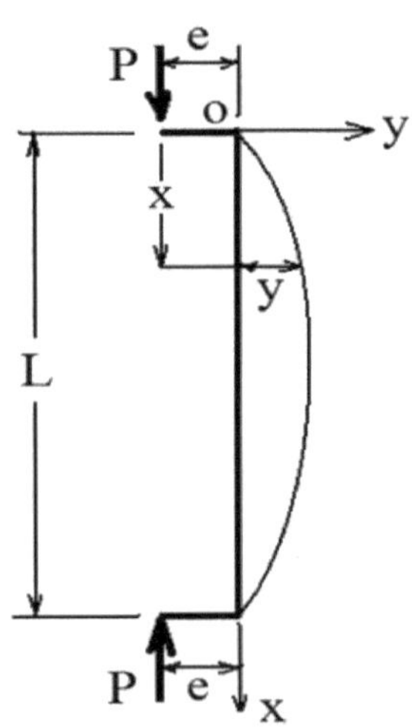

Fig. 10.10 An eccentrically loaded column

Assume a column hinged at both ends and subjected to a compressive force P at a distance “e” from the column axis, see Fig. (10.10)

$$Thus,\ EI\frac{d^2y}{dx^2} + \mu^2 y = -\frac{Pe}{EI} \tag{3}$$

$$Where{:}\mu = \frac{P}{EI}$$

The solution of equation (3) is given by:

$$y = A.\sin \mu x + B\cdot\cos \mu x$$

The constants A and B are determined from the conditions at the end supports as follows:

At x = 0, y = 0

Hence: B = e

At x = L, y = 0

$$Hence\ : 0 = A.\sin \mu L + e\cdot\cos \mu L - e$$

$$Thus\ : A = e(1-\cos \mu L)/\ \sin \mu L$$

Hence the equation of the deflection curve is given by:

$$y = e\left(\frac{1-\cos \mu L}{\sin \mu L}\right)\sin \mu x + e\cos \mu L - e$$

$$= e\left(\frac{1-\cos \mu L}{\sin \mu L}\right)\sin \mu x + e(\cos \mu L - 1) \tag{4}$$

The maximum deflection of the column occurs at $x = L/2$ and is given by:

$$y_{\max} = e\left(\frac{1-\cos \mu L}{\sin \mu L}\right)\sin\frac{\mu L}{2} + e\left(\cos\frac{\mu L}{2} - 1\right) \tag{5}$$

Using the following transformations:

$$1-\cos x = 2\sin^2\frac{x}{2}$$

$$\sin x = 2\sin^2\frac{x}{2}\cos\frac{x}{2}$$

$$\cos x = 2\cos^2\frac{x}{2} - 1$$

Substituting these transformations in equation (5), we get:

$$Y_{max} = e\left(\sec\frac{\mu L}{2} - 1\right) \tag{6}$$

The maximum bending moment occurs at x = L/2 and is given by:

$$\begin{aligned} M_{max} &= P\left(e + y_{max}\right) \\ &= P.e\left(1 + \sec\frac{\mu L}{2} - 1\right) \\ &= P.e.\sec\frac{\mu L}{2} \end{aligned} \tag{7}$$

The maximum stress at mid length of the column is given by:

$$\begin{aligned} \sigma_{max} &= \frac{P}{A} + \frac{M.c}{I} \\ &= \sigma + \frac{P.e.c}{I}\sec\frac{\mu L}{2} \\ &= \sigma\left[1 + \frac{e.c}{k^2}\sec\frac{\mu L}{2}\right] \end{aligned} \tag{8}$$

Where: A = sectional area of column
$k = I / A$ = radius of gyration of column section

The maximum stress should satisfy the following condition:

$$\sigma_{max} \leq \sigma_y \tag{9}$$

Where: σ_y = yield stress of the material

Substituting equation (8) in condition (9), we get:

$$\frac{\sigma}{\sigma_y} \leq \frac{1}{1 + \frac{e.c}{k^2}\sec\frac{\mu L}{2}}$$

Let: $ec / k^2 = r$ = eccentricity ratio.

The variation of σ/σ_y with L_e/k for different values of ec/k^2 is shown in fig. (10.11). The variation of the ratio of ec/k^2 with L_e/k is given in Fig. (10.11) for the yield stress $\sigma_y = 2.4$ t/cm^2 and elastic modulus E = 2.1 x 10^6 kg/cm^2

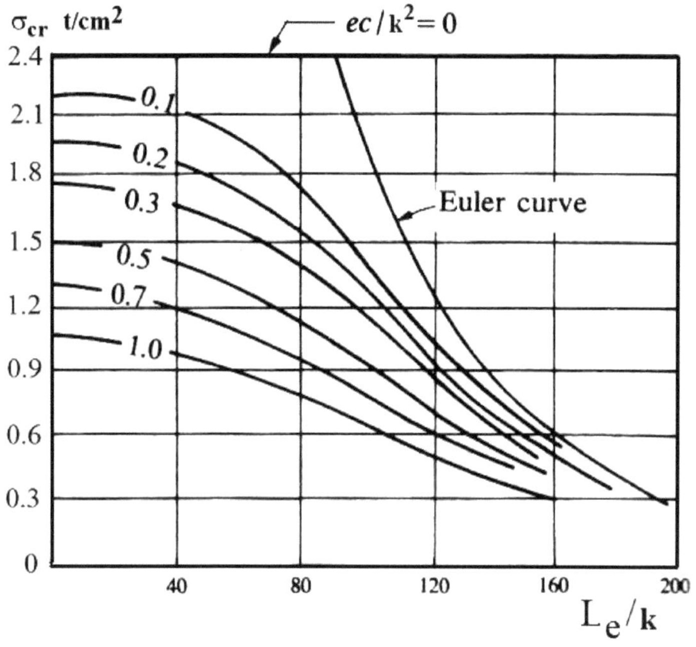

Fig. 10.11 Variation of the ratio of ec/r^2 with L_e/k

5 Beam Columns

Beam columns normally refer to beams subjected to an axial force and a lateral load. The lateral load could be either concentrated, end moments or uniformly/linearly distributed. A beam column subjected to an axial force at both ends and a uniformly distributed lateral load is shown in Fig. (10.12). A horizontal beam column subjected to an axial load P and a concentrated force W acting at mid-span is shown in Fig. (10.13).

A beam column subjected to an axial force and end moments at both ends of the beam is shown in Fig. (10.14).

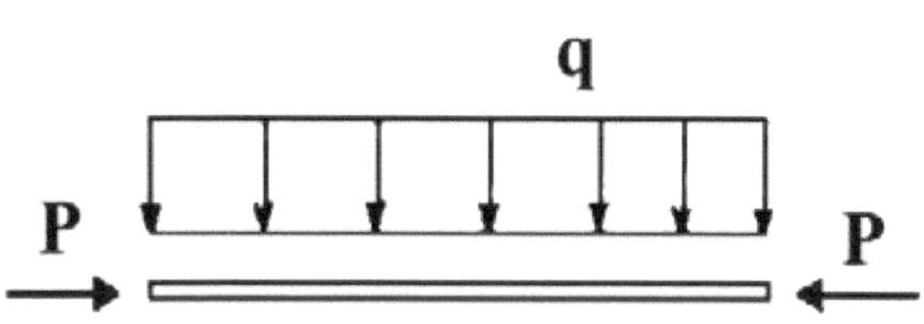

Fig. 10.12 A beam column subjected to an axial force and a uniformly distributed transverse load

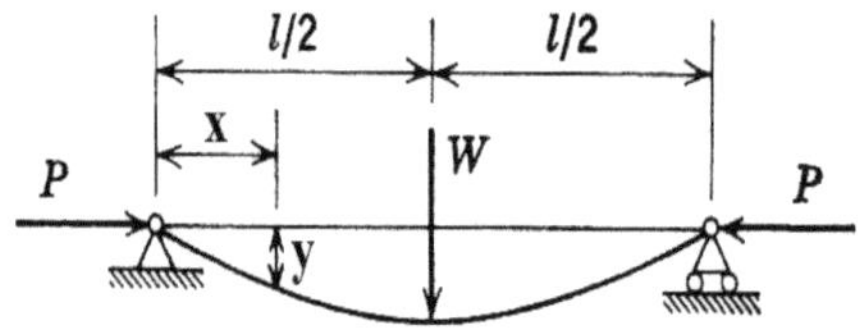

Fig. 10.13 Basic configuration of a horizontal beam column subjected to an axial force and a concentrated lateral force at midspan

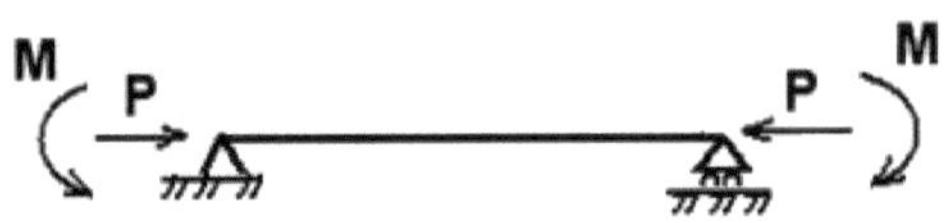

Fig. 10.14 A horizontal beam column subjected to an axial force and end moments at both ends of the beam

5.1 Load-Deflection Relationship of Beam Columns

Consider a beam uniformly loaded transversely and subjected to an axial force P. The load defection curve of the beam column is shown in Fig. (10.15) for the following axial loading cases:

a) The axial force P is a compressive force
b) The axial force P is a tensile force
c) The beam is subjected only to the transversely uniform load

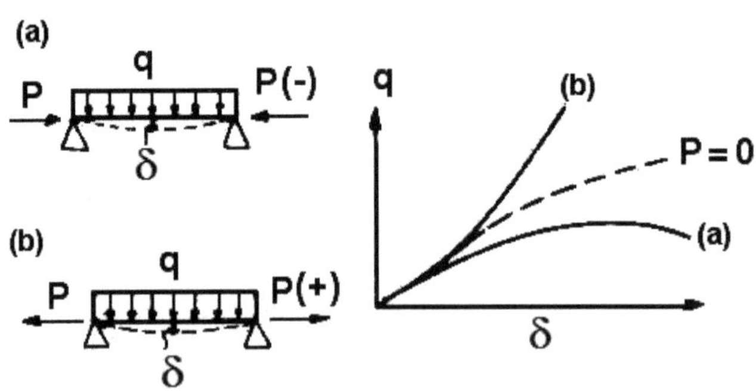

Fig. 10.15 Load-deflection curves of beam columns

The bending moment at any section along the beam column is composed of the primary moment resulting from the lateral load and the secondary moment resulting from the axial load and the deflection of the beam, see Fig. (10.16). For the case of a compressive axial load, the secondary moment is added to the primary moment. For the case of a tensile axial load, the secondary moment is subtracted from the primary moment, see Fig. (10.17).

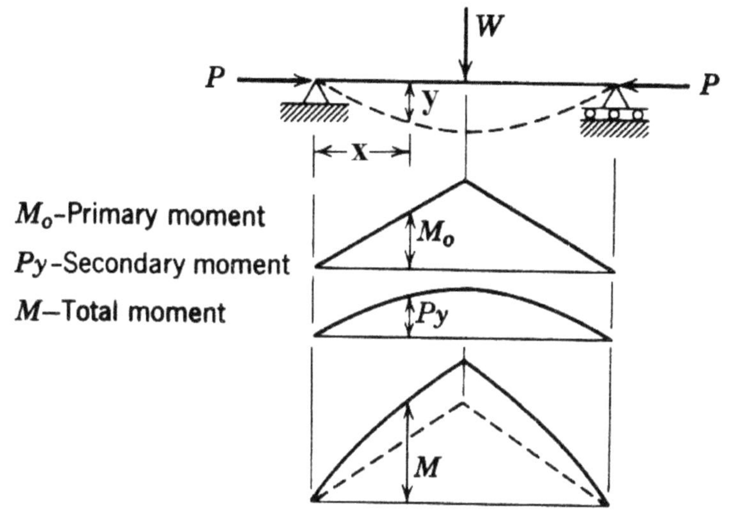

Fig. 10.16 A beam column subjected to an axial compressive force and a concentrated lateral force

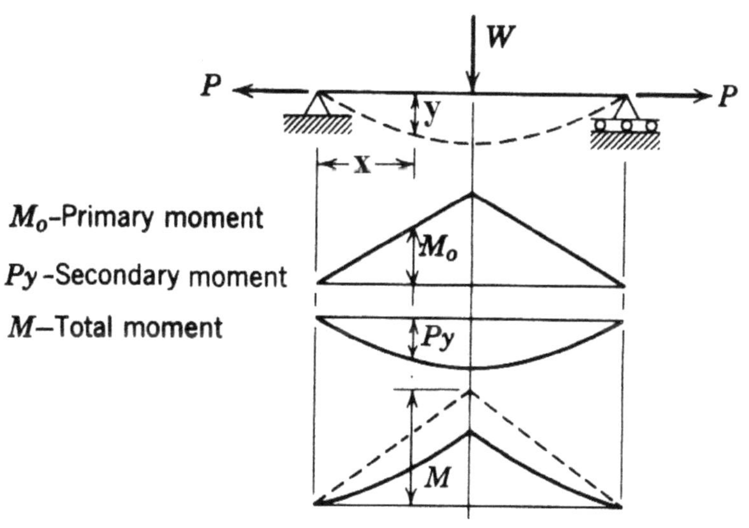

Fig. 10.17 A beam subjected to an axial tensile force and a concentrated lateral force

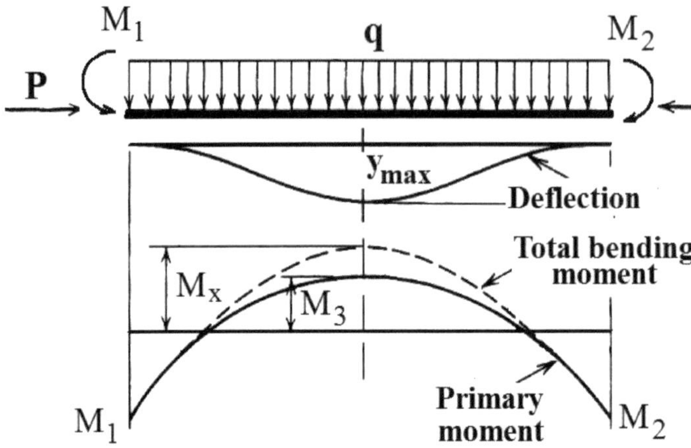

Fig. 10.18 Primary and total bending moment for a beam column fixed at both ends and subjected to a transverse uniform loading

Fig. (10.18) shows a beam column subjected to an axial force P and a transverse uniform load q. The total bending moment curve is composed of the primary bending moment and the additional bending moment induced by the axial force P and the presence of the deflection y. The primary bending moment at both fixed ends of the beam is given by:

$$m_1 = m_2 = qL^2 / 12$$

It is clear from Fig. (10.18) that the presence of a compressive axial load on a transversely loaded beam reduces its load carrying capacity significantly.

6 Stresses in Beam Columns

The calculation of stresses in beam columns requires the determination of the equation of the deflection curve and the calculation of the maximum bending moment. Consider the following cases of a beam column:

Case 1: Beam column under uniform loading and fixed at both Ends

The bending moment at a distance x from the left end of the beam is given by, see Fig. (10.19):

$$M = P \cdot y + \frac{qL}{2} \cdot x - \frac{q \cdot x^2}{2} - M_o$$

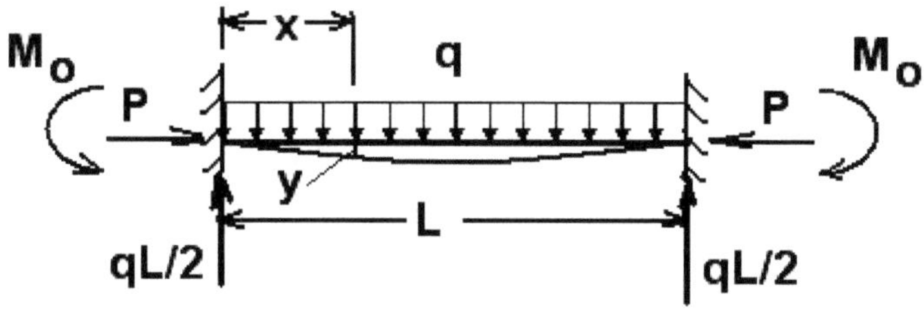

Fig. 10.19 A beam column subjected to axial force and bending moments at both ends and a transverse uniform loading

Thus: $EI\dfrac{d^2y}{dx^2} = -P\cdot y - \dfrac{qL}{2}\cdot x + \dfrac{q\cdot x^2}{2} + M_o$

Let: $\mu^2 = P/EI$

Hence: $\dfrac{d^2y}{dx^2} + \mu^2 y = \dfrac{-q}{2EI}\left(xL - x^2\right) + \dfrac{M_o}{EI}$

The solution of equation ( ) is given by:

$$y = A\cdot\sin\mu x + B\cdot\cos\mu x + \frac{M_o}{P} - \frac{q}{2P}\left(xL - x^2 + \frac{2}{\mu^2}\right)$$

The constants A and B are determined from the end conditions as follows:

At x = 0 y = 0

Then: $B = \dfrac{q - \mu^2 M_o}{P\cdot\mu^2}$

At x = L y = 0

Then: $A = \dfrac{q \quad \mu^2 M_o}{P\cdot\mu^2}\left(\dfrac{1-\cos\mu L}{\sin\mu L}\right)$

Hence the deflection equation of the beam column is given by:

$$y = \frac{q-\mu^2 M_o}{P\cdot\mu^2}\left(\frac{1-\cos\mu L}{\sin\mu L}\right)\cdot\sin\mu x + \frac{q-\mu^2 M_o}{P\cdot\mu^2}\cdot\cos\mu x$$

$$+\frac{\mu^2 M_o - q}{P\cdot\mu^2} - \frac{q}{2P}\left(Lx - x^2\right)$$

The bending moments at both ends of the beam column could be determined using the following end conditions:

$$At \quad x = 0, \; y' = \frac{dy}{dx} = o$$

Substituting for these conditions, the end moment is given by:

$$M_o = qL^2 \left(\frac{2 \tan \frac{\mu L}{2} - \mu L}{2 \mu^2 L^2 \cdot \tan \frac{\mu L}{2}} \right)$$

The maximum deflection occurs at midspan of the beam and is given by:

$$y_{\max} = \frac{q - \mu M_o}{P \cdot \mu^2} \left[\cos \frac{\mu L}{2} + \frac{1 - \cos \mu L}{2 \cos \frac{\mu L}{2}} - 1 \right] - \frac{qL^2}{8P}$$

$$= \frac{q - \mu M_o}{P \cdot \mu^2} \left[\frac{1}{\cos \frac{\mu L}{2} - 1} \right] - \frac{qL^2}{8P}$$

The equation of the maximum deflection could be simplified by using the following transformations:

$$cos \frac{\mu L}{2} = 1 - \frac{\mu^2 L^2}{8} + \ldots.$$

$$\tan \frac{\mu L}{2} = \frac{\mu L}{2} + \frac{\mu^3 L^3}{3} + \ldots$$

Substituting for $\cos \frac{\mu L}{2}$ and $\tan \frac{\mu L}{2}$, the maximum deflection is given by:

$$y_{\max} \approx \frac{y_q}{1 - P/P_E}$$

The bending moment at both ends of the beam column is given by:

$$M_O \cong \frac{qL^2/12}{1 - P/P_E}$$

The maximum bending moment at midspan is given by:

$$M_{\max} = P \cdot y_{\max} + \frac{qL^2}{8} - M_o$$

$$\approx \frac{1}{1-P/P_E}\left[P\cdot y_q - \frac{qL^2}{12}\right] + \frac{qL^2}{8}$$

The maximum stress is therefore given by:

$$\sigma_{\max} = \frac{P}{A} + \frac{1}{Z}\left[\frac{1}{1-P/P_E}\left(P\cdot y_q - \frac{qL^2}{12}\right) + \frac{qL^2}{8}\right]$$

<u>Summary of case 1:</u>

The bending moment at a distance x from the left end of the beam is given by:

$$M = P\cdot y + \frac{qL}{2}\cdot x - \frac{qx^2}{2} - M_o$$

$$y_{\max} = \frac{y_q}{1-P/P_E},$$

$$y_q = \frac{qL^4}{384\,EI}$$

$$M_o = \frac{qL^2/12}{1-P/P_E}$$

Where : $\frac{1}{1-P/P_E}$ = *Amplificat ion factor*

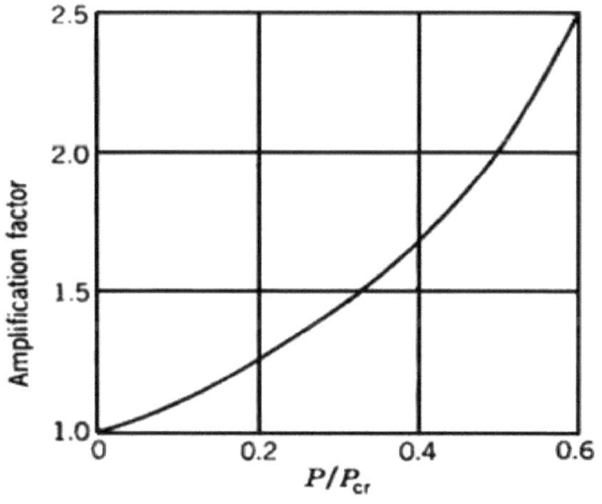

Fig. 10.20 Variation of amplification factor with P/P_{cr}

The variation of the amplification factor with P/P_{cr} is shown in Fig. (10.20). The maximum bending moment at midspan is given by:

$$M_{max} = P \cdot y_{max} + \frac{qL^2}{8} - M_o$$

$$\sigma_{max} = \frac{P}{A} + \frac{M_{max}}{Z}$$

Case 2: Beam column subjected to an axial force and end moments

The bending moment at any distance x along the beam column is given by, see Fig. (10.21):

$$M = -P \cdot y - M_o$$

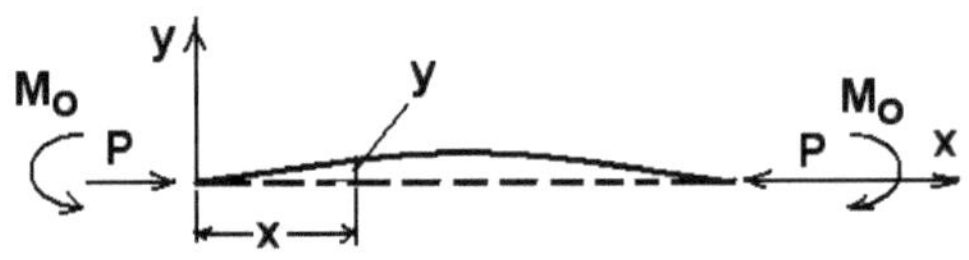

Fig. 10.21 A beam column subjected to an axial force and end moments

Hence: $EI \dfrac{d^2 y}{dx^2} = -P \cdot y - M_o$

Let: $\dfrac{P}{EI} = \mu^2$

Then: $\dfrac{d^2 y}{dx^2} + \mu^2 y = -\dfrac{M_o}{EI}$

The solution of equation ( ) is given by:

$$y = A \cdot \sin \mu x + B \cdot \cos \mu x - \frac{M_O}{P}$$

The constants A and B are determined from the conditions of deflections at both end of the beam column as follows:

At $x = 0$ $y = 0$

Then: $B = \dfrac{M_O}{P}$

At $x = L$ $y = 0$

Then: $A = \dfrac{M_O}{P}(1 - \cos \mu L)/\sin \mu L$

Hence, the deflection equation is given by:

$$y = \frac{M_O}{P}\left(\frac{1-\cos\mu L}{\sin\mu L}\right)\sin\mu x + \frac{M_O}{P}(\cos\mu x - 1)$$

The maximum bending moment M_{max} occurs at x = L/2

$$M_{max} = P \cdot y_{max} + M_o = \frac{M_o}{1-\frac{P}{P_E}},$$

Where: $P_E = \pi^2 EI/L^2$

The maximum deflection is given by:

$$y_{max} = y = \frac{M_O}{P}\left(\frac{1-\cos\mu L}{\sin\mu L}\right)\sin\frac{\mu L}{2} + \frac{M_O}{P}\left(\cos\frac{\mu L}{2} - 1\right)$$

Consider an arbitrary cross section of the beam column, see Fig. (10.22).

The maximum stress over the column cross section is given by:

$$\sigma_{max} = \frac{P}{A} + M_{max} \cdot \frac{c}{I}$$

$$= \sigma + \frac{M_o}{1-\frac{P}{P_E}} \cdot \frac{c}{I}$$

Where: A = cross section of the assumed beam section

c = distance of the top extreme fibers from the neutral axis of the section

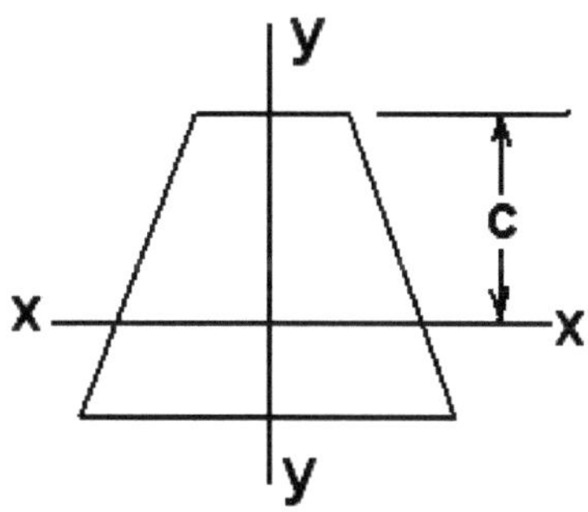

Fig. 10.22 An arbitrary cross section of the beam column

The bending stress due to the bending moment at the beam end is given by:

$$\sigma_o = \frac{M_o \cdot c}{I}$$

Then the total maximum stress should satisfy the following condition:

$$\sigma\left[1 + \frac{\sigma_o}{\sigma\left(1 - P/P_E\right)}\right] \leq \sigma_y$$

Case 3: Beam column hinged at both ends and subjected to uniform loading

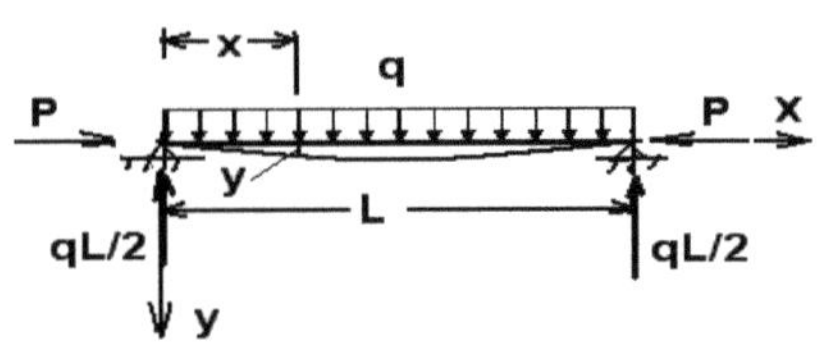

Fig. 10.23 A beam column subjected to an axial force and a uniformly distributed transverse load

The bending moment at a distance ‘x” from the left support is given by, see Fig. (10.23):

$$M = P \cdot y + \frac{qL}{2} \cdot x - \frac{qx^2}{2}$$

The maximum deflection at midspan is given by:

$$y_{\max} = \frac{y_q}{1 - P/P_E}$$

The transverse maximum deflection resulting from the transverse uniformly distributed load is given by:

$$y_q = \frac{5qL^4}{384\,EI}$$

The maximum bending moment at midspan is given by:

$$M_{\max} = P \cdot y_{\max} + \frac{qL^2}{8}$$

The maximum stress over the cross section of the beam is given by:

$$\sigma_{\max} = \frac{P}{A} + \frac{M_{\max}}{Z}$$

Where: $Z = I/c$

Summary of beam column formulae

Case 1: Beam hinged at both ends and subjected to a concentrated load

The maximum deflection occurs at midspan of the beam and is given by, see Fig. (10.24):

$$y_{\max} = \frac{WL^3}{48\,EI}\left[\frac{3(\tan\mu - \mu)}{\mu^3}\right] = \frac{WL^3}{48\,EI}\cdot\varphi_1$$

The maximum bending moment occurs at midspan of the beam and is given by:

$$M_{\max} = \frac{WL}{4}\left[\frac{3(\tan\mu)}{\mu}\right] = \frac{WL}{4}\cdot\varphi_2$$

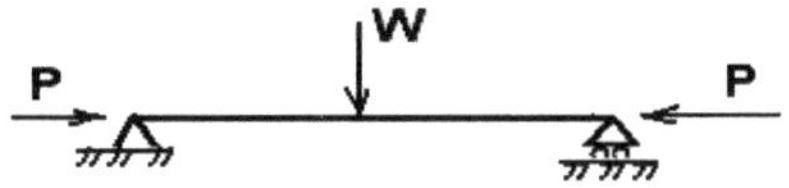

Fig. 10.24 A beam column subjected to an axial force and a concentrated force at midspan of the beam.

Case 2: Beam hinged at both ends and subjected to uniform loading

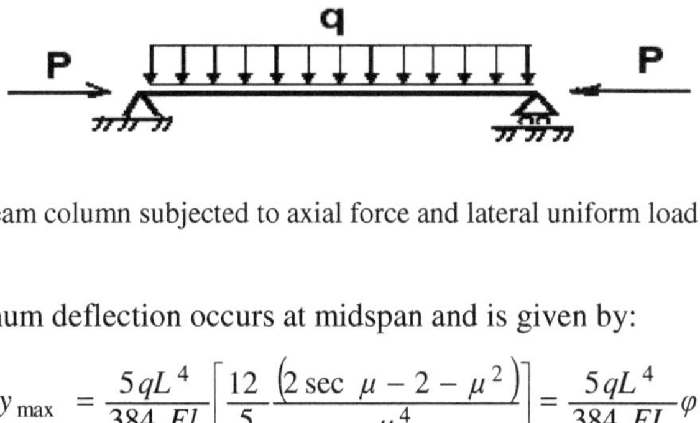

Fig. 10.25 A beam column subjected to axial force and lateral uniform load

The maximum deflection occurs at midspan and is given by:

$$y_{\max} = \frac{5qL^4}{384\,El}\left[\frac{12}{5}\frac{\left(2\sec\mu - 2 - \mu^2\right)}{\mu^4}\right] = \frac{5qL^4}{384\,EI}\varphi_3$$

The maximum bending moment occurs at midspan and is given by:

$$M_{\max} = \frac{qL^2}{8}\left[\frac{2(\sec\mu - 1)}{\mu^2}\right] = \frac{qL^2}{8}\varphi_4$$

Case 3: Beam hinged at both ends and subjected to end moments

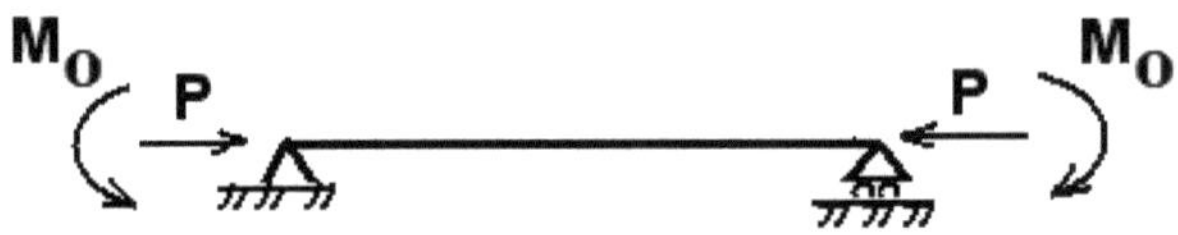

Fig. 10.26 A beam column subjected to an axial force at both ends and a concentrated moment at both ends of the beam

Following the same procedure given before, the maximum deflection at mid span could be determined as follows:

$$y_{\max} = \frac{M_o L^2}{8EI}\left[\frac{2(\sec\mu - 1)}{\mu^2}\right] = \frac{M_o L^2}{8EI}\varphi_4$$

The maximum bending moment at midspan is given by:

$$M_{\max} = M_o[\sec\mu] = M_o\varphi_5$$

The different "ϕ" values are given in table (10.1).

Table 10.1 Different values of "φ"

P/P_E	$1/(1-P/P_E)$	φ_1	φ_2	φ_3	φ_4	φ_5
0	1	1	1	1	1	1
0.1	1.111	1.109	1.091	1.107	1.114	1.137
0.2	1.25	1.247	1.205	1.253	1.258	1.31
0.3	1.429	1.42	1.35	1.43	1.441	1.533
0.4	1.667	1.657	1.545	1.663	1.684	1.831
0.5	2	1.982	1.815	2	2.029	2.252
0.6	2.5	2.477	2.223	2.502	2.544	2.884
0.7	3.333	3.303	2.901	3.347	3.407	3.942
0.8	5	4.943	4.253	5.013	5.124	6.057
0.9	10	9.876	8.308	10.04	10.29	12.42

Example. Calculate the maximum compressive stress induced in a bottom longitudinal of an oil tanker. The scantling of the bottom longitudinal is shown in Fig. (10.27).

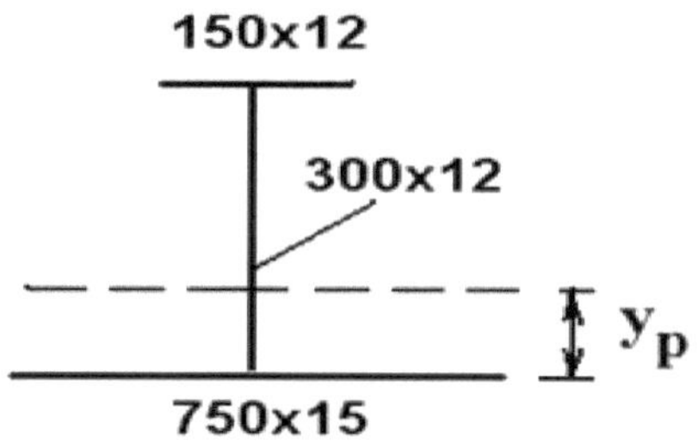

Fig. 10.27 Scantlings of the bottom longitudinal

Assume the following data:

The ship section is subjected to a hogging moment of 300 MNm, The section modulus of the tanker at the keel = 8m^3 . E = 210 GN/m^2. (1 ton = 9.964 kN), The design draught T = 10 m. Longitudinal spacing b = 750 mm

Transverse spacing L = 3 m. Assume longitudinal hinged at both ends.

Solution

The maximum stress in the section of the longitudinal is given by:

$$\sigma_{\max} = \frac{P}{A} + \frac{1}{Z}\left[\frac{1}{1-\frac{P}{P_e}}\left(P \cdot y_q - \frac{qL^2}{12}\right) + \frac{qL^2}{8}\right]$$

The longitudinal is subjected to a uniformly distributed load q induced by the water pressure and is given by:

$$q = b \times T \times \gamma = 0.75 \times 10 \times 1.025 = 7.6875 \quad t/m$$

The length of the longitudinal L = 3 m

The cross sectional area of the longitudinal is given by:

$$A = 0.75 \times 0.015 + 0.15 \times 0.012 + 0.3 \times 0.012 = 0.01665 \quad m^2$$

The bending stress induced in the bottom longitudinal due to bending of the hull girder is given by:

$$\sigma = M/Z = \frac{300}{8} = 37.5 \quad MN/m^2$$

The axial force acting on the longitudinal section is therefore given by:

$$P = \sigma \cdot A = 37.5 \times 0.01665 = 0.624 \quad MN$$

The critical buckling load is given by:

$$P_e = \frac{4\pi^2 EI}{L^2}$$

The moment of inertia of the longitudinal section about the base line of the section is given by:

$$I = I_o - A \cdot y_P^2$$

Where: I = Moment of inertia of the longitudinal section about the section neutral axis
I_o = Moment of inertia of the longitudinal section about the section base line
A = the cross sectional area of the longitudinal
y_P = Distance of the neutral axis from the base line of the longitudinal section

The moment of inertia of the longitudinal section about the section neutral axis is given by:

$$I_o = 0.15 \times 0.012 \times 0.3^2 + 0.3 \times 0.012 \times \left(\frac{0.3}{2}\right)^2 + \frac{0.012 \times 0.3^3}{12}$$

$$= 0.00027\, m^4$$

The distance of the neutral axis of the longitudinal section from the base line is given by:

$$y_P = \frac{0.15 \times 0.012 \times 0.3 + 0.3 \times 0.012 \times 0.15}{A} = 0.0648 \quad m$$

Hence the moment of inertia of the section about the neutral axis is given by:

$$I = 0.00027 - 0.01665 \times 0.0648^2 = 0.0002 \quad m^4$$

The Euler buckling load is given by:

$$P_e = \frac{4\pi^2 \times 210 \times 10^3 \times 0.0002}{3^2} = 184.0 \ MN$$

The deflection at midspan of the longitudinal is given by:

$$y_q = \frac{qL^4}{384\ EI} = \frac{3^4 \times 7.6875 \times 9.964}{384 \times 210 \times 10^6 \times 0.0002} = 0.00038 \quad m$$

The maximum stress is given by:

$$\sigma_{\max} = 37.5 + \frac{1}{8}\left[\frac{1}{1-\frac{0.624}{184}}\left(0.624\times 0.031 - \frac{7.6875\times 9.964\times 3^2}{12}\right) + \frac{7.6875\times 9.964\times 3^2}{8}\right]$$

$$= 41.068 \quad MN/m^2$$

Chapter 11
Buckling of Stiffened Panels

1 Introduction

This chapter presents the basic configurations of stiffened panels commonly used in ship construction. A comprehensive analysis of buckling of stiffened panels is introduced. Global and local buckling modes of deformation of stiffened panels are given. The commonly used idealizations of boundary support conditions of ship plating are clarified. The basic equations of plate buckling for various boundary conditions and different loading combinations are presented. Interaction equations of buckling of plating subjected to a variety of combined loading systems are given. The non-uniform in-plane compressive loadings are idealized by the combined loadings of uniform compression and pure bending. The various boundary conditions commonly assumed for girders and plating are given. The general modes of buckling of face plates and web plates of girders are illustrated. General modes of buckling deformation of longitudinals and plating are illustrated. Post–buckling strength of plating is introduced. Ultimate stress of simply supported plate panels is given. The concept of effective width of plating is clarified.

2 Basic Configurations of Stiffened Panels

The basic structural configuration of stiffened panels is shown in Fig.(11.1). In longitudinally stiffened deck structures, the stiffeners could be the deck longitudinals and the girders are the heavy beam transverses. In transversely stiffened deck structures, the stiffeners are the deck beams and the transverses are the heavy deck girders.

The most common in-plane loading conditions of stiffened panels are, see Fig. (11.2):

- Loading orthogonal to the direction of stiffeners
- Loading in the same direction of stiffeners

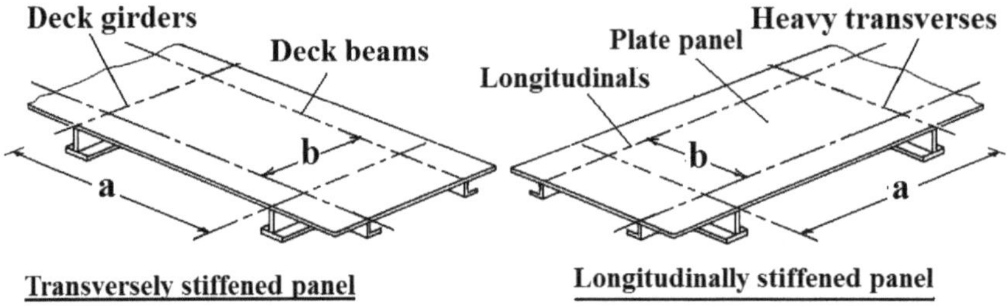

Fig 11.1 Longitudinally and transversely stiffened plate panels

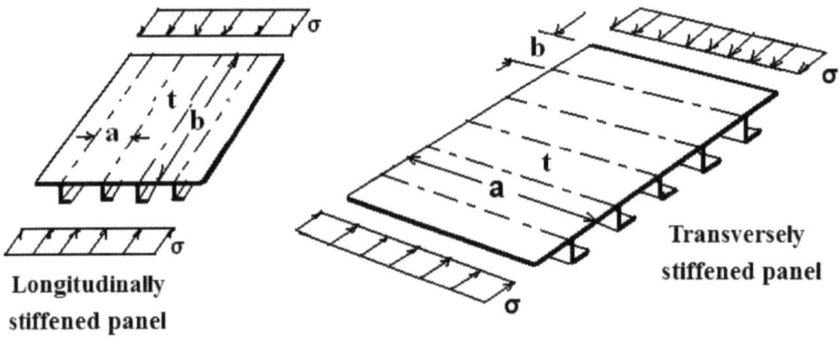

Fig 11.2 In-plane loading conditions on stiffened panels

3 Modes of Deformation of Transversely Stiffened Plate Panels

The general buckling modes of deformation of transversely stiffened plate panels under in-plane loading orthogonal to the direction of stiffeners are:

- Buckling of plate panels between stiffeners
- Buckling of stiffeners
- Flexural buckling of plate-stiffener combination (global buckling of stiffened panel)

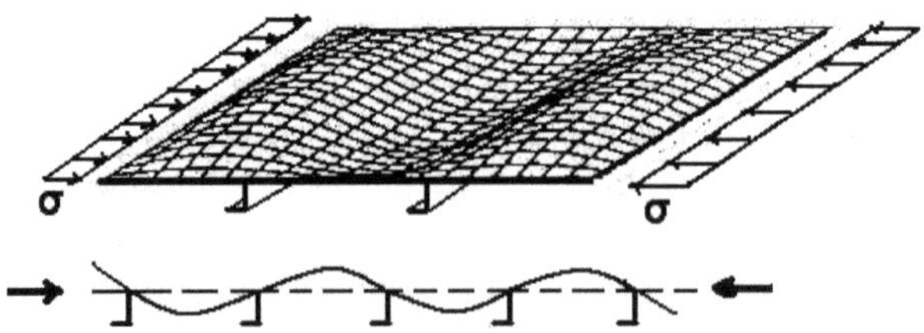

Fig. 11.3 Mode of deformation of a stiffened plate panel subjected to in-plane loading orthogonal to the direction of stiffeners

The torsional rigidity of the stiffening members has significant effect on the mode of deformation of transversely stiffened plate panels. The different modes of deformation of a transversely stiffened plate panel subjected to in-plane compressive loading orthogonal to the stiffening members are shown in Figs. (11.3, 11.4).

Fig. 11.4 Mode of deformation of a transversely stiffened panel under in-plane loading, the stiffeners has low torsional rigidity

The mode of deformation when the stiffeners have torsional rigidity much greater than the flexural rigidity of the plating is shown in Fig. (11.3). The mode of deformation when the stiffeners have relatively small torsional rigidity is shown in Fig. (11.4). Global flexural buckling of a stiffened panel under in-plane compressive stresses is shown in Fig. (11.5).

An example of the global mode of deformation of a transversely stiffened deck structure is shown in Fig. (11.6).

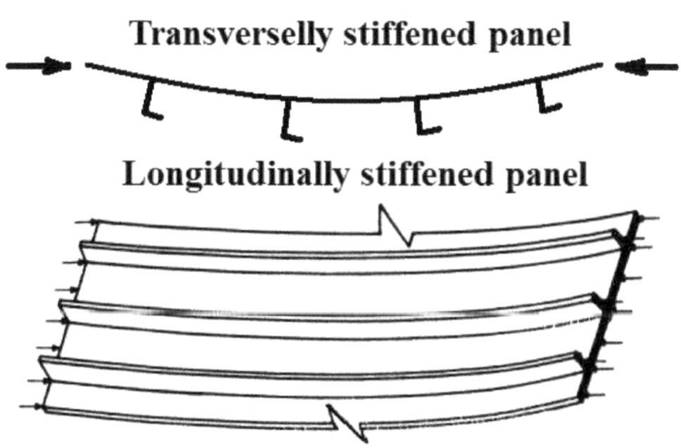

Fig.11.5. Global flexural buckling of a stiffened panel under in-plane compressive stresses orthogonal to direction of stiffeners

3.1 Modes of Buckling Deformation of Stiffeners

The various modes of buckling deformation of stiffeners subjected to in-plane loading in the same direction of the stiffeners are:

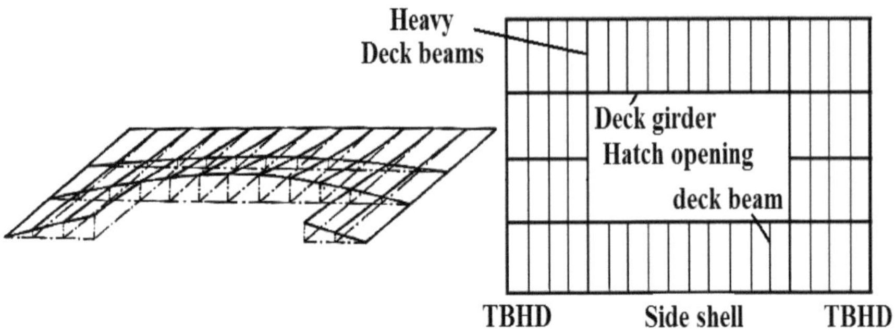

Fig.11.6. Flexural buckling of a deck structure

- Flexural buckling of stiffeners
- Torsional buckling of stiffeners
- Lateral buckling of stiffeners
- Global buckling of stiffener with attached plating
- Local buckling of web plate of the stiffeners
- Local buckling of the face plate of the stiffeners

The mode of deformation of flexural buckling of stiffeners is shown in Fig. (11.7). In this case the stiffeners behave like a beam-column. When the supporting heavy members at both ends of the stiffeners have large torsional rigidity, the mode of deformation of the beam column is as shown in Fig. (11.8).

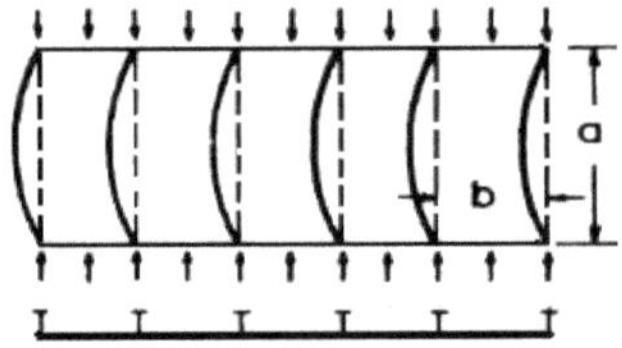

Fig. 11.7 Mode of deformation of a longitudinally stiffened panel under in-plane compressive stresses (column failure)

On the other hand, when the torsional rigidity of the supporting members is relatively low, the mode of deformation of the beam column is as shown in Fig. (11.9). When the torsional rigidity of the stiffeners is relatively low, the mode of deformation of the stiffeners is shown in Fig. (11.10). When the flexural rigidity of the plating and the torsional rigidity of the stiffeners are of the same order of magnitude, the mode of deformation of the stiffener with attached plating is as shown in Fig. (11.11).

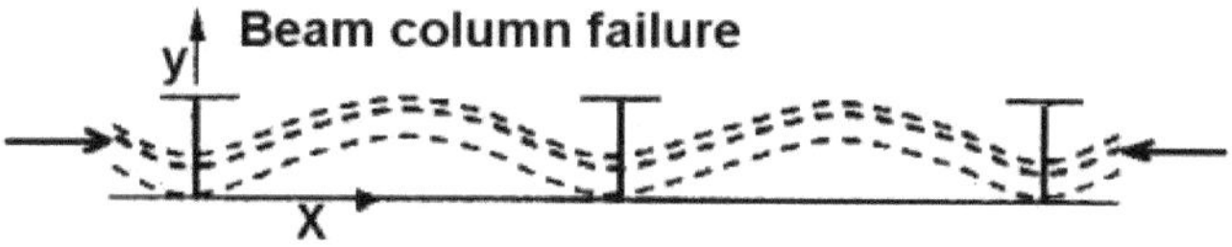

Fig. 11.8 Mode of deformation of a stiffener under in-plane compressive stresses, (beam column failure)

The mode of deformation of the web plate of a stiffener could take the shape shown in Fig. (11.12). The face plate of stiffeners could experience the mode of deformation shown in Fig. (11.13) whether symmetrical or asymmetrical face plate is used.

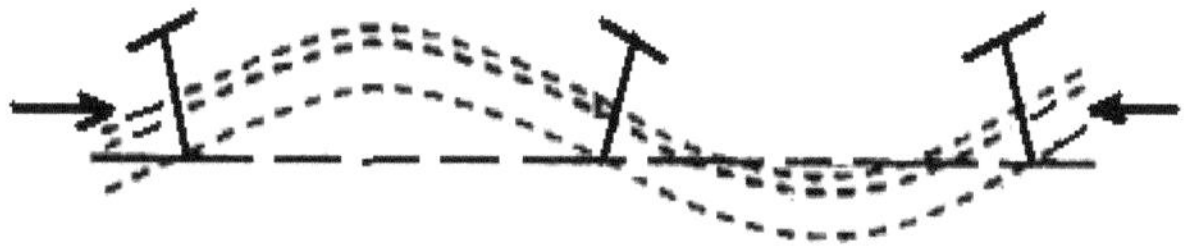

Fig. 11.9 Mode of deformation of a stiffener under in-plane compressive stresses, (low torsional rigidity of the supporting members)

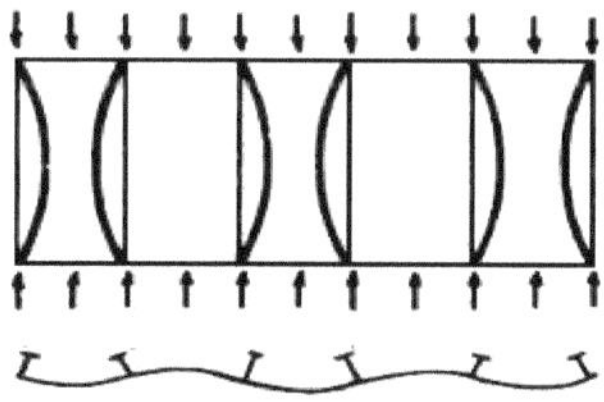

Fig. 11.10 Mode of deformation of a longitudinally stiffened panel under in-plane compressive stresses,(Tripping failure)

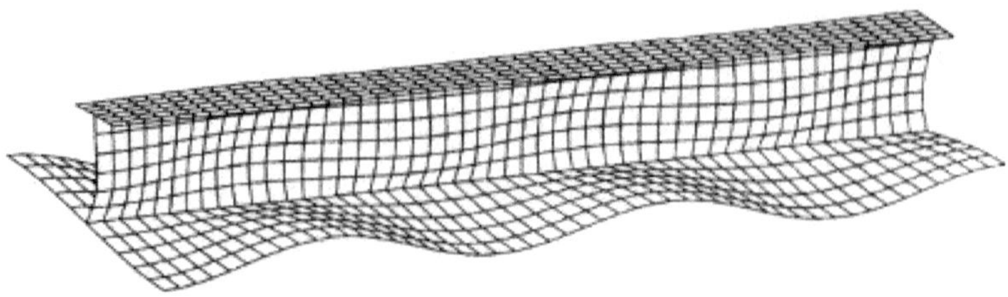

Fig. 11.11 Mode of deformation of an isolated stiffener with attached plating under in-plane compressive stresses

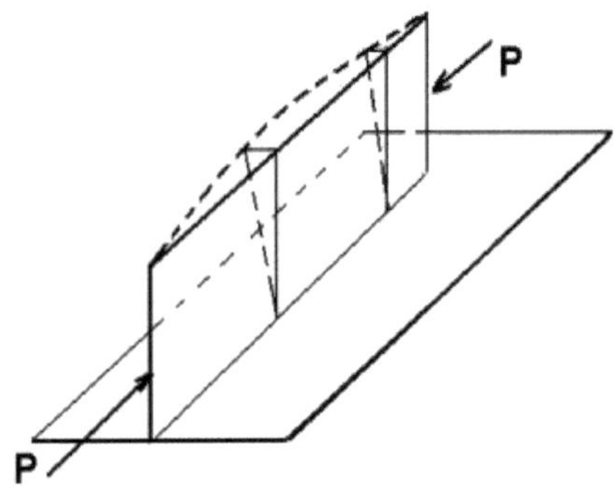

Fig. 11.12 Mode of deformation of the web plate of a stiffener under in-plane compressive stresses

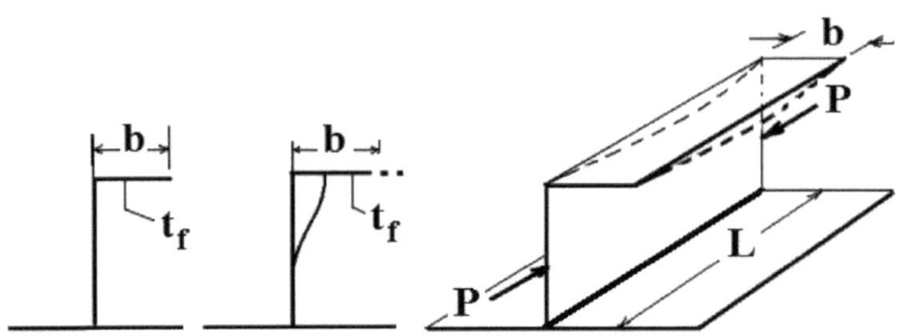

Fig. 11.13 Mode of deformation of the face plate of a stiffener under in-plane compressive stresses

3.2 Global Mode of Deformation of Stiffened Panels

The mode of deformation of a stiffened panel depends on the direction of the in-plane loading with respect to the direction of the stiffeners.

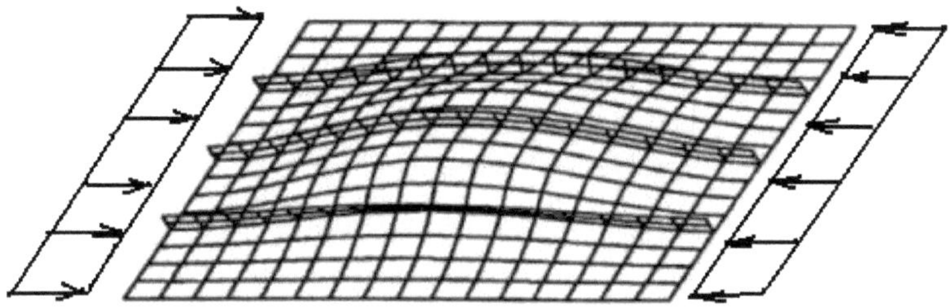

Fig. 11.14 Mode of deformation of a stiffened panel under in-plane compressive stresses in-line with stiffeners

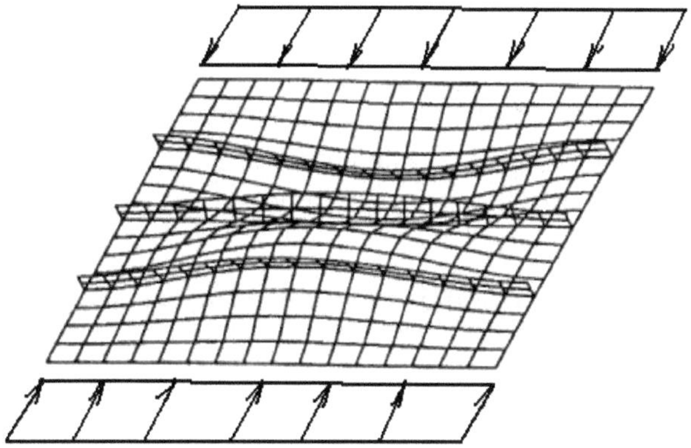

Fig. 11.15 Mode of deformation of a stiffened panel under in-plane compressive stresses orthogonal to the stiffeners

The mode of deformation when the in-plane loading is in the same direction of the stiffening members is shown in Fig. (11.14) and when the in-plane loading is orthogonal to the stiffening members the mode of deformation is shown in Fig. (11.15).

4 Assessment of Buckling Strength of Plating

The general Euler buckling equation of a plate panel is given by, see Fig. (11.16):

$$\sigma_E = \frac{k.\pi^2 E}{12\left(1-\upsilon^2\right)}\left(\frac{t}{b}\right)^2$$

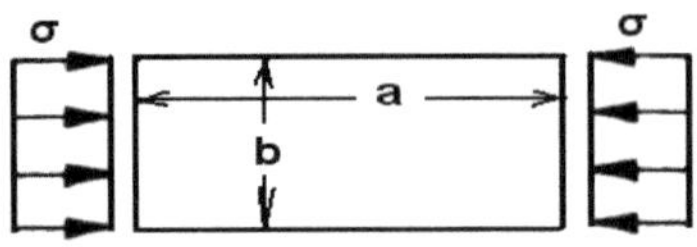

Fig. 11.16 Plate panel under in-plane loading on the smaller width of plate

Where: σ_E = Euler buckling stress

b = length of smaller width of plate

a = length of the larger width of plate

t = plate thickness

ν = Poisson's ratio

k = a factor depending on the following main parameters:

- Plate geometry and dimensions (plate aspect ratio)
- Plate boundary conditions
- Type and distribution of in-plane loadings

4.1 Commonly Used Idealized Plate Boundary Support Conditions

The plate boundary support conditions commonly used in the analysis of buckling strength of plates are usually a combination of the following standard idealized support conditions: fixed, simple support or free edge, see Fig. (11.17).

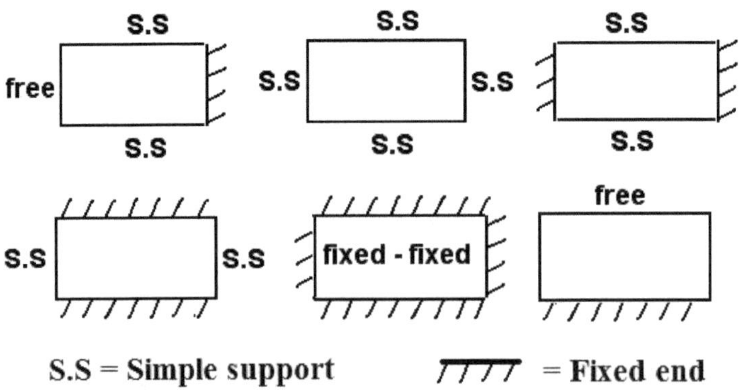

Fig. 11.17 Different plate boundary support conditions

4.2 General In-Plane Loading Conditions

The general in-plane loading condition on all edges of a plate panel is composed of compression, bending and shear stresses, see Fig. (11.18).

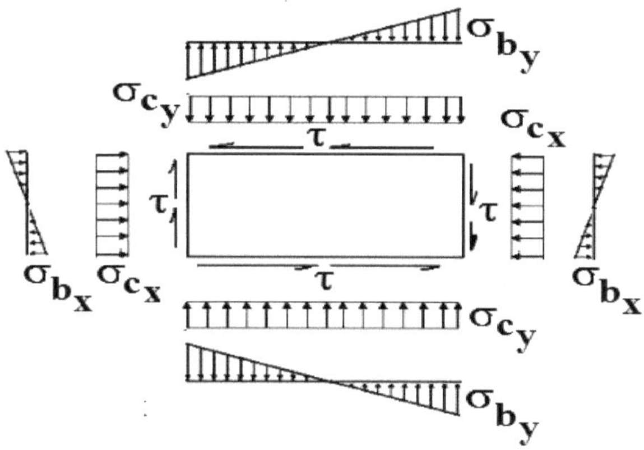

Fig. 11.18 General in-plane loading condition

4.2.1 Single Loading Conditions

The single in-plane loading conditions are as follows:

- Uniform compressive stresses on either the short or long edges, see Fig. (11.19)
- Non-uniform compressive stresses on either short or long edges, see Fig. (11.20)
- Pure bending stresses on either the short or long edges, see Fig. (11.21)
- Pure shear stress, see Fig. (11.22)

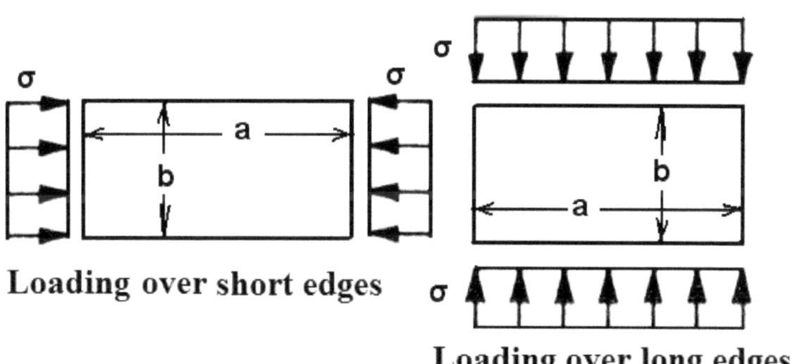

Fig. 11.19 In-plane uniform compression on the long/short edges of the plate

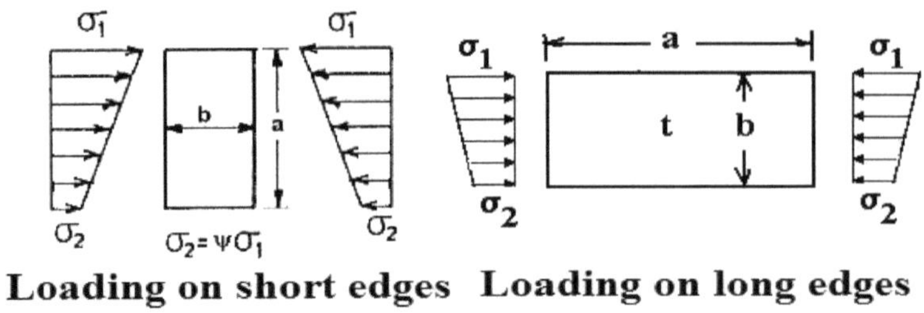

Fig. 11.20 In-plane non-uniform compression on long/short edges

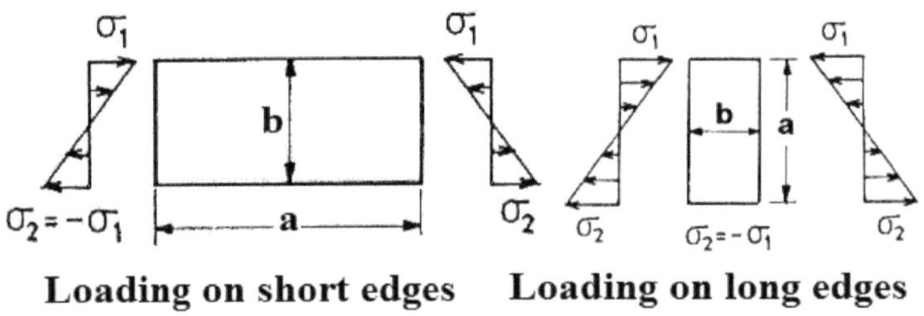

Fig. 11.21 In-plane pure bending stresses on the long/short edges of the plate

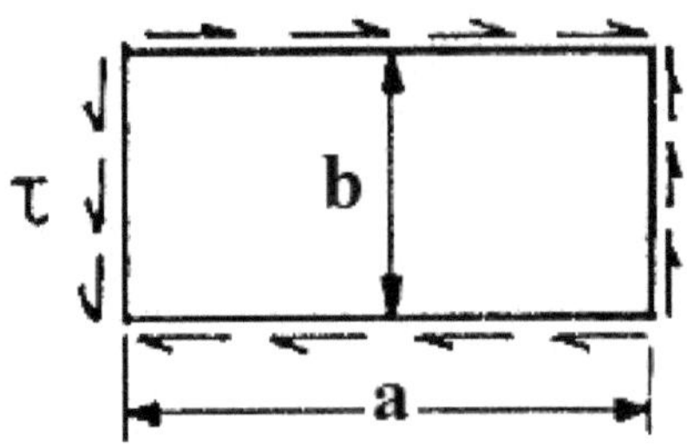

Fig. 11.22 Pure shear stresses on all edges of the plate

4.2.2 Combined Loading Conditions

The combined loading conditions are composed of a combination of two or more of the basic single loading conditions as follows:

- Uniform compressive stresses on the four edges of the plate, see Fig. (11.23)
- Shear stresses and uniform compressive stresses on long/short edges, see Fig. (11.24)

- Shear stresses and uniform compressive stresses on all edges, see Fig. (11.25)
- Pure bending and shear stresses on short/long edges, see Fig. (11.26)

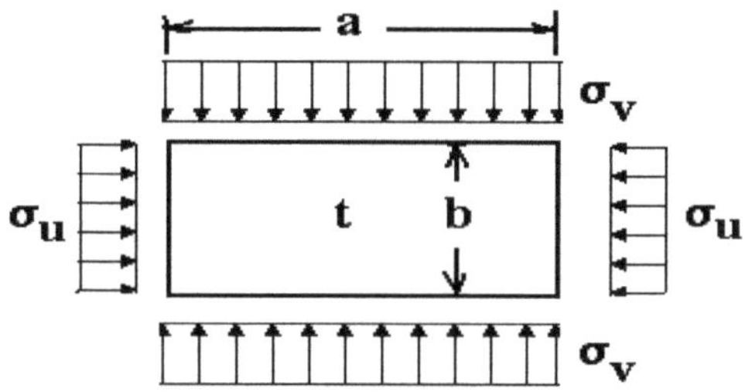

Fig. 11.23 Unequal uniform compression on all edges of the plate

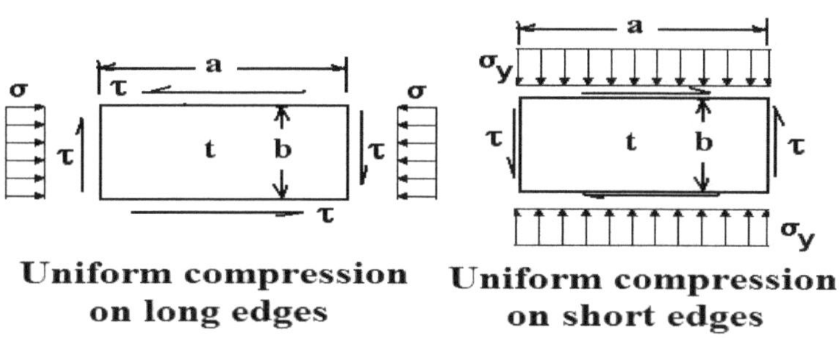

Fig. 11.24 Shear stresses and uniform compression on all edges

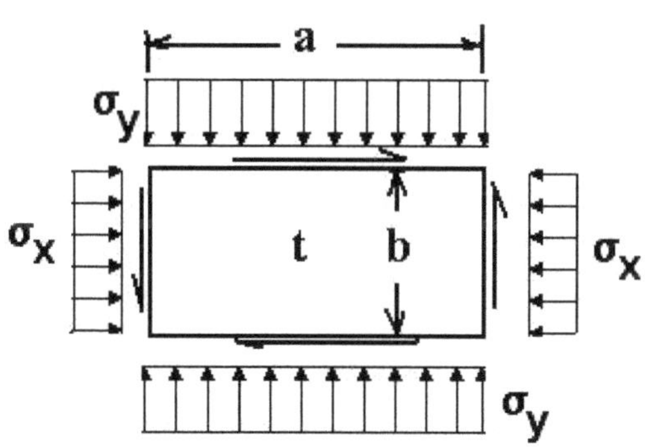

Fig. 11.25 Shear stresses and uniform compression on all edges of the plate

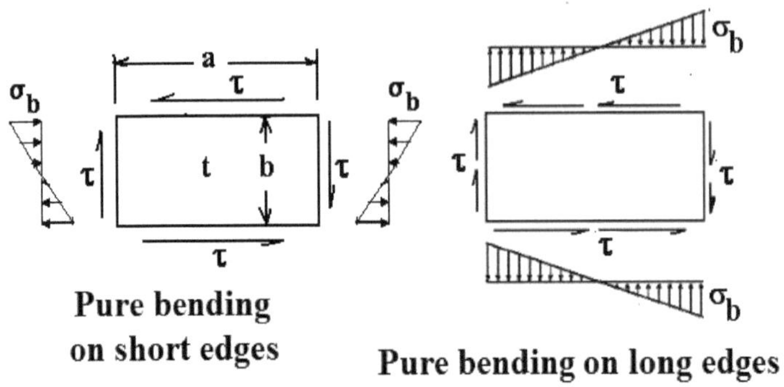

Fig. 11.26 Shear stresses and pure bending on the long/short edges

4.3 Modes of Buckling Deformation

The mode of buckling deformation of a plate subjected to in-plane loadings depends on:

- Type of loading
- Type of end supports
- Aspect ratio of plate

Consider the following cases:

4.3.1 Mode of Buckling Deformation of a Long Plate for the following Conditions

- Plate aspect ratio = 3
- All edges are simply supported
- Plate subjected to compressive stresses over the short edges

A long rectangular plate of length "a" supported at all edges will buckle into a number of half waves so that the length of each half wave approximately equals the width of the plate "b", see Fig. (11.27). The magnitude of the buckling stress depends, among other things, on the aspect ratio a/b of the plate.

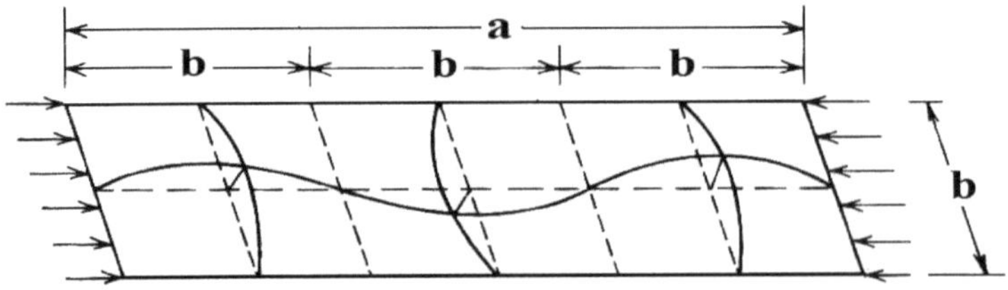

Fig. 11.27 Shape of deformation of a plate having an aspect ratio = 3 and subjected to in-plane compressive stresses

4.3.2 Mode of Buckling Deformation of a Plate Fixed at the Long Edges is as Shown in Fig. (11,28)

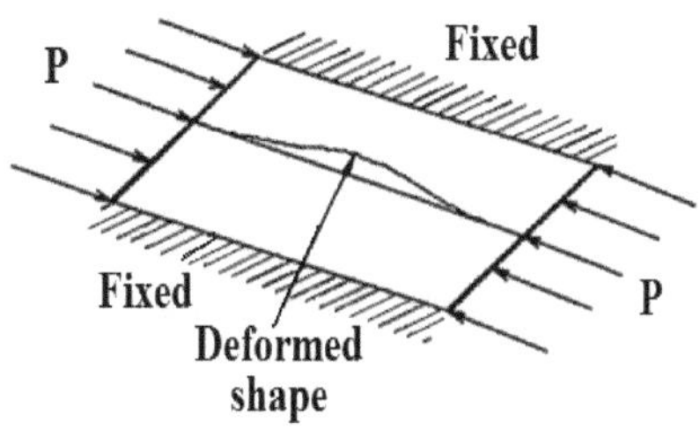

Fig. 11.28 Buckling deformation of a plate fixed at the long edges

4.3.3 Mode of Buckling Deformation of a Long Plate for the following Edge Supports and Loading Conditions is shown in Fig. (11.29):

- Plate aspect ratio = 3
- All edges are simply supported
- Plate subjected to pure shear loading

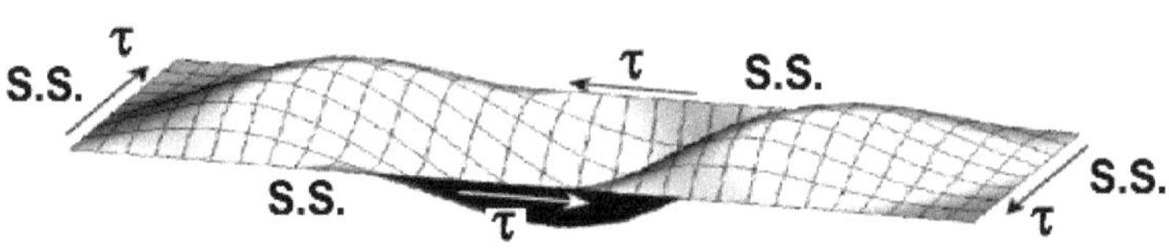

Fig.11.29.Plate deformation under shear stresses, all edges simply supported

4.3.4 Mode of Buckling Deformation of a Long Plate for the following Edge Supports and Loading Conditions is shown in Fig. (11.30):

- Plate aspect ratio = 4
- All edges are clamped
- Plate subjected to pure shear loading

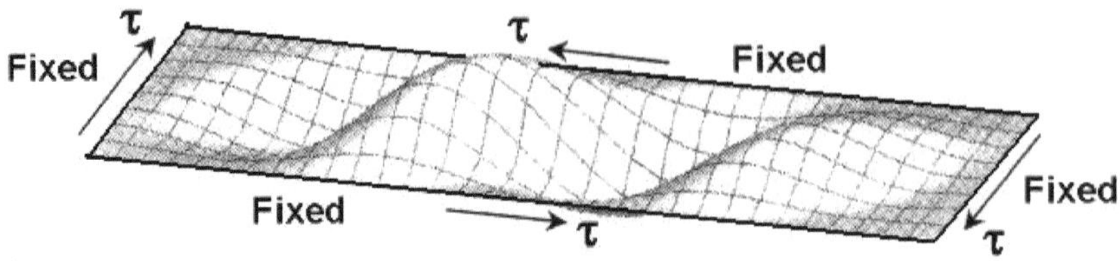

Fig.11.30 Plate deformation under shear stresses, all edges are clamped

The mode of buckling deformation of a plate panel under pure shear loading is as shown in Fig. (11.31).

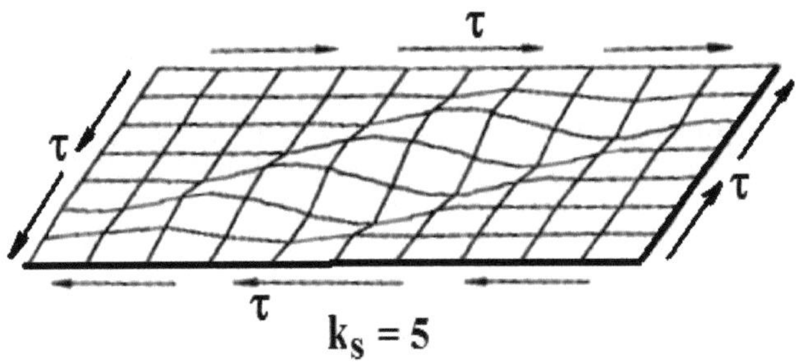

Fig.(11.31).Plate deformation under shear stresses

4.3.5 Mode of Buckling Deformation of a Long Plate for the following Edge Supports and Loading Conditions is Shown in Fig. (11.32):

- Plate aspect ratio = 4
- All edges are simply supported
- Plate subjected to compressive stresses over the short edges

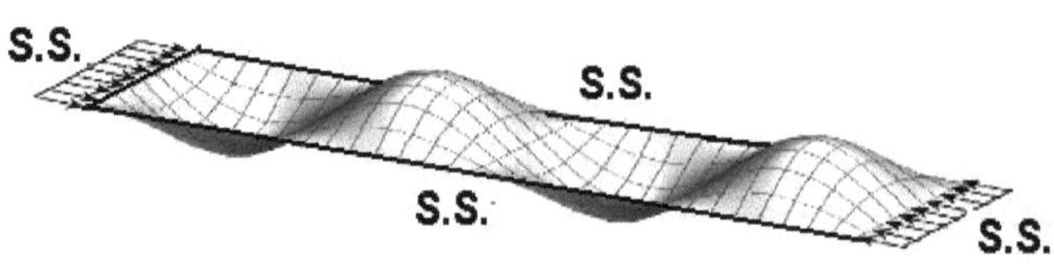

Fig.11.32 Plate deformation under end compressive stresses, all edges simply supported

4.3.6 Mode of Buckling Deformation of a Long Plate for the following Edge Supports and Loading Conditions is Shown in Fig. (11.33)

- Plate aspect ratio = 5
- All edges are simply supported
- Plate subjected to compressive stresses over the short edges

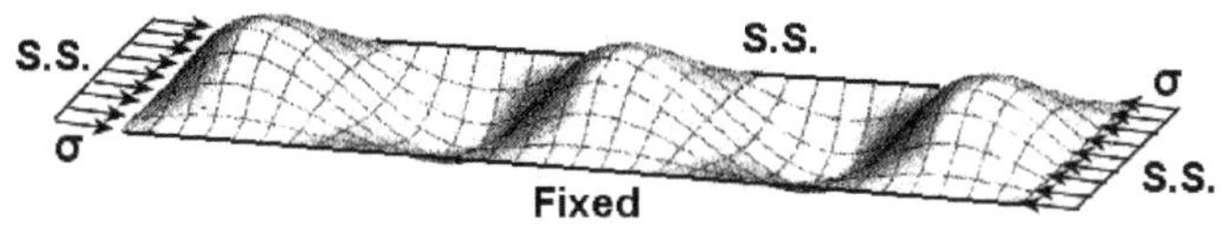

Fig.11.33 Plate deformation under end compressive stresses, one edge fixed and the other free

4.3.7 Mode of Buckling Deformation of a Long Plate for the following Edge Supports and Loading conditions is Shown in Fig. (11.34)

- Plate aspect ratio = 6
- All edges are simply supported
- Plate subjected to compressive stresses over the short edges

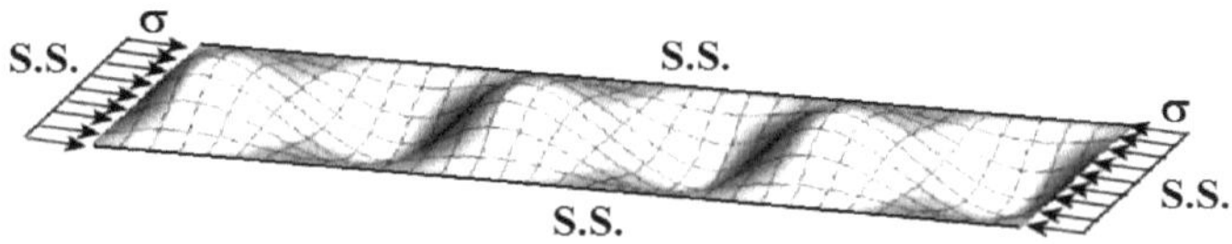

Fig.11.34 Plate deformation under end compressive stresses, all edges are fixed

4.3.8 Mode of Buckling Deformation of a Long Plate for the following Edge Supports and Loading Conditions is Shown in Fig. (11.35)

- Plate aspect ratio = 2
- Short edges are simply supported
- One long edge is fixed and the other is free
- Plate subjected to compressive stresses over the short edges

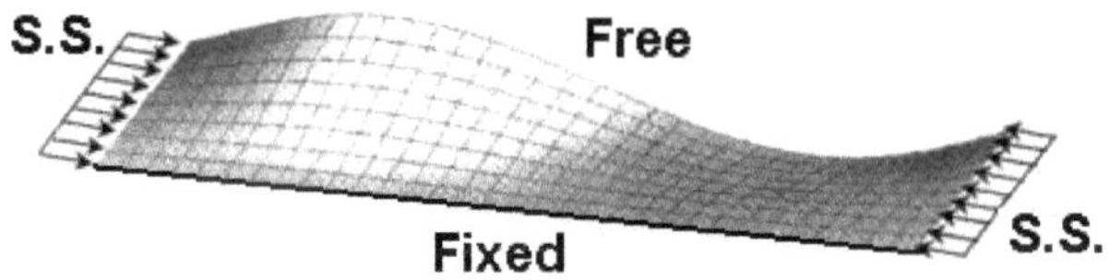

Fig.11.35 Plate deformation under end compressive stresses and one edge free and one edge fixed

The mode of buckling deformation of a plate panel fixed at one edge and free at the other edge is as shown in Fig.(11.36)

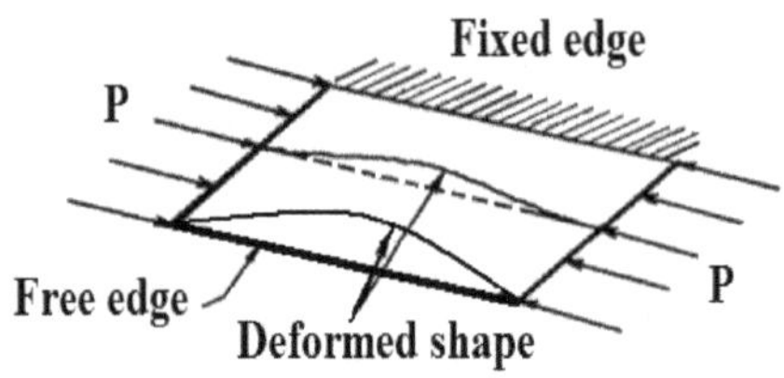

Fig.11.36 Buckling deformation of a plate fixed at one edge and free on the other edge

4.3.9 Mode of Buckling Deformation of a Long Plate for the following Edge Supports and Loading conditions is Shown in Fig. (11.37)

- Short edges are simply supported
- One long edge is simply supported and the other is free
- Plate subjected to compressive stresses over the short edges

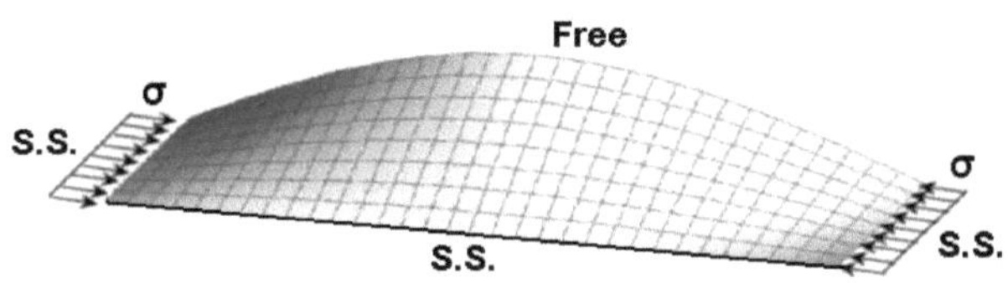

Fig.11.37 Plate deformation under end compressive stresses and one edge free and one edge simply supported

5 Basic Equations of Plate Buckling for Various Boundary Conditions and Different Loading Combinations

5.1 Plate Fixed at all Edges and Subjected to In-Plane Compressive Stresses over the Short Edges

Assuming the plate is clamped at all edges, the mode of buckling deformation is shown in Figs. (11.38).

The Euler buckling equation is given by:

$$\sigma_E = \frac{\pi^2 E}{12\left(1-\upsilon^2\right)}\left(\frac{t}{b}\right)^2\left[\left(\frac{b}{a}\right)^2+1\right]^2$$

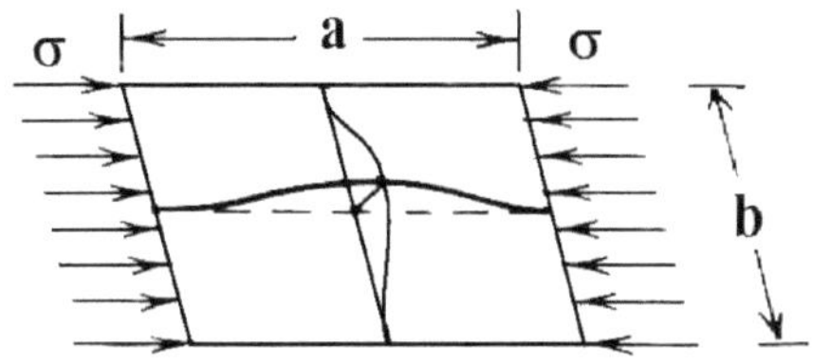

Fig.11.38 Deformed shape of plate clamped at all edges

\Where: σ_E = Euler buckling equation
b = length of shorter edge of plate
a = length of the long edge of plate
t = plate thickness
ν = Poisson's ratio

For a = b and substituting for ν = 0.3, the critical buckling stress is given by, see Fig.(11.38):

$$\sigma_{cr} = \frac{\pi^2 E}{3}\left(\frac{t}{b}\right)^2 = 3.6E\left(\frac{t}{b}\right)^2$$

5.2 Plate Simply Supported at All Edges and Subjected to In-Plane Compressive Stresses over the Long Edges

The mode of buckling deformation of a simply supported plate under in-plane loading over the long edges is shown in Figs.(11.39).

When the length "a" is much greater than the width "b", see Fig.(11.40), the Euler buckling stress of the strip is given by,:

$$\sigma_{ec} = \frac{\pi^2 E}{12\left(1-\upsilon^2\right)}\left(\frac{t}{b}\right)^2 \qquad a >> b$$

$$\cong 0.9E\left(\frac{t}{b}\right)^2$$

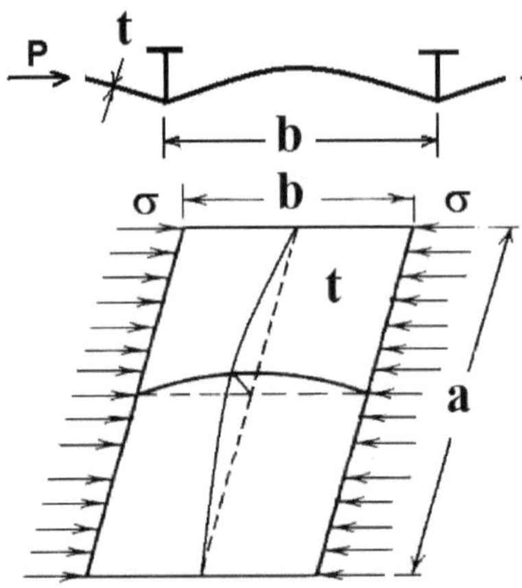

Fig. 11.39 Configuration and mode of buckling deformation of a simply supported plate panel

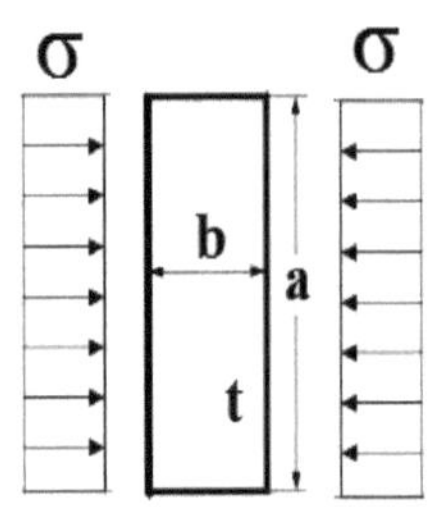

Fig. 11.40 A plate having the length "a" much greater than the width "b" and subjected to in-plane loading over the long edges

Note: It is clear that the critical buckling stress for a longitudinally stiffened panel is four times the critical buckling stress for transversely stiffened panels.

5.3 Plate Has One Long Edge Free and All Other Edges Simply Supported

The Euler buckling equation for a plate having one long edge free and all other edges simply supported and the plate is subjected to end compressive stresses over the short edges is given by, see Fig. (11.41):

$$\sigma_{ec} = \frac{k.\pi^2 E}{12\left(1\text{-}\upsilon^2\right)}\left(\frac{t}{b}\right)^2, k = 0.43 + \left(\frac{b}{a}\right)^2$$

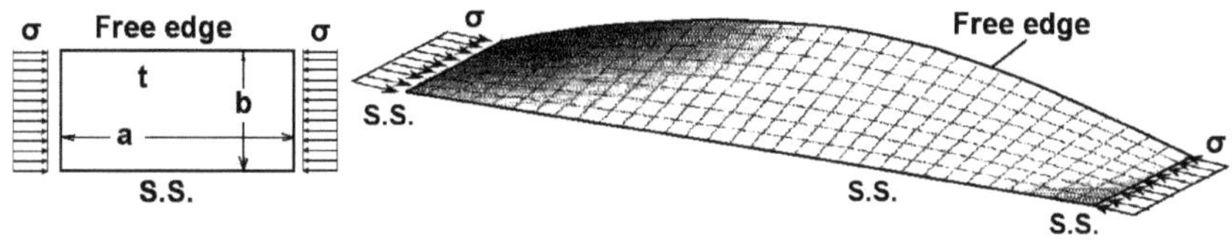

Fig.11.41 Buckling of a plate with one edge free and the other edges simply supported

5.4 Plate has One Long Edge Free and the Other Long Edge Clamped

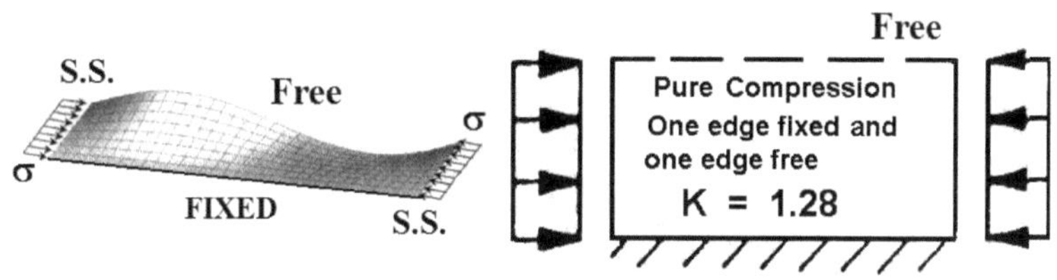

Fig.11.42 Buckling of a plate with the short edges simply supported and one long edge free and the other fixed

6 Buckling of Simply Supported Plate under Various in Plane Loading Conditions

6.1 Plate Subjected to Pure Bending Stresses

The Euler buckling equation for a plate subjected to in-plane bending stresses is given by, see Fig. (11.43):

$$\sigma_{eb} = \frac{\pi^2 E}{12(1-\upsilon^2)}\left(\frac{t}{b}\right)^2 k_b$$

σ_{eb} = Euler buckling stress under pure bending

The factor k_b depends on a/b ratio. For typical panels of ship structure, the constant k_b= 24.0.

The mode of deformation of a plate panel subjected to pure bending stresses is shown in Figs. (11.44, 45).

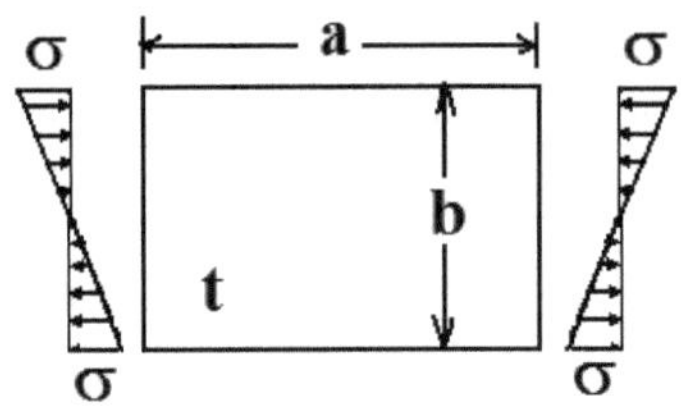

Fig. 11.43 Plate subjected to pure bending stresses

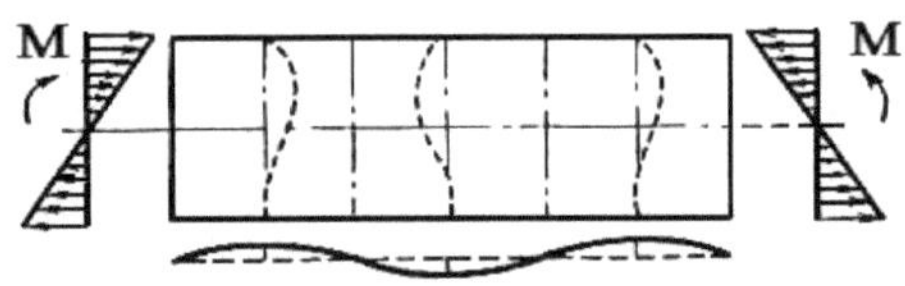

Fig. 11.44 Mode of plate deformation under the action of in-plane pure bending on the short edges

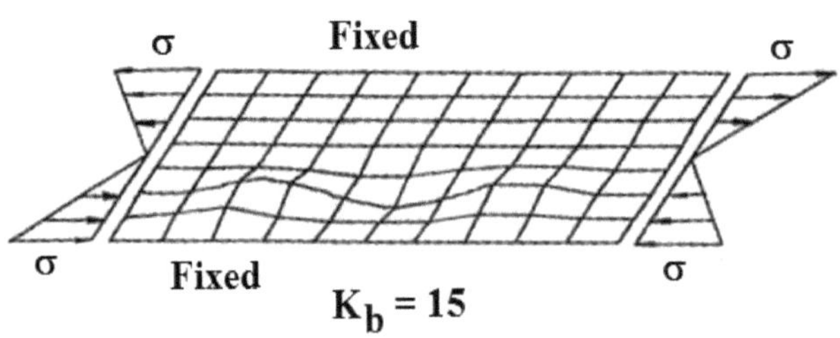

Fig.11.45 Mode of deformation of a plate under pure bending stresses

6.2 Plate Subjected to Pure Shear Stresses

The mode of deformation of a plate panel subjected to pure shear stresses over all edges is shown in Fig. (11.46). The pure shear loading causes wrinkling of the plate at 45 degree, see Fig. (11.46).

The Euler buckling equation for a plate subjected to in-plane shear stress is given by:

$$\tau_{es} = \frac{\pi^2 E}{12\left(1-\upsilon^2\right)}\left(\frac{t}{b}\right)^2 k_S$$

τ_{es}= Euler buckling stress for in-plane pure shear loading

The constant k_s is given by:

$k_s = 4.8 + 3.6(b/a)^2$ for simply supported edges

$k_s = 8.1 + 5.1(b/a)^2$ for clamped edges

For typical ship structure plate panels, we have:

$$\frac{b}{a} = \frac{1}{3} - \frac{1}{5}$$

Hence $\left(\frac{b}{a}\right)^2 = 0.04 - 0.1$ with a mean value = 0.07

Then for a simply supported plate, k = 5.05

And for a fixed-fixed plate. k = 8.45

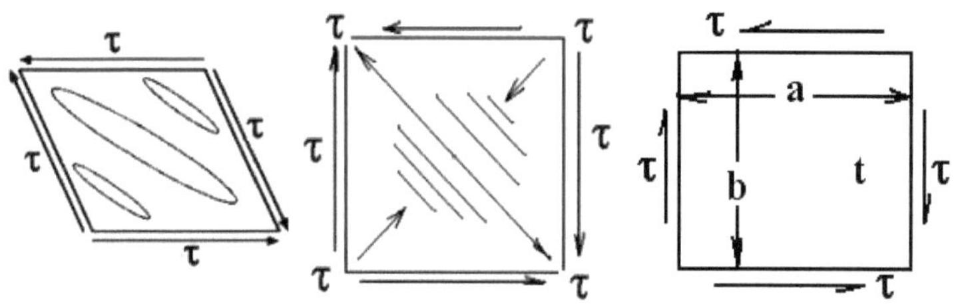

Fig.11.46 A plate panel subjected to pure shear loading along all edges

The values of the constant k for different boundary conditions is given in Fig.(11.47).

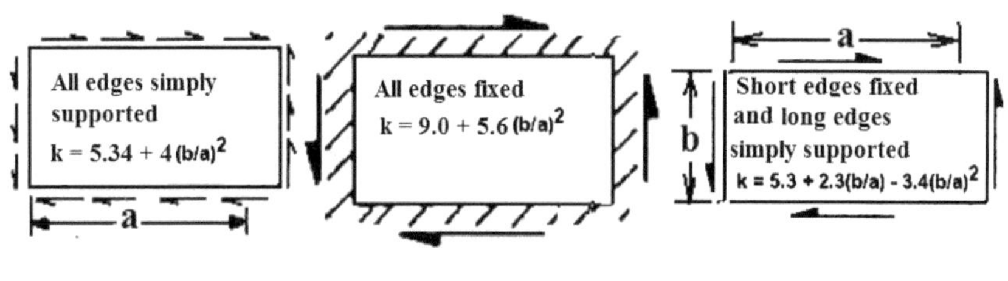

Fig.11.47. k factors of shear buckling for different boundary conditions

7 In-Elastic Bucking

When the Euler buckling stress for a plate panel subjected to in-plane compressive stress is less than half the yield stress of the material:

$$\sigma_E \leq \sigma_y / 2$$

The critical buckling stress σ_{cr} is given by:

$$\sigma_{cr} = \sigma_E = \frac{k\pi^2 E}{12\left(1 - \upsilon^2\right)}\left(\frac{t}{b}\right)^2$$

Where: σ_E = Euler buckling stress
σ_y = yield stress

On the other hand, when the Euler buckling stress is greater than half the yield stress of the material:

$$\sigma_E > \frac{\sigma_y}{2}$$

The critical buckling stress σ_{cr} is given by the Johnson-Ostenfeld formula:

$$\sigma_{cr} = \sigma_y \left[1 - \frac{\sigma_y}{4\sigma_e} \right]$$

7.1 Inelastic Buckling due Shear Loading

Elastic buckling due to shear loading occurs when:

$$\tau_{es} \leq 0.5\tau_y$$

Where: τ_{es}= Euler buckling stress for in-plane pure shear loading

$$\tau_y = \frac{\sigma_y}{\sqrt{3}}$$

Hence, the elastic critical shear buckling stress is given by:

$$\tau_{cr} = \tau_{es}$$

Inelastic buckling due to shear loading occurs when:

$$\tau_{es} \geq 0.5\tau_y$$

In this case, the critical shear buckling stress is given by:

$$\tau_{cr} = \frac{\sigma_y}{\sqrt{3}} \left[1 - \frac{\sigma_y}{4\sqrt{3}.\tau_{es}} \right]$$

8 Assessment of Buckling Strength of Plating Subjected to Combined Loading

The following interaction formulae are used to evaluate the critical buckling stresses for plates subjected to a combination of in plane compressive, bending and shear stresses.

8.1 Combined Shear and Compressive Stresses on the Short Edges

The plate buckling equation is given by, see Fig. (11.48):

$$\left(\frac{\tau}{\tau_{es}}\right)^2 + \frac{\sigma}{\sigma_{ec}} \leq 1.0$$

τ = applied stress

τ_{es} = Euler critical shear buckling stress for pure shear loading

When $\tau_{es} \leq 0.5\tau_y$, the critical buckling stress for pure shear loading is given by:

$$\tau_{cr} = \frac{\pi^2 E}{12\left(1-\upsilon^2\right)}\left(\frac{t}{b}\right)^2 .k_s$$

Where: k_s = buckling factor depending on the boundary conditions

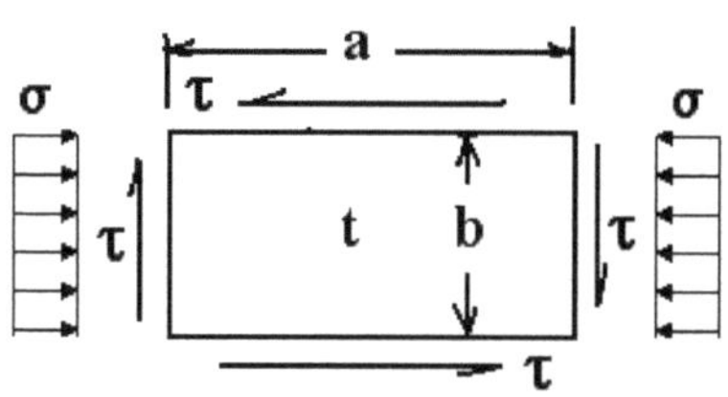

Fig. 11.48 Combined shear and pure compression on the short edges

When $\tau_{es} \geq 0.5\tau_y$, the critical buckling stress for pure shear loading is given by:

$$\tau_{cr} = \tau_y\left(1-\frac{\tau_y}{4\tau_{es}}\right)$$

σ = applied stress

When $\sigma_{ec} \leq 0.5\ \sigma_y$, the critical buckling stress is given by:

$$\sigma_{ec} = \frac{\pi^2 E}{12\left(1-\upsilon^2\right)}\left(\frac{t}{b}\right)^2 \cdot k_c$$

When $\sigma_{ec} \geq 0.5\,\sigma_y$, the critical buckling stress is given by:

$$\sigma_{cr} = \sigma_y \left(1 - \frac{\sigma_y}{4\sigma_{ec}} \right)$$

Where: k_c = buckling factor depending on the boundary conditions

The interaction equation for the combined in plane compression and shear is represented by the buckling curve shown in Fig.(11.49)

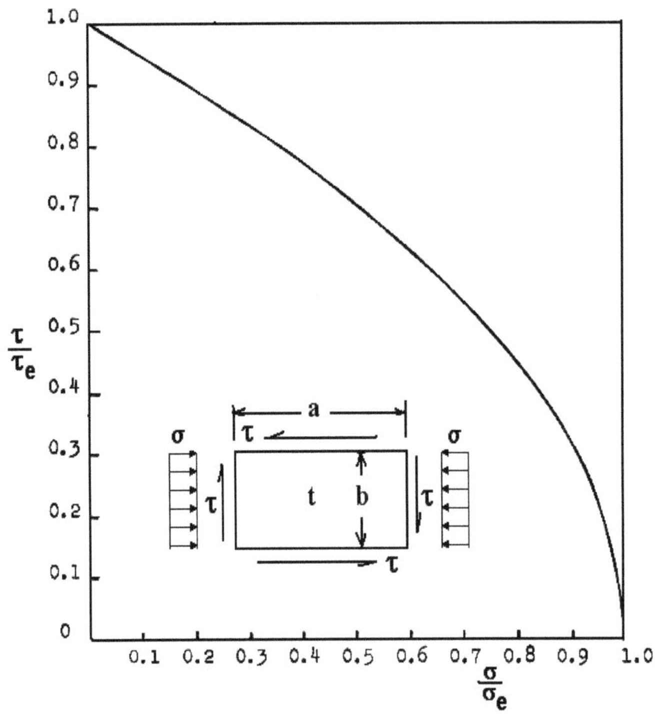

Fig.11.49. Combined shear and in plane compression on short edges

8.2 *Combined Shear and In-Plane Bending on the Short Edge*

The plate buckling equation for in-plane shear and pure bending on the short edges of a plate panel is given by the following interaction equation, see Fig. (11.50):

$$\left(\frac{\tau}{\tau_{es}} \right)^2 + \left(\frac{\sigma_b}{\sigma_{eb}} \right)^2 \leq 1.0$$

Where: σ = applied bending stress

$$\sigma_{eb} = \frac{\pi^2 E}{12(1-\upsilon^2)}\left(\frac{t}{b}\right)^2 \cdot k_b \quad \text{for } \sigma_{eb} \le 0.5\,\sigma_y$$

$$= \sigma_y\left(1 - \frac{\sigma_y}{4\sigma_{eb}}\right) \quad \text{for } \sigma_{eb} \ge 0.5\,\sigma_y$$

The interaction equation for the combined in plane pure bending and shear is represented by the buckling curve shown in Fig.(11.50).

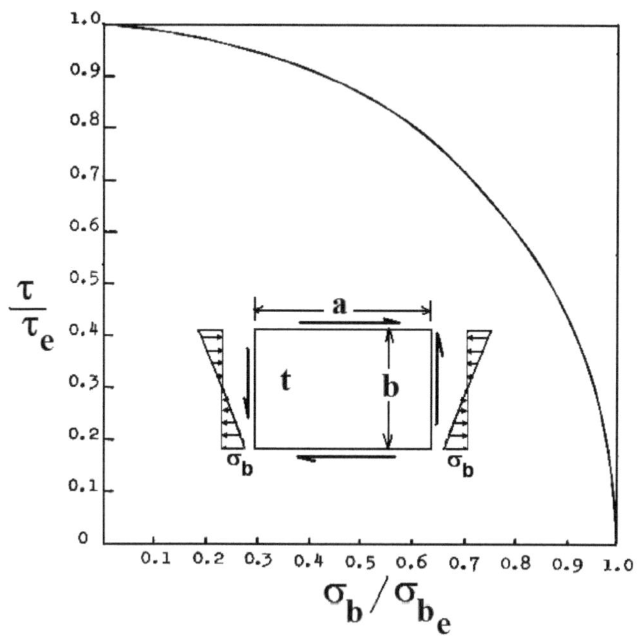

Fig. 11.50 Combined shear and in plane bending on short edges

8.3 Combined In-Plane Bending and Compression

The plate buckling equation for in-plane compression and pure bending on the short edges of a plate panel is given by the following interaction equation, see Fig. (11.51):

$$\frac{\sigma}{\sigma_{ec}} + \left(\frac{\sigma_b}{\sigma_{eb}}\right)^2 \le 1.0$$

Where: $\sigma = \dfrac{\sigma_1+\sigma_2}{2}$,

$$\sigma_b = \frac{\sigma_1-\sigma_2}{2}$$

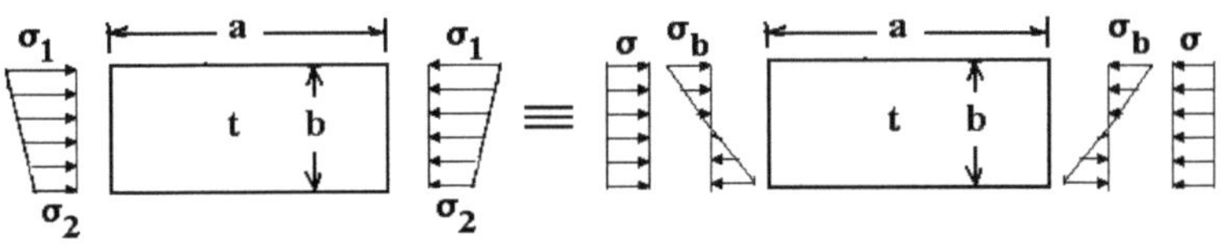

Fig.11.51 Combined in-plane bending and compression on short edges

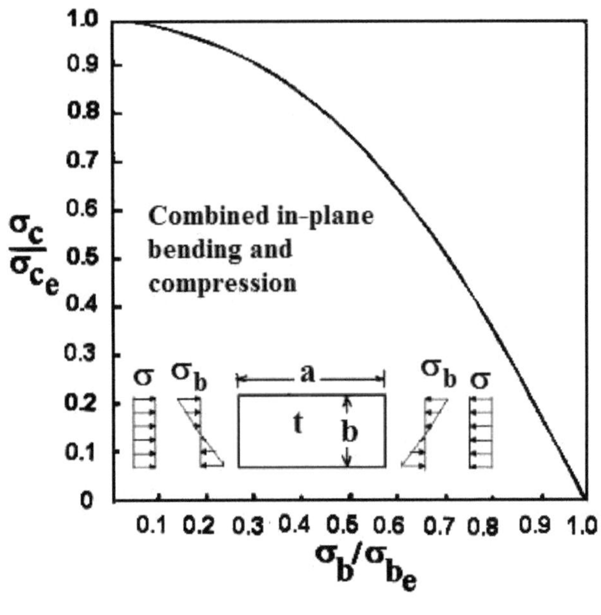

Fig.11.52 Plate subjected to in-plane compression and bending on short edges

The interaction equation for the combined in plane pure bending and compression over the short edges is represented by the buckling curve shown in Fig.(11.52).

8.4 In-Plane Compression in Two Orthogonal Directions

Buckling of plate panels subjected to bi-axial in-plane compression is given by the following interaction equation, see Fig. (11.53):

$$\frac{\sigma_u}{\sigma_{cu}} + \frac{\sigma_v}{\sigma_{cv}} \leq 1.0$$

Where: σ_u = in-plane compressive stress over the short edges
σ_v = in-plane compressive stress over the long edges
σ_{cv} = critical buckling stress for loading on the long edges
σ_{cu} = critical buckling stress for loading on the short edges

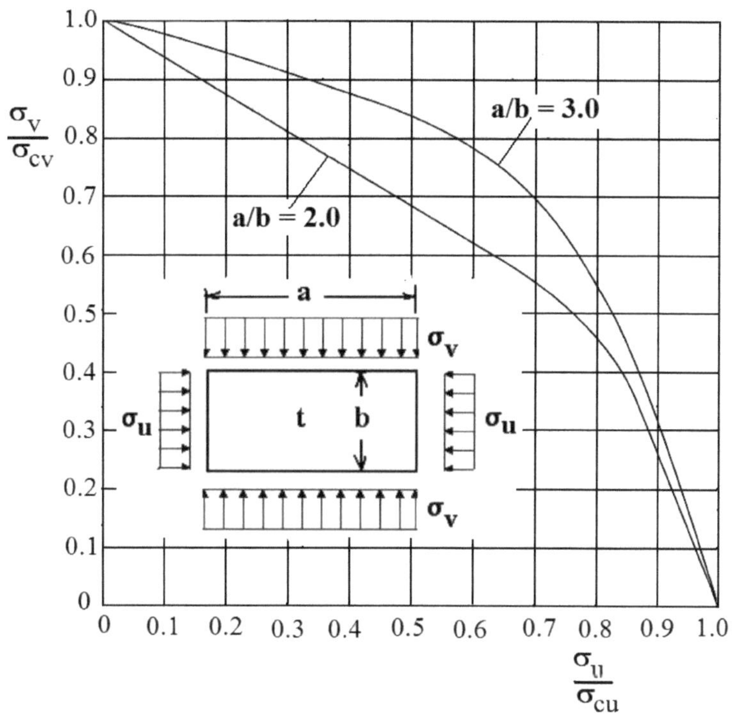

Fig. 11.53 Interaction curves of plate subjected to bi-axial compression

The interaction curves of plate panels subjected to bi-axial compression is shown in Fig. (11.53).

8.5 Combined Shear, In-Plane Bending and Compression

The general loading condition of combined shear, in-plane bending and compression is shown in Fig. (11.54).

The interaction buckling equation for a plate panel subjected to combined shear, in-plane compression and bending on the short edges is given by, see Fig.(11.54):

$$\frac{\sigma}{\sigma_{ec}} + \left(\frac{\sigma_b}{\sigma_{eb}}\right)^2 + \left(\frac{\tau}{\tau_{es}}\right)^2 \leq 1.0$$

The definitions of stresses given in the interaction equation are as given in the cases given before.

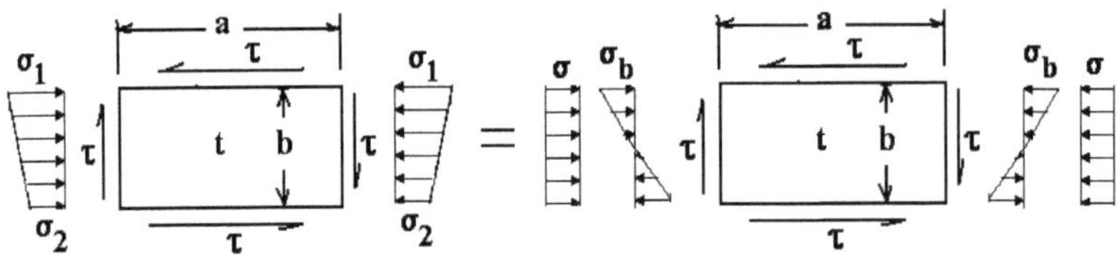

Fig 11.54 Combined in plane shear, bending and uniform compression on short edges

9 Critical Buckling Stress of Plating of Stiffened Panels

9.1 Longitudinally Stiffened Panels

Consider the longitudinally stiffened panel shown in Fig. (11.55).

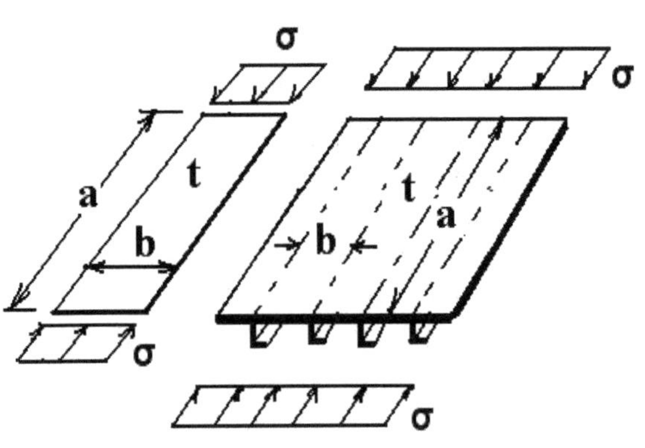

Fig.11.55 Configuration of plating of a longitudinally stiffened panel

The plate panel is assumed clamped at both short edges and subjected to in-plane compressive stresses. The Euler buckling stress of the panel is given by, see Fig. (11.55):

$$\sigma_E = \frac{4\pi^2 D}{b^2 t} = 3.6E\left(\frac{t}{b}\right)^2$$

$$Where \ : D = \frac{Et^3}{12\left(1-\upsilon^2\right)}$$

The Euler buckling stress for a panel simply supported at both ends is given by:

$$\sigma_E = \frac{\pi^2 D}{b^2 t} = 0.9E\left(\frac{t}{b}\right)^2$$

The critical buckling stress σ_{cr} could be determined as follows:

When $\sigma_E \leq 0.5\,\sigma_y$

The critical buckling stress is given by:

$$\sigma_{cr} = \sigma_E$$

When $\sigma_E \geq 0.5\,\sigma_y$

The critical buckling stress is given by:

$$\sigma_{cr} = \sigma_y\left(1-\frac{\sigma_y}{4\sigma_e}\right)$$

Where: σ_y = yield stress of the material.

9.2 *Transversely Stiffened Panels*

The Euler buckling stress for a simply supported panel subjected to in-plane compressive stresses along the long edges is given by, see Fig.(11.56):

$$\sigma_E = \frac{\pi^2 D}{b^2 t}\left(1+\frac{b^2}{a^2}\right)^2$$

$$D = \frac{Et^3}{12\left(1-\upsilon^2\right)}$$

Where: b = beam spacing

a = length of plate

For ship structure, we have:

$$\frac{b}{a} \approx \frac{1}{3} - \frac{1}{5},$$

Then $\sigma_E = \dfrac{\pi^2 D}{b^2 t}$

The critical buckling stress is given by:

When: $\sigma_E \leq \dfrac{\sigma_y}{2}$

$$\sigma_{cr} = \sigma_E$$

When: $\sigma_E \geq 0.5\sigma_y$

$$\sigma_{cr} = \sigma_y \left(1 - \frac{\sigma_y}{4\sigma_e}\right)$$

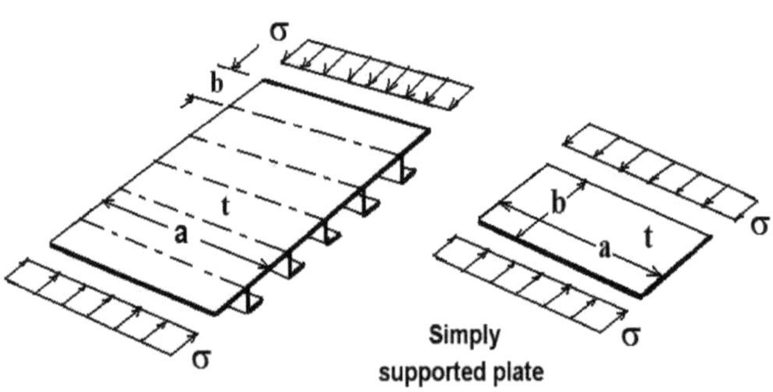

Fig. 11.56 Transversely stiffened plate panel

Example: Determine the critical buckling stress for the simply supported plate shown in Fig. (11.57) using the Euler buckling equation for columns.

<u>Solution</u>

The Euler buckling load for a column simply supported at both ends is given by:

$$P_E = \frac{\pi^2 EI}{L^2}$$

Consider a strip of width 1.0 m and length "b". The strip is simply supported at both ends.

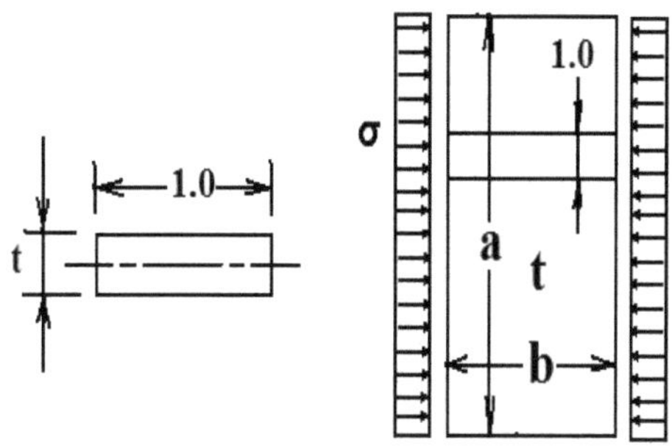

Fig. 11.57 Plate panel under compressive stress over the long edges

The critical buckling load is given by:

$$P_E = \frac{\pi^2 EI}{L^2} = \frac{\pi^2 Et^3}{12b^2}$$

The moment of inertia of section of the strip about its neutral axis is given by:

$$I = \frac{1.0 \times t^3}{12}$$

The critical buckling stress is given by:

$$\sigma_{cr} = \frac{P_{cr}}{A} = \frac{\pi^2 Et^3}{12b^2 t} = \frac{\pi^2 E}{12}\left(\frac{t}{b}\right)^2$$

The actual Euler critical buckling stress for a plate is given by:

$$\sigma_E = \frac{\pi^2 E}{12(1-\upsilon^2)}\left(\frac{t}{b}\right)^2 \cdot k$$

It is evident that $\sigma_e(plate) = \sigma_e(column) \times K$

The difference between the two equations is represented by the following factor:

$$K = \frac{k}{(1-\upsilon^2)}$$

The factor K takes account of the bending of the plate in the orthogonal direction, boundary conditions, type of loading, etc.

10 Post–buckling Strength of Plating

Initial buckling of plate panels supported over the plate edges does not lead to collapse of the plate as the plate is constrained over its edges, see Fig. (11.58). Lateral deflection due to buckling of the plate is partially restrained by transverse bending of the plate between the edge supports.

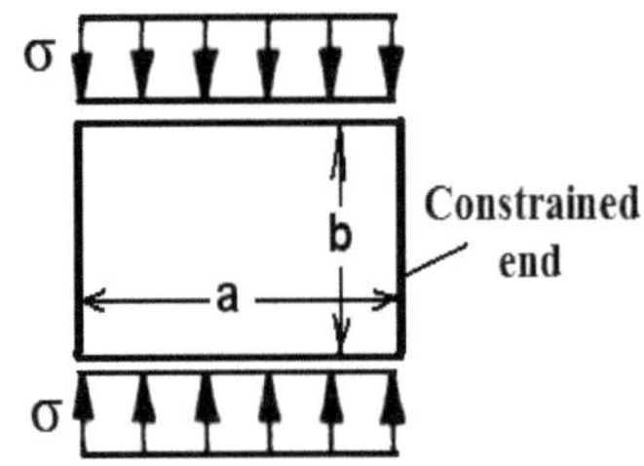

Fig.11.58 Plate panel supported over the short edges of the plate

In order to determine the post buckling strength, assume a simply supported rectangular plate subjected to uniformly distributed load along the short edges of the plate. When the applied stress exceeds the critical buckling stress, the additional load will be concentrated towards the edges of the plate and the stresses will increase in those regions, see Fig. (11.59). The pattern of stress distribution over the edge of the plate panel before and after buckling occurs is as shown in Fig.(11.59). The definition of stresses in Fig. (11.59) is given by:

σ_{cr} = critical buckling stress
σ_{max} = maximum stress along the side-edges
σ_{mean} = mean stress in plating

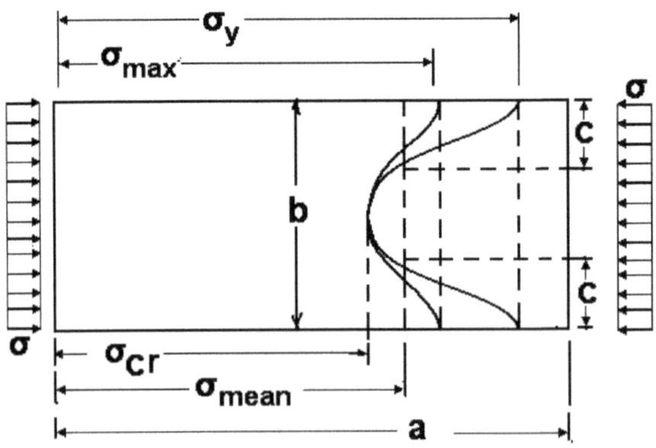

Fig. 11.59 Shape of stress distribution for a plate panel under compressive stresses before and after buckling of the plate

This effective width of the plate is represented by a width C from each side edge of the plate panel and is given by, see Fig.(11.60):

$$b_e = 2C$$

The effective width of plating is defined as the width of plating which, when taking the full load, has a uniformly distributed stress equal to the maximum stress, σ_{max}, see Fig. (11.60).

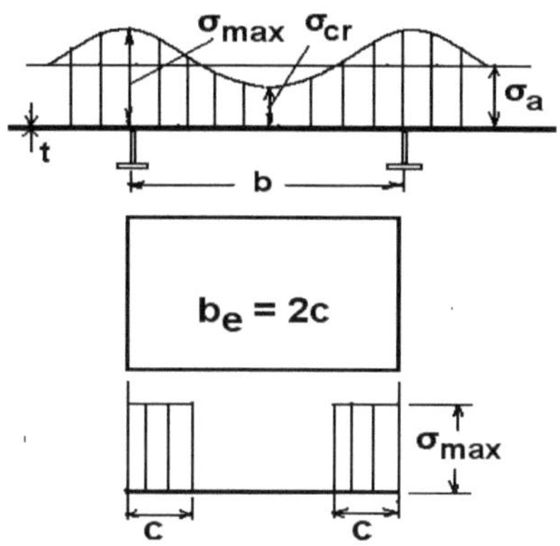

Fig. 11.60 The effective width of plating

The middle sector of the plate can be disregarded and the two strips can be handled as a long simply supported rectangular plate of total width 2C. The Euler buckling stress for such a plate is given by:

$$\sigma_{ec} = \frac{4\pi^2 E}{12(1-\upsilon^2)}\left(\frac{t}{2C}\right)^2$$

Assuming that the ultimate load is reached when σ_{ec} becomes equal to the yield stress of the material σ_y, then we obtain:

$$C = \frac{\pi t}{\sqrt{12(1-\upsilon^2)}}\sqrt{\frac{E}{\sigma_y}}$$

Hence, the ultimate load is given by:

$$P_u = 2Ct\sigma_y = \frac{\pi t^2}{\sqrt{3(1-\upsilon^2)}}\sqrt{E\sigma_y}$$

Where: $\nu = 0.3$,

$$P_u = 1.9t^2\sqrt{E\sigma_y}$$

An approximate expression for "C" is given by:

$$C = 1.9\left(1 - 0.55\frac{t}{b}\sqrt{\frac{E}{\sigma_y}}\right)$$

Alternatively, an approximate expression for the effective width is given by :

$$\frac{b_e}{t} = 1.9\sqrt{\frac{E}{\sigma_e}}\left[1 - 0.475(t/b)\sqrt{\frac{E}{\sigma_e}}\right]$$

For ship building steel, $\sigma_y = 2.4$ t/cm^2 and E = 2.1 x 10^3 t/cm^2

Hence: $\sqrt{\frac{E}{\sigma_y}} = \sqrt{\frac{2100}{2.4}} = 29.58$

$$\frac{b_e}{t} = 1.9 \times 29.58[1 - 0.475(t/b) \times 29.58]$$
$$= 56.2 \times (1 - 14.05t/b)$$

For t/b = 1/40

$b_e/t = 36.46$

The post–buckling reserve of strength varies with the geometry of the plate, the type of loading and the boundary support conditions and can be disregarded in the following conditions:

- When b/a < ¼
- When one edge is free
- When $\sigma_e \geq \sigma_y$

10.1 Ultimate Stress of Simply Supported Plate Panels

10.1.1 Long Edges Loaded

To determine the ultimate strength of a plate panel having $a/b > 3.3$, assume a strip of thickness "t", width 1.0 m and length "b'. Treating the strip as a column, the ultimate buckling strength of the plate is given by, see Fig. (11.61):

$$\sigma_u = \frac{\pi^2 E}{12(1-\upsilon^2)}\left(\frac{t}{b}\right)^2$$

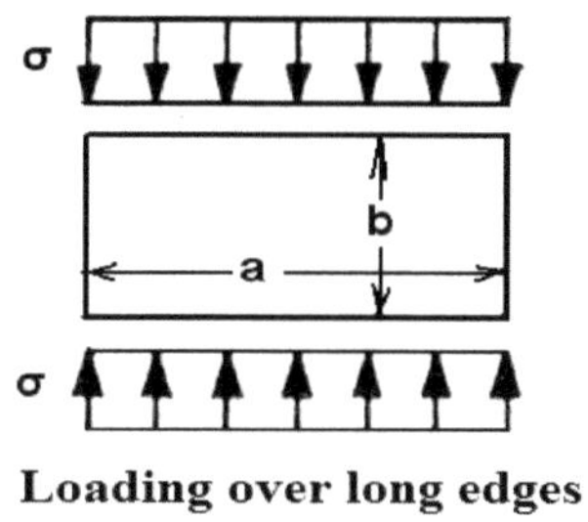

Fig. 11.61 A plate panel under uniform compression over the long edges

10.1.2 Short Edges Loaded

The ultimate strength depends on the effective width of the plate. The effective width of plating for plate panels under compressive stresses over the short edges could be obtained from the following equation, see Fig.(11.62):

$$\frac{b_e}{b} = \frac{2}{\beta} - \frac{1}{\beta^2}$$

Where: a = long dimension of plate
b = short dimension of plate
β =slenderness parameter

$$\beta = \frac{b}{t}\sqrt{\frac{\sigma_y}{E}}$$

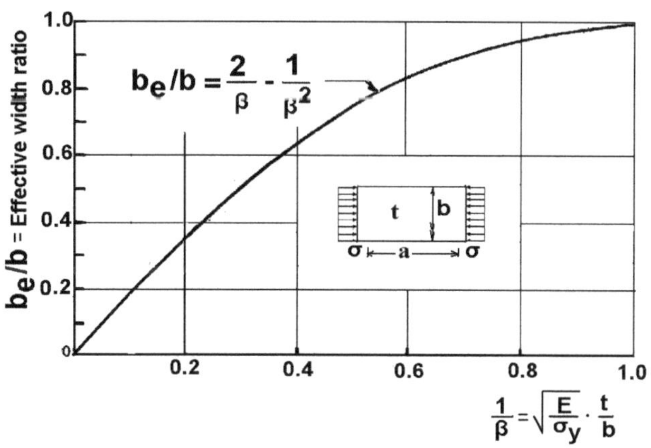

Fig.11.62 Effective width of plating

Example: Determine the effective width of a plate panel having the following data:

$\sigma_y = 2.4$ t/cm^2 ,E = 2.1 x 10^3 t/cm^2, a x b = 200 x 60 cm and t = 10 mm

Solution:

b/t = 60

$\beta = 2.04$

$b_e/b = 0.985 - 0.243 = 0.742$

Hence $b_e = 44.5$ cm

11 Buckling Limit State of Plate Panels

The buckling limit state of a transversely stiffened deck structure is given by the following equation:

$$M_f \leq M_{cr}$$

Where: M_{cr} = Limit state for buckling of deck plating
M_f = sagging yield moment at deck

An example showing the yielding of deck structure of a ship hull girder as a result of the induced high compressive stress by sagging bending moment exceeding the ultimate buckling strength of deck plating, see Fig.(11.63).

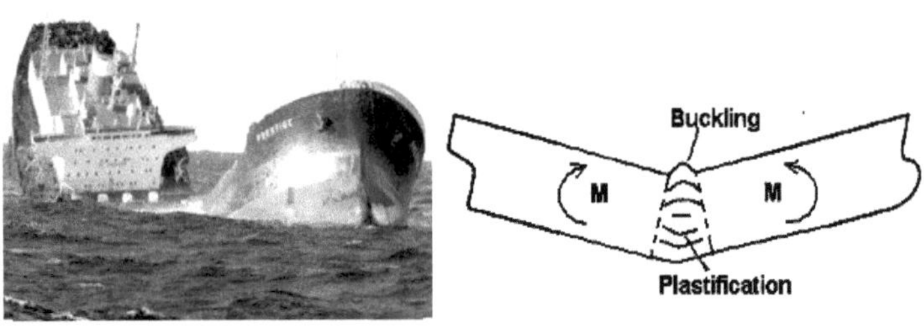

Fig.11.63 Yielding of deck structure of a ship hull girder

The safety margin against buckling of the deck plating in the mid-ship region is given by:

$$S = M_{cr} - M_f > 0$$

Where: S = safety margin against buckling failure of deck structure

$$M_{cr} = Z_{eD} \times f_{cr}$$

Where: f_{cr} = critical buckling stress of deck plating

Z_{eD} = elastic modulus of ship section at deck

The limiting state for instability of deck plating could be given by the following failure equation:

$$Z_{eD} \times f_{cr} - M_f \geq 0$$

When $f_e \leq 05 f_y$

The critical buckling stress of the deck plating is given by:

$$f_{cr} = f_e = \frac{k\pi^2 E}{12(1-v^2)} \cdot \left(\frac{t}{b}\right)^2$$

Where: f_e = Euler buckling stress

f_y = yield stress of the material

f_{cr} = critical buckling stress

When $f_e \geq 05 f_y$

The critical buckling stress of the deck plating is given by:

$$f_{cr} = f_y \left(1 - \frac{f_e}{4 f_y}\right)$$

11.1 Uncertainty Modeling of Buckling Safety Margin

Because the main parameters of the ultimate strength of the deck plating are subject to several sources of uncertainties, it is necessary to determine their statistical characteristics. The uncertainties of the ultimate strength of the deck plating are associated with the uncertainties of the yield stress of the material as well as the geometrical and flexural properties of the deck plate panels.

To determine the statistical characteristics of the safety margin of a deck plate panel under compressive stresses, assume that the variables involved have normal distribution functions.

Hence: $f_{cr} \equiv N\left(u_{fcr}, \sigma_{fcr}\right)$,

$$Z_e \equiv N\left(u_{Z_e}, \sigma_{Z_e}\right)$$

$$M_f \equiv N\left(u_{Mf}, \sigma_{Mf}\right)$$

Where u_X = mean value of X, X = f_{cr}, Z_e and M_f

σ_X = standard deviation of X

Since the deck capability against buckling and the buckling load induced by sagging moments are statistically independent, the statistical parameters of the safety margin are calculated as follows:

$$S \equiv N(u_S, \sigma_S)$$

$$u_S = u_{M_{cr}} - u_{Mf}$$

Where: $$u_{M_{cr}} = u_{Z_e} \times u_{f_{cr}}$$

$$\sigma^2_{M_{cr}} = \sigma^2_{Z_e} + \sigma^2_{f_{cr}}$$

$$\sigma^2_s = \sigma^2_{Mcr} + \sigma^2_{Mf}$$

Chapter 12
Assessment of Buckling of Ship Structure

1 Introduction

This chapter presents a simplified procedure for the assessment of buckling of ship strength members. The main strength members of longitudinally and transversely stiffened bottom and deck structures sustaining compressive forces are identified. The compressive loadings are the compound in-plane stresses induced by the hull girder, secondary, and tertiary stresses. The assessment of buckling of strength members of bottom structure is carried out when the ship is in hogging condition and for strength members of deck structure when the ship is in sagging condition. For both cases, the compounding of stresses is carried out when the secondary and tertiary loadings are inducing compressive stresses. The assessment of buckling of web plates and face plates of deck and bottom girders is presented. The assessment of buckling of side shell plating for the various induced in-plane loading conditions is introduced. Assessment of buckling strength of plating for different end support conditions and for a variety of loading patterns is given. The importance of providing ship strength members sustaining compressive forces with adequate strength against buckling failure is stressed.

2 Ship Strength Members Sustaining Compressive Forces

Ship strength members sustaining compressive forces should have adequate buckling strength to ensure safety against buckling failure. The assessment of buckling strength of strength members of bottom structure is to be carried out when the ship is in hogging condition and for strength members of deck structure when the ship is in sagging condition. For both cases, the compounding of stresses is to be carried out when the secondary and tertiary loadings are inducing compressive stresses. A simplified procedure for the assessment of buckling of ship strength members is given Fig. (12.1).

For the assessment of buckling and ultimate strength of plates and stiffened panels subjected to combined stress fields, the buckling assessment method should consider the following effects when deriving the buckling capacity:

(a) nonlinear geometrical behaviour
(b) inelastic material behaviour
(c) initial imperfections (geometrical out-of flatness of plate and stiffeners)
(d) welding residual stresses
(e) Interactions between structural elements; plates, stiffeners, girders etc.
(f) simultaneous acting loads; bi-axial compression/tension, shear and lateral pressure
(g) boundary conditions

The main strength members of longitudinally and transversely stiffened bottom and deck structures sustaining compressive forces and need to be considered for the assessment of buckling strength are:

- Girders
- Longitudinals
- Plating

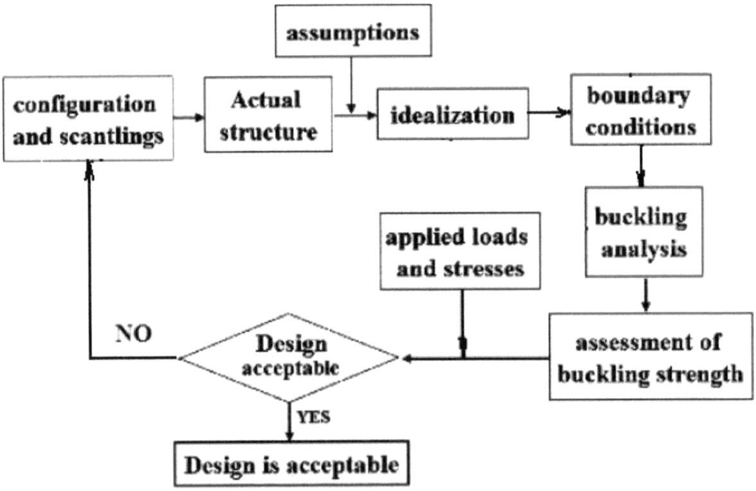

Fig. 12.1 Basic approach for the assessment of buckling strength

The assessment of buckling requires the knowledge of:

- Geometry and scantlings (dimensions and thicknesses) of each component of the strength member (web plate and face plate).
- Boundary conditions of each member
- Type, distribution and magnitude of the compressive stresses exerted on each member

- Realistic modeling of each strength member so as not to depart significantly from the actual structure
- Buckling of girders and longitudinals should cover both general and local modes of buckling failure.

3 Basic Equations of Buckling of Plate Panels Subjected to Non-uniform In-Plane Compression

3.1 Idealization of the In-Plane Compressive Loadings

The buckling of a plate panel subjected to the general case of non-uniform in-plane compressive stresses over the short edges of the plate is as shown in Fig. (12.2).

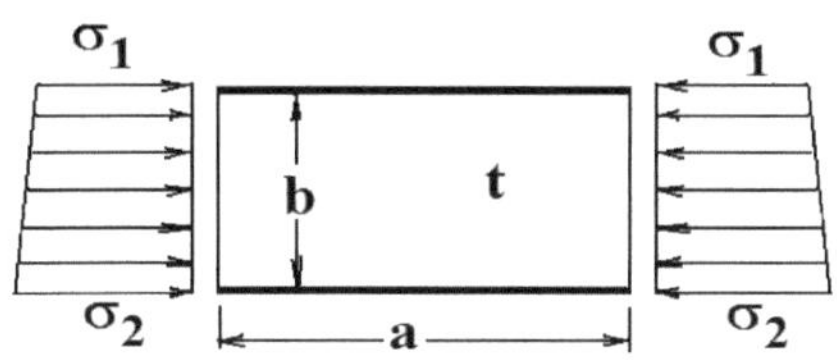

Fig. 12.2 Non-uniform in-plane compression over the short edges of plate

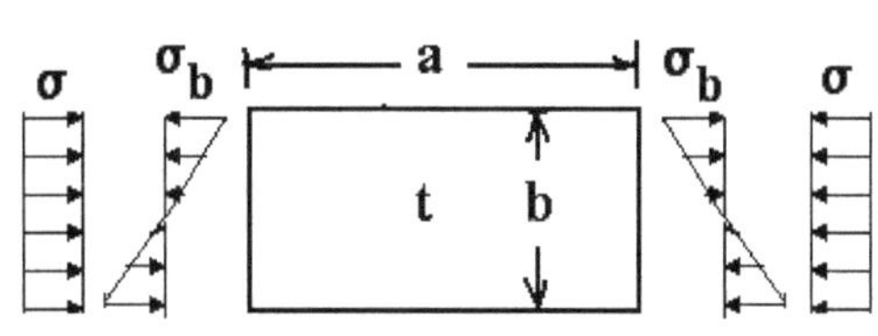

Fig. 12.3 Modeled in-plane compressive stress over the face plate

This general loading condition of non-uniform in-plane compression could be transformed into the basic stress components of uniform compression and pure bending stress, see Fig. (12.3).

The values of the modeled in–plane uniform compression and bending stresses could be determined as follows:

$$\sigma = \frac{\sigma_1 + \sigma_2}{2} \quad \text{and} \quad \sigma_b = \frac{\sigma_2 - \sigma_1}{2}$$

Where: σ = uniform in-plane compressive stress
σ_b = pure in-plane bending stress

The assessment of buckling could be carried out using the following interaction equation of combined in-plane uniform and bending stresses:

$$\frac{\sigma}{\sigma_{ec}} + \left(\frac{\sigma_b}{\sigma_{eb}}\right)^2 \leq 1.0$$

Where: σ_{ec} = critical buckling stress for pure compressive stress
σ_{eb} = critical buckling stress for pure bending stress

Deck, side and bottom plate panels are subjected to significant values of shear stresses. The assessment of buckling strength of these plate panels could be evaluated using the following equation:

$$\frac{\sigma}{\sigma_{ec}} + \left(\frac{\sigma_b}{\sigma_{eb}}\right)^2 + \left(\frac{\tau}{\tau_{es}}\right)^2 \leq 1.0$$

Where: σ_{ec} = Euler buckling stress for pure in-plane compression
σ_{eb} = Euler buckling stress for pure in-plane bending stress
τ_{es} = Euler buckling stress for pure shear stresses

The Euler buckling equations are given by:

For uniform compression:

$$\sigma_{ec} = \frac{\pi^2 E}{12(1-\upsilon^2)}\left(\frac{t}{b}\right)^2 \cdot k_C$$

For pure bending:

$$\sigma_{eb} = \frac{\pi^2 E}{12(1-\upsilon^2)}\left(\frac{t}{b}\right)^2 \cdot k_b$$

For shear:

$$\tau_{es} = \frac{\pi^2 \cdot E}{12(1-\upsilon^2)}\left(\frac{t}{b}\right)^2 .k_s$$

The values of the constant "k" in the above equations depend upon the boundary condition of the plate panel.

3.2 *Critical Buckling Stress*

The buckling strength of any strength member of ship structure could be evaluated using the concept of the critical buckling stress. The magnitude of the critical buckling stress is related to the classical Euler buckling equation and depends on whether buckling occurs in the elastic or in the in-elastic condition.

Elastic buckling occurs when:

$$\sigma_E \leq 0.5\sigma y$$

Where: σ_y = yield stress of the plate material

σ_E = Euler buckling stress

The Euler buckling stress of a flat plate is given by:

$$\sigma_E = \frac{k \cdot \pi^2 \cdot E}{12(1-\upsilon^2)}\left(\frac{t}{b}\right)^2$$

Where: ν = Poisson's ratio, $\nu = 0.3$

E = Modulus of elasticity, $\left(E = 2.1 \times 10^3 \ t/cm^2\right)$

t = plate thickness

b = length of short side of plate

k = Constant depending on the boundary conditions of the plate and the aspect ratio (a/b ratio)

For $a/b >> 1.0,\ k = 4.0$

Inelastic buckling occurs when:

$$\sigma_E \geq 0.5\sigma y$$

In this case, the critical buckling stress is given by:

$$\sigma_{cr} = \sigma_y\left(1 - \frac{\sigma_y}{4\sigma_E}\right)$$

3.3 *Boundary Conditions*

3.3.1 Boundary Conditions of Girders

Girders and longitudinals could be assumed clamped at the connection with transverse bulkheads and heavy transverse members, see Fig. (12.4).

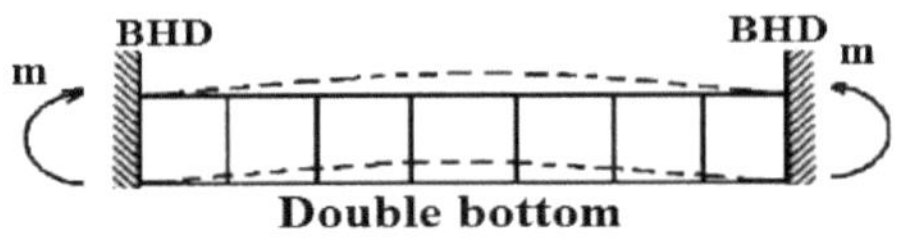

Fig. 12.4 Girders assumed clamped at transverse bulkheads

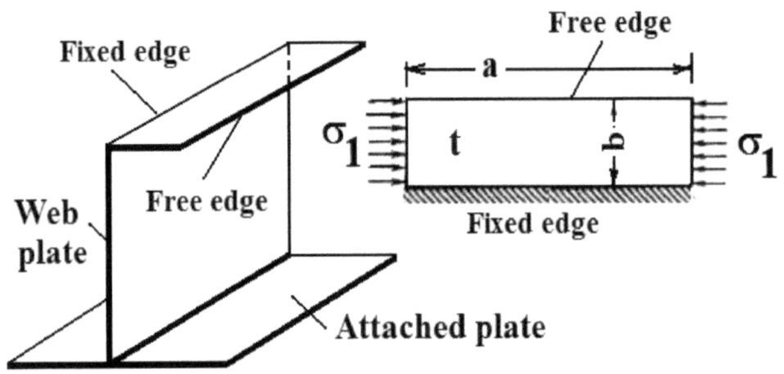

Fig. 12.5 Boundary conditions of face plate of a girder having an angle section

The boundary conditions for the face plate of a girder having an angle section can be modeled as shown in Fig. (12.5). The face plate is assumed fixed with the web plate along the line of connection between them. On the other hand, the web plate could be assumed either simply supported or fixed along the line of connection with the face plate. The web plate could be assumed fixed with the attached plating.

3.3.2 Boundary Conditions of Plating

The boundary condition of any plate edge could be modeled as being clamped or simply supported depending on the flexural and torsional rigidities of the stiffening members at the edges of the plate panel.

For stiffened plate panels supported by heavy transverse members over the short edges, the connection along the short edges of the plate could be assumed clamped. Similarly, stiffened plate panels loaded with in-plane compressive stresses orthogonal to the long edges of the plate, the plate connection with the stiffener along the long edges of the plate could be also assumed clamped because of the flexural behavior of the plating, see Fig. (12.6).

Also when the torsional rigidity of the stiffening members are relatively high, the connection of the plating with the stiffeners are assumed clamped, see Fig. (12.7). On the other hand, when the torsional rigidity of the stiffeners are relatively low, the connection of the plating with the stiffeners are assumed simply supported, see Fig. (12.8).

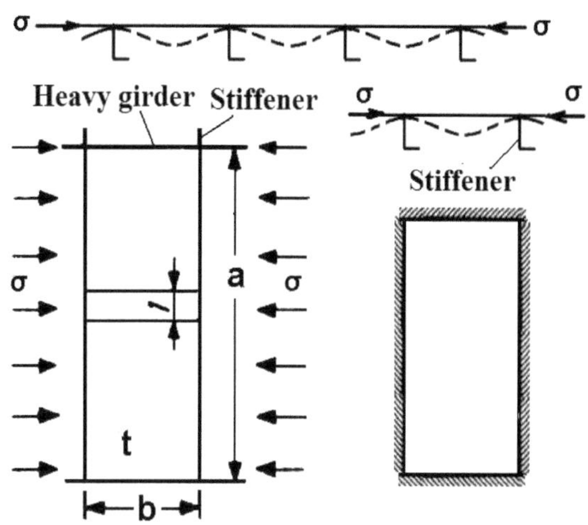

Fig. 12.6 Plate panel loaded with in-plane compressive stresses orthogonal to the long edges

Fig. 12.7 Plate panel stiffened with high torsional rigidity stiffeners loaded with in-plane compressive stresses orthogonal to the longitudinal

Fig. 12.8 Plate panel stiffened with low torsional rigidity stiffeners loaded with in-plane compressive stresses orthogonal to the longitudinals

4 Modes of Buckling

4.1 General Modes of Buckling of Girders

In the assessment of buckling strength of girders and longitudinals, both global and local buckling failure should be considered. The general mode of failure of T-section girder is illustrated by the deformation of the web plating longitudinally and transversely and by the waviness of the attached plating as shown in Fig. (12.9).

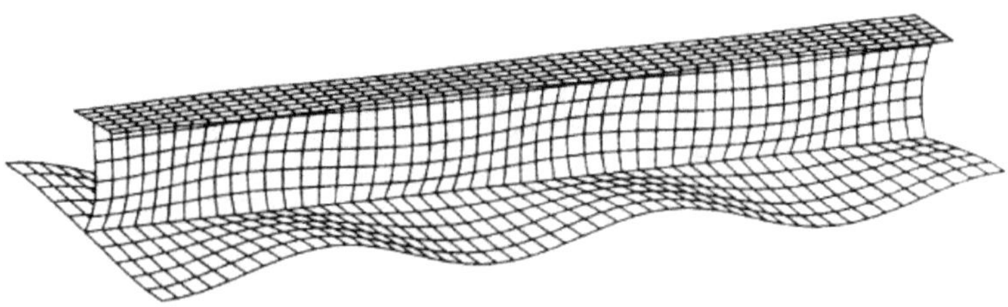

Fig. 12.9 General buckling of a T-section girder with attached plating

4.2 Modes of Buckling of Face Plates

The mode of buckling failure of the face plate could be either lateral deformation, see Fig. (12.10) for asymmetrical face plates and Fig. (12.11) for symmetrical face plate.

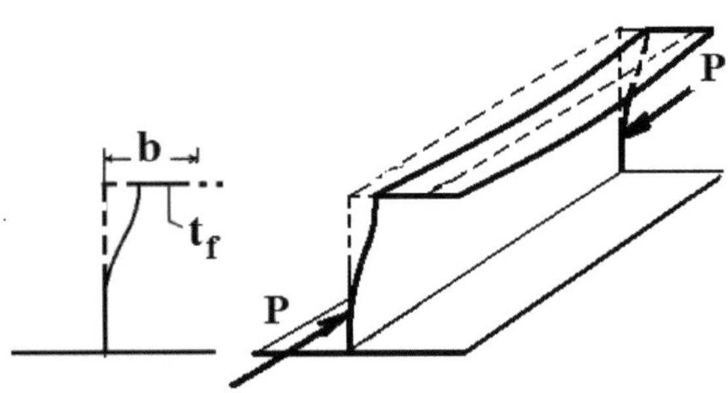

Fig. 12.10 Lateral deformation of an asymmetrical face plate

The face plate could be also buckled and deformed vertically. The mode of buckling of an asymmetrical face plate is shown in Fig. (12.12) and for symmetrical face plate is shown in Fig. (12.13).

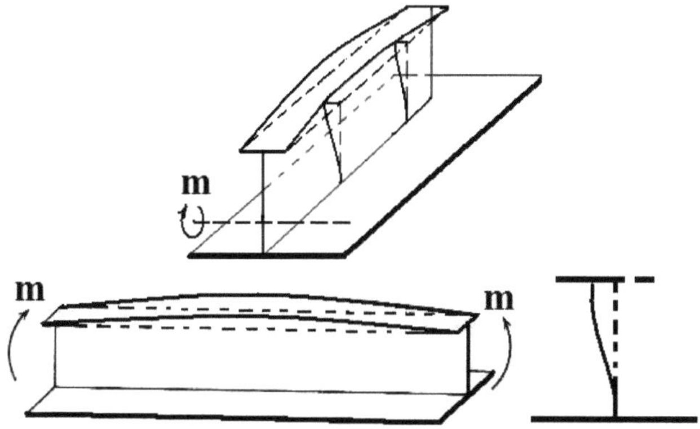

Fig. 12.11 Lateral deformation of symmetrical face plate under bending stresses

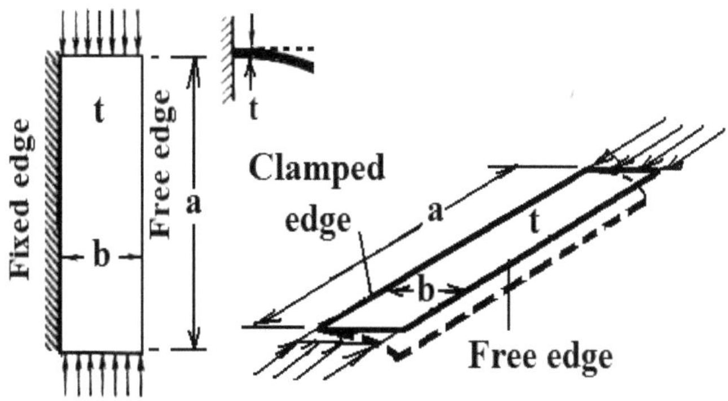

Fig. 12.12 Vertical deformation of asymmetrical face plate under compression

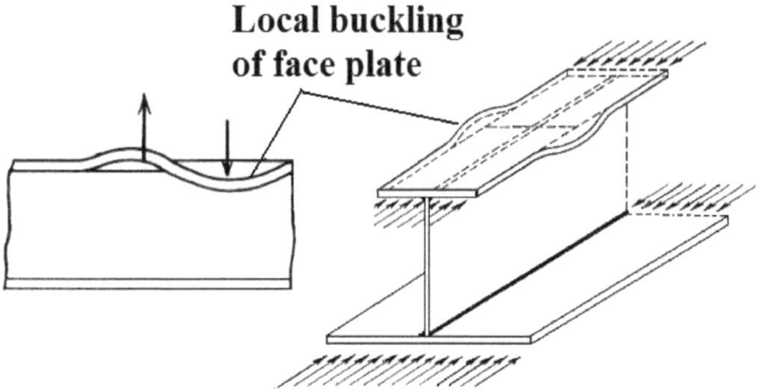

Fig. 12.13 Local buckling of the girder face plate

4.3 Modes of Buckling of the Web Plate

The mode of buckling of the web plate depends on the type of the in-plane loading on the web plate. When the web plate is subjected to in-plane bending stresses, the mode of buckling is as shown in Fig. (12.14).

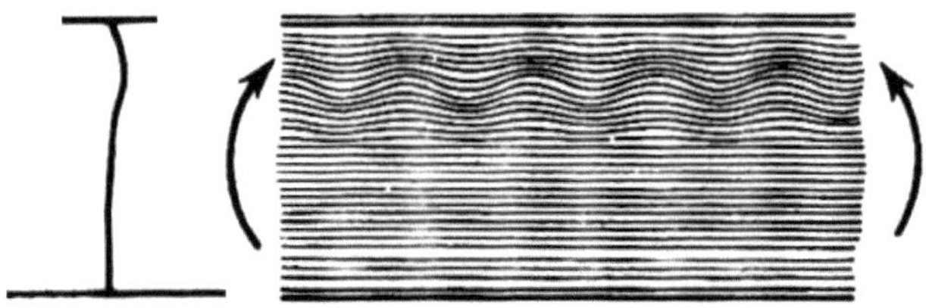

Fig. 12.14 Longitudinal buckling of the web plate of girder

When the web plate is subjected to high shear stresses, the mode of buckling is as shown in Fig. (12.15).

Fig. 12.15 Diagonal buckling of the web plate of girder

5 General Mode of Buckling of Longitudinals

The mode of buckling of stiffeners supported by heavy transverse members having torsional rigidity lower than the flexural rigidity of the stiffener with the attached plating is shown in Fig. (12.16). The longitudinal behaves as a beam column clamped at both ends.

The mode of buckling of stiffeners/longitudinals supported by heavy transverses having torsional rigidity much larger than the flexural rigidity of the longitudinals with the attached plating is shown in Fig. (12.17). The longitudinal behave as a beam fixed at both ends.

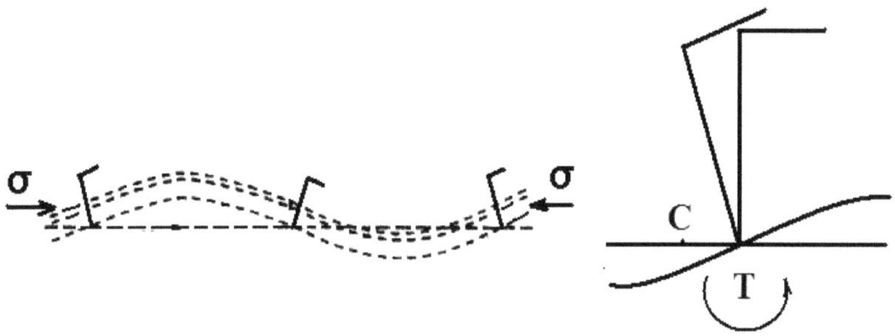

Fig. 12.16 Mode of buckling of longitudinals supported by transverses having low torsional rigidity

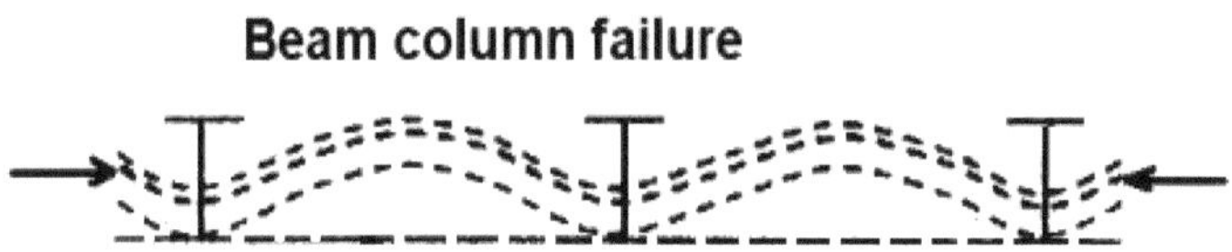

Fig. 12.17 Mode of buckling of stiffener supported by heavy transverses

6 General Mode of Buckling of Plating

In a stiffened panel subjected to in-plane compressive stresses orthogonal to the direction of the stiffeners, the mode of buckling of the plating is as shown in Fig. (12.18) when the stiffeners have high torsional rigidity and in Fig. (12.19) when the stiffeners have low torsional rigidity.

Fig. 12.18 Mode of buckling of transversely stiffened plate panel when the stiffeners have large torsional rigidity

Fig. 12.19 Mode of buckling of transversely stiffened plate panel when the stiffeners have low torsional rigidity

7 Assessment of Buckling of Girders and Longitudinals

7.1 Torsion Buckling

The Euler stress for torsion buckling of a beam of length L is given by:

$$\sigma_e = \beta \frac{C \cdot \frac{\pi^2}{L^2} \cdot E\Gamma + GJ}{I_\rho}$$

G = shear modulus = E/2.6
I_P = polar m.o.i, see Fig. (12.20)

$$I_p = I_x + I_y$$

I_y = m.o.i of section about y-y axis without effective plate flange
I_x = m.o.i of cross-section about x-x axis without effective plate flange
C = constant depending on the support condition of the stiffener.
C = 1.0 for simply supported ends
= 4.0 for fixed ends
β = 1.0 = constant depending on the support conditions along the plate

Where: $\Gamma = I_y \left(h + \frac{t_f}{2} \right)^2$

$$J = \frac{1}{3}\left(f \cdot t_f^3 + h \times t_h^3 \right)$$

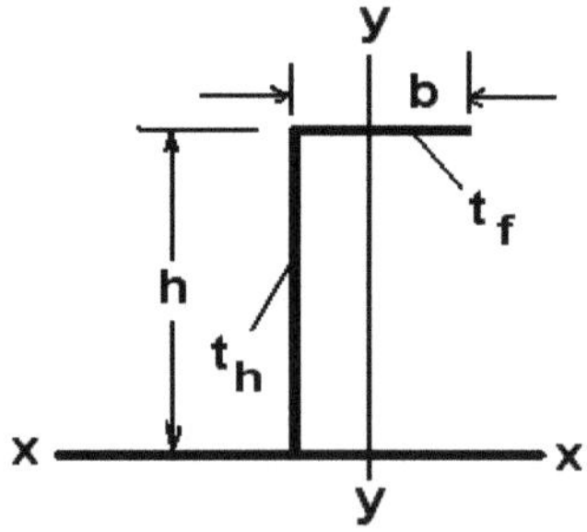

Fig. 12.20 Geometry and scantlings of angle section

The torsion buckling stress of the top flange of a section is given by:

$$\sigma_e = \frac{GJ}{I_y}$$

Where: J = torsion constant of section,

$$J = bt_f^3/3$$

I_y = m.o.i of flange about y-y axis

For offset flange, see Fig. (12.21):

$$I_y = t_f \cdot \frac{b^3}{3}$$

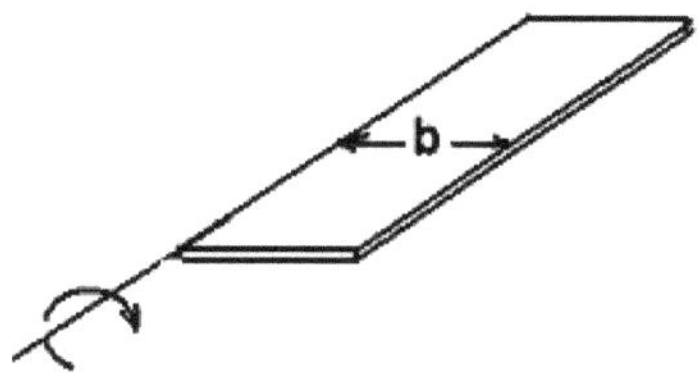

Fig. 12.21 Offset flange

Hence, the torsion buckling stress for an offset flange is given by:

$$\sigma_e = G \cdot \frac{bt_f^3/3}{t_f \cdot b^3/3} = G \cdot \left(\frac{t_f}{b}\right)^2$$

For T-stiffener, see Fig. (12.22):

$$I_y = t_f \cdot b^3/12$$

The torsion buckling stress for the flange of a T-stiffener is given by:

$$\sigma_e = G \cdot \frac{bt_f^3/3}{t_f \cdot b^3/12} = 4G \cdot \left(\frac{t_f}{b}\right)^2$$

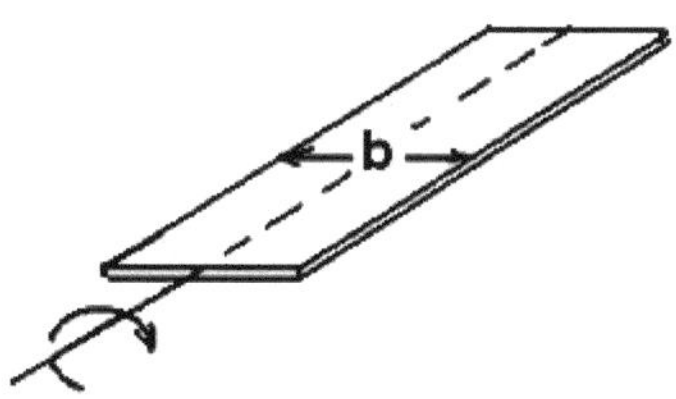

Fig. 12.22 Symmetrical flange

7.2 Lateral Buckling of Flanges

The lateral buckling stress of the flange of the section is given by:

$$\sigma_E = \frac{\pi^2 E I_y}{L^2 A}$$

Where: A = cross–section area of flange

For T-stiffeners

The Euler buckling stress for a symmetrical flange is given by, see Fig. (12.23):

$$\sigma_E = \frac{\pi^2 E}{12} \cdot \left(\frac{b}{L}\right)^2$$

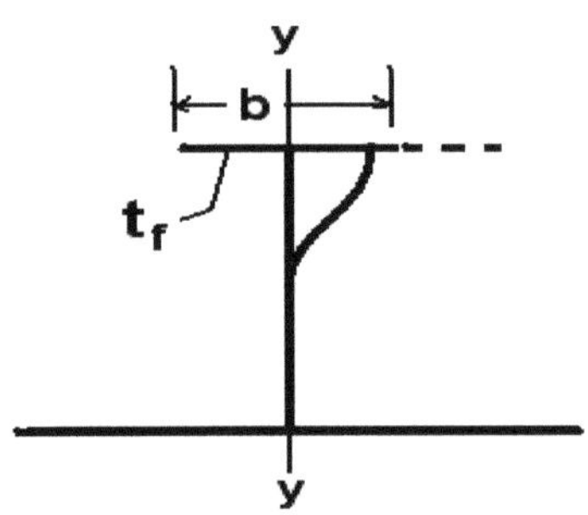

Fig. 12.23 Lateral deformation of symmetrical flange

For offset flange

The lateral buckling of offset flange is given by, see Fig. (12.24):

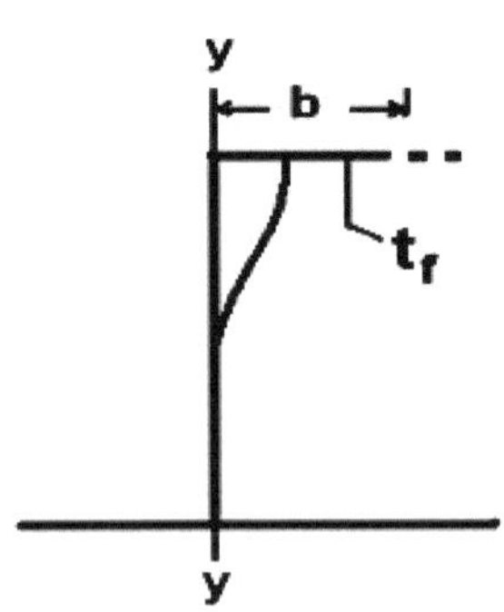

Fig. 12.24 Geometry and scantlings of an

$$\sigma_E = \frac{\pi^2 E I_y}{L^2 A}$$

$$= \frac{\pi^2 E}{3} \cdot \left(\frac{b^2}{L^2} \right)$$

8 Assessment of Buckling of Deck Girders

The compound stress in the section of a deck girder could be modeled as shown in Fig. (12.25). The compound stress is composed of the primary and secondary stresses when the vessel is in a sagging condition.

The compound stress in the web plate could be modeled as shown in Fig. (12.25). The stress components shown in Fig. (12.25) are as follows:

σ_{p1} = primary hull girder stress at the top edge of the web plate
σ_{p2} = primary hull girder stress at the bottom edge of the web plate
σ_{s1} = secondary stress at the top edge of the web plate
σ_{s2} = secondary stress at the bottom edge of the web plate
σ_{t1} = total stress at the top edge of the web plate
σ_{t2} = total stress at the bottom edge of the web plate

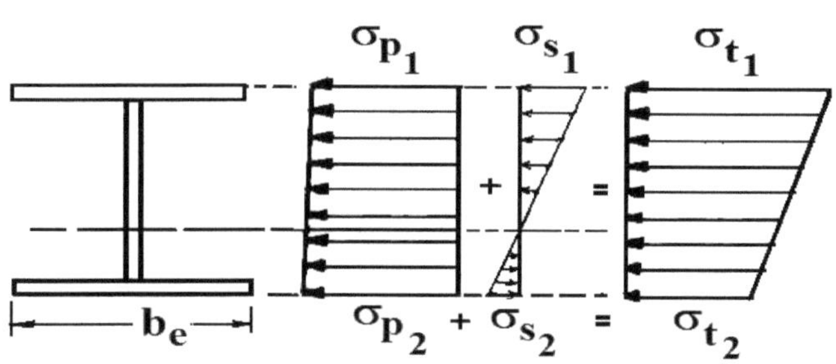

Fig. 12.25 Primary, secondary and compound stresses on a bottom girder

8.1 Assessment of Buckling of the Web Plate

Because the depth of the bottom girder is much smaller than the depth of the ship, the primary stress is assumed uniformly distributed over the girder section, see Fig. (12.26). The distribution of the in-plane compound stress over the web depth is as shown in Fig. (12.26). The values of the compound stress at the top and bottom edges of the web plate are given by:

$$\sigma_1 = \sigma_p + \sigma_{s1}$$

$$\sigma_2 = \sigma_p - \sigma_{s2}$$

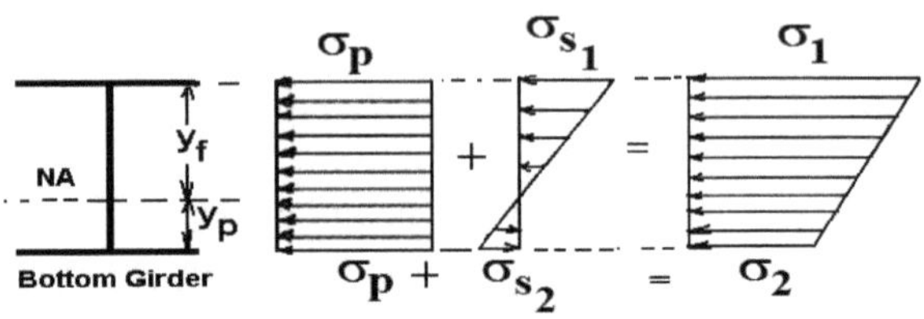

Fig. 12.26 Compound in-plane compressive stress over bottom girder section

The web plate is therefore subjected to non-uniform compressive in-plane stress as shown in Fig. (12.27). The non-uniform distribution of the compressive stresses can be decomposed into uniform and pure bending stresses as shown in Fig. (12.28).

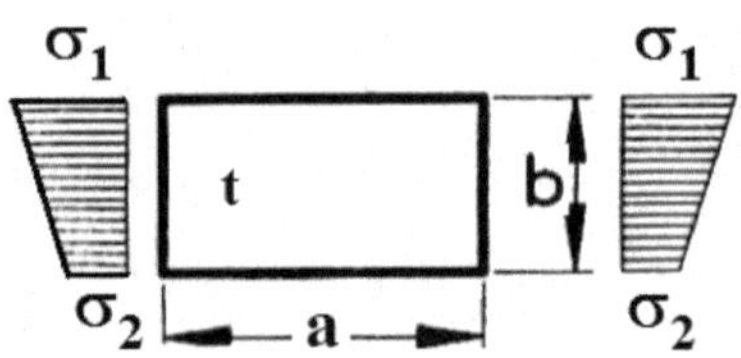

Fig. 12.27 Non-uniform compressive stress over the short edges of the web plate

The values of the pure compressive and pure bending stresses are given by, see Fig. (12.28):

$$\sigma = \frac{\sigma_1 + \sigma_2}{2}$$

$$\sigma_b = \frac{\sigma_1 - \sigma_2}{2}$$

The assessment of buckling of the web plate of the girder could be based on the following basic equation:

$$\frac{\sigma}{\sigma_{ec}} + \left(\frac{\sigma_b}{\sigma_{eb}}\right)^2 \leq 1.0$$

Where: σ_{ec} = critical buckling stress for pure compressive stress
σ_{eb} = critical buckling stress for pure bending stress
σ = uniform compressive stress
σ_b= pure bending stress

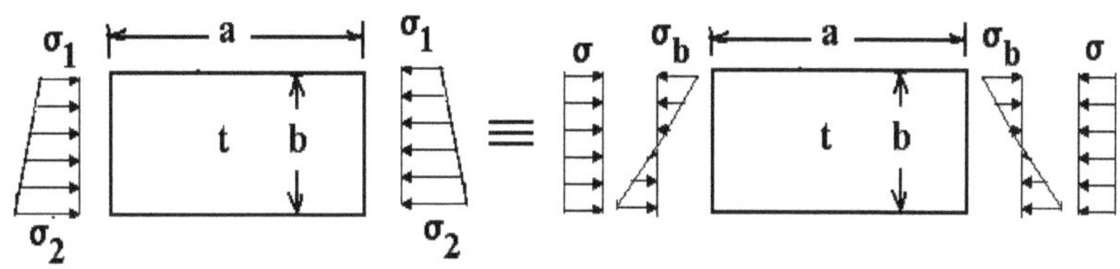

Fig. 12.28 In-plane non-uniform compressive stress over the web plate

8.2 *Assessment of Buckling of the Face Plate*

For girders having asymmetrical face plates, the compound stress loading in the face plate is uniformly distributed over the width of the face plate, see Fig. (12.29).

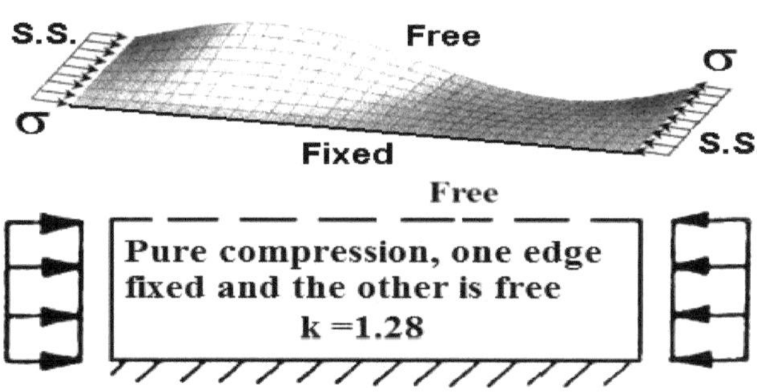

Fig. 12.29 Mode of buckling of half width of a face plate

The Euler buckling equation of a face plate under pure compression on the short edges and having one long edge fixed and the other free is given by, see Fig. (12.29):

$$\sigma_E = \frac{k \cdot \pi^2 \cdot E}{12(1-\upsilon^2)} \left(\frac{t}{b}\right)^2$$

Where: k = 1.28, see Fig. (12.29).

In general, the assessment of buckling of the face plate is necessary when: $(2b/t_f) > 30$, where 2b = width of face plate, see Fig. (12.30). For girders having asymmetrical face plate, the in-plane compound compressive stress is not uniformly distributed over the width of the face plate, see Fig. (12.31). This non-uniform stress distribution could be decomposed into uniform and pure bending distributions and the same approach for the assessment of buckling could be used.

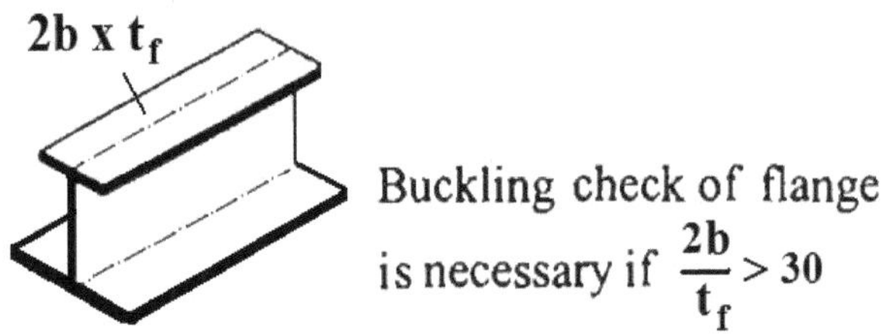

Fig. 12.30 Stress components induced in bottom longitudinals

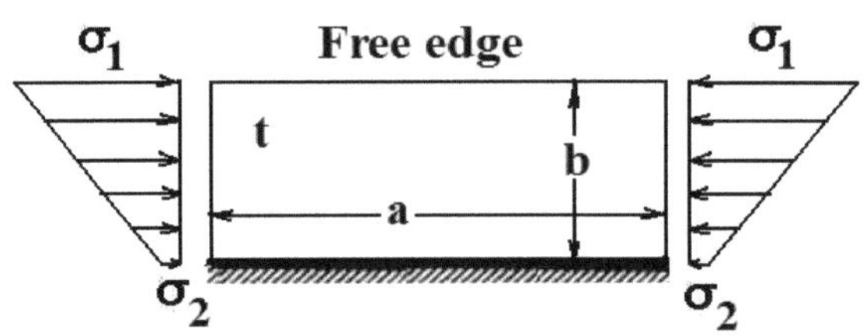

Fig. 12.31 In-plane non-uniform compressive stress over asymmetrical face plate

9 Assessment of Buckling of Longitudinals

For a ship in a hogging condition, the compound compressive stress for bottom longitudinals is given by, see Fig. (12.32).

$$\sigma_1 = \sigma_{p_1} + \sigma_{s_1} + \sigma_{t_1}$$

$$\sigma_2 = \sigma_{p_2} + \sigma_{s_2} - \sigma_{t_2}$$

Where: σ_p = Primary hull girder stress

σ_s = Secondary stress

σ_t = Tertiary stress

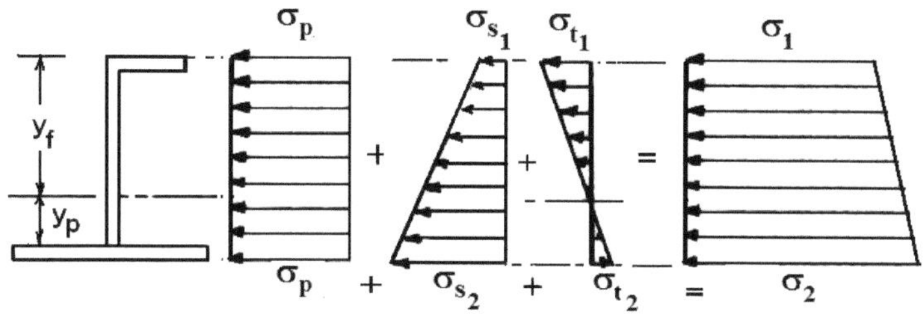

Fig. 12.32 Primary, secondary, tertiary and compound stresses in bottom longitudinals

The compound stress in the face plate and web plate of a bottom longitudinal is shown in Fig. (12.33). The primary in-plane compressive stress induced by the hull girder bending is assumed uniformly distributed over the longitudinal section, and is given by:

$$\sigma_p = \frac{M_V \cdot y_B}{I_V}$$

Where: M_V = hull girder vertical bending moment

y_B = distance of bottom plating from ship section N.A

σ_p = primary hull girder stress

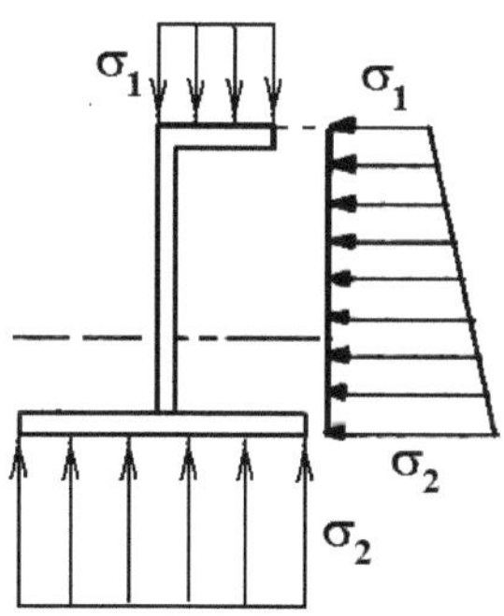

Fig.12.33 Compound stresses on a bottom longitudinal

The secondary stress induced by the secondary loadings is linearly distributed over the longitudinal section, see Fig. (12.34).

$$\sigma_{s_2} = \frac{m_2 . y_p}{i_s}$$

$$\sigma_{s_1} = \frac{m_2 . y_f}{i_s}$$

Where: i_s = second moment of area of the tank top and bottom longitudinal sections with attached plating about the neutral axis

m_s = secondary bending moment

The tertiary stress is the bending stress induced by the local loadings and is linearly distributed over the longitudinal section, see Fig. (12.35).

The tertiary stress σ_3 at the lower edge of the longitudinal is given by:

$$\sigma_{t_2} = \frac{m_3 \cdot y_P}{i}$$

The tertiary stress at the flange of the longitudinal is given by:

$$\sigma_{t_1} = \frac{m_3 \cdot y_f}{i}$$

Where: m_3 = tertiary bending moment

i = moment of inertia of longitudinal section

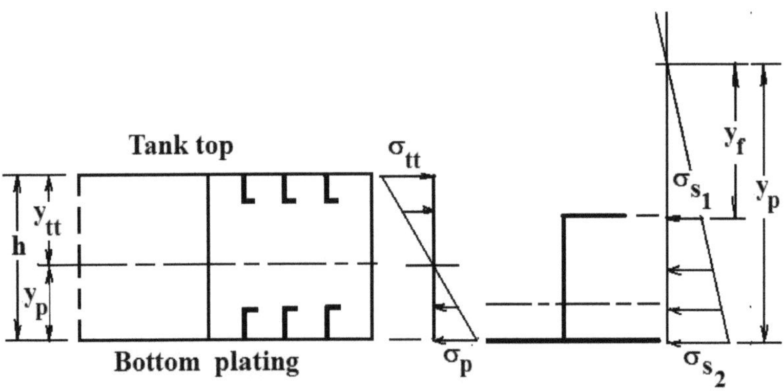

Fig. 12.34 Secondary stress over longitudinal section

The compound non-uniform compressive stress over the short edges of the web plate of longitudinal is shown in Fig. (12.36).

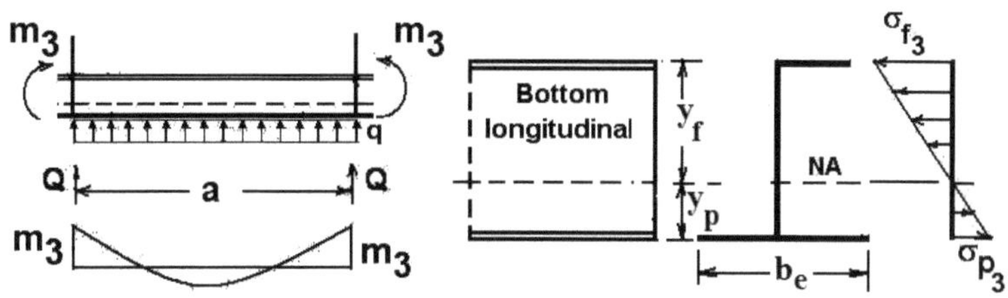

Fig. 12.35 Tertiary stress distribution over bottom longitudinal section

The assessment of buckling of the web plate and face plate follows the same approach used for girders. The non-uniform in-plane compressive stresses over the short edges of the web plate of the longitudinal could be modeled by uniform and pure bending stresses as shown in Fig. (12.37).

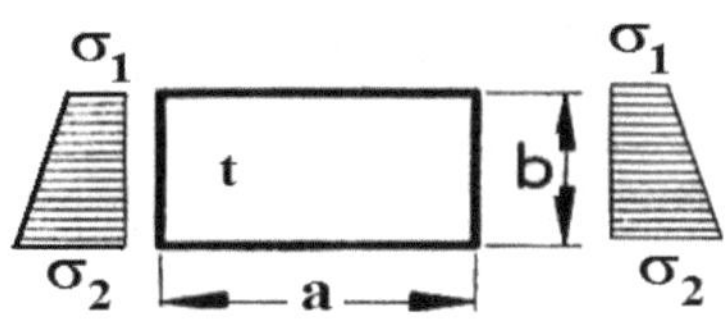

Fig. 12.36 Compound non-uniform compressive stress over the short edges of the web plate of longitudinal

The values of the uniform and bending stress components could be determined as follows:

$$\sigma = \frac{\sigma_1 + \sigma_2}{2} \text{ and } \sigma_b = \frac{\sigma_2 - \sigma_1}{2}$$

In the absence of high values of shear stresses, the assessment of buckling of the web plate could be carried out using the following equation:

$$\frac{\sigma}{\sigma_{ec}} + \left(\frac{\sigma_b}{\sigma_{eb}}\right)^2 \leq 1.0$$

σ_{ec} = critical buckling stress for pure compressive stress

σ_{eb} = critical buckling stress for pure bending stress

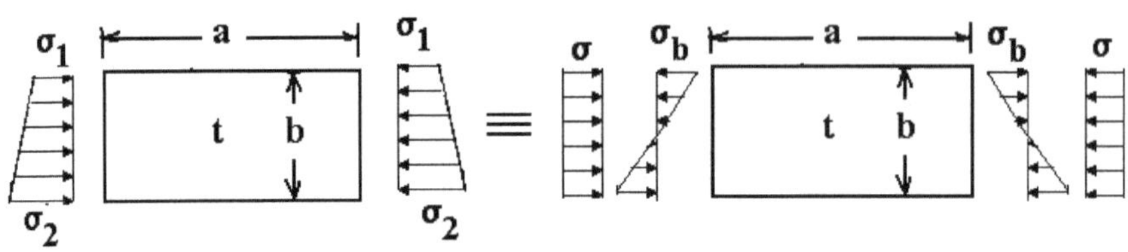

Fig. 12.37 In-plane compressive stress over the face plate

10 Assessment of Buckling of Plating

10.1 Buckling of Bottom Plating

The various stress components in a plate panel of the bottom plating of a transversely stiffened bottom structure of a ship is shown in Fig. (12.38). The compound in-plane stress to be used in the assessment of plate buckling is given by:

$$\sigma = \sigma_p + \sigma_s + \sigma_{local}$$

Where: σ_p = hull girder stress

σ_s = secondary stress

σ_{local} = local stress

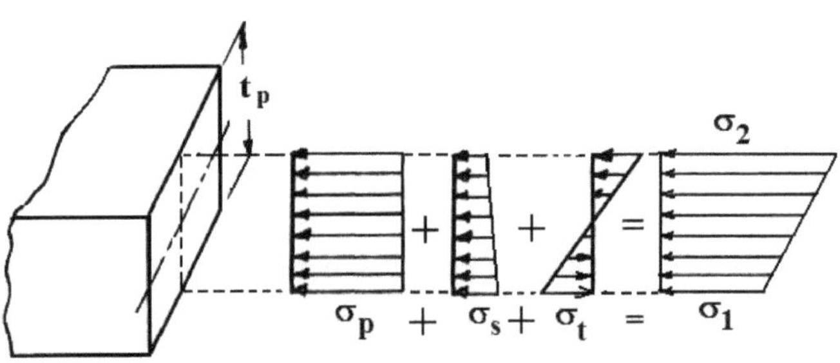

Fig. 12.38 Various stress components affecting bottom plating

In order to determine the magnitude of the local stress in the same direction as the primary and secondary stresses, consider a strip of width one meter in a bottom plate panel, see Fig. (12.39). The local bending stress at the clamped edge of the strip in the longitudinal direction of the ship is given by, see Fig. (12.40):

$$\sigma_A = \frac{m}{z} = \frac{qb^2}{12} \Big/ \frac{t^2}{6} = 50q\left(\frac{b}{t}\right)^2$$

Where:

t = plate thickness

m = bending moment at the fixed end of the strip, see Fig. (12.39)

$$m = \frac{qL^2}{12}$$

The local bending stress at mid-span of the plate panel is given by:

$$\sigma_B = 25q\left(\frac{b}{t}\right)^2$$

For a longitudinally stiffened bottom structure, the compound stress in the bottom plating is given by:

$$\sigma_{tp} = \sigma_1 + \sigma_2 + \sigma_3 \pm \sigma_4$$

σ_1 = hull girder stress at the bottom plating
σ_2 = secondary stress at the bottom plating
σ_3 = tertiary stress at the bottom plating
σ_4 = local stress in the bottom plating
σ_t = total compressive stress

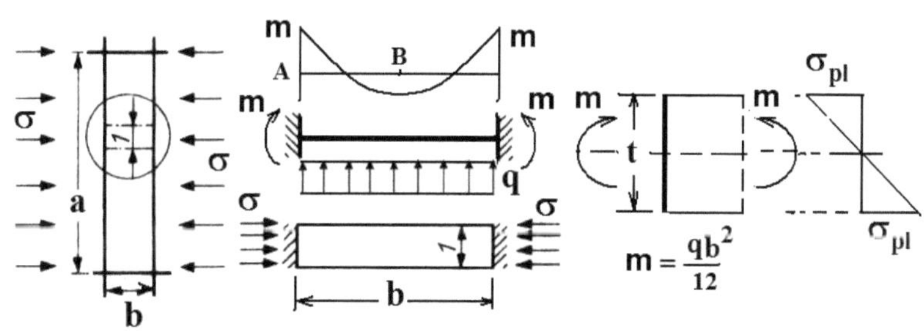

Fig. 12.39 Local bending stress in bottom plating in the transverse direction

For longitudinally stiffened bottom plating, the local stress in the longitudinal direction of the ship at the clamped edge of a plate panel is given by, see Fig. (12.41):

$$\sigma_4 = \nu . \sigma_{pl}$$

$$= 50\nu q\left(\frac{b}{t}\right)^2$$

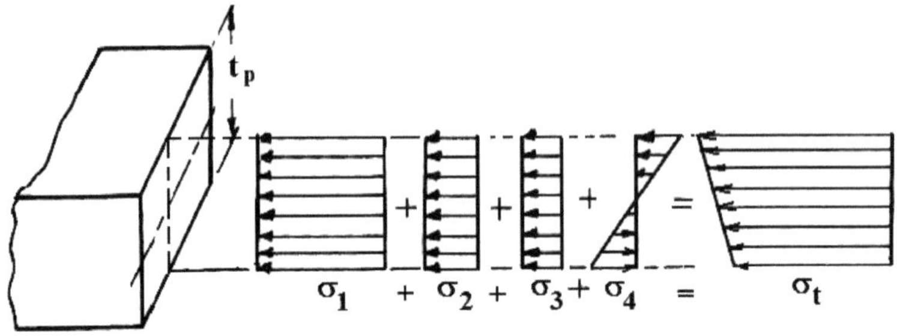

Fig. 12.40 Various stress components acting on the bottom plating

Where: σ_ℓ = Local bending stress

σ_4 = local plate stress in the longitudinal direction

σ_{pl} = local plate stress in the transverse direction

ν = Poisson ratio = 0.3

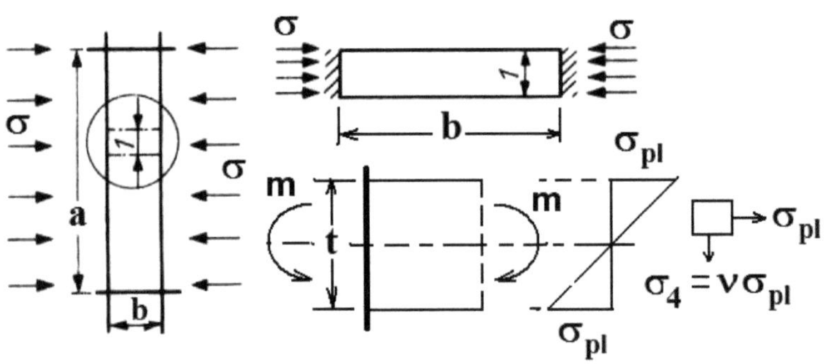

Fig. 12.41 Local bending stress in bottom plating in the transverse direction

10.2 Buckling of Deck Plating

The in-plane stress induced by hull girder horizontal bending moment in a deck plate is shown in Fig. (12.42).

The assessment of buckling of a deck plate panel subjected to in-plane compressive stresses could be carried out as follows using the elastic Euler equation, see Fig. (12.43):

$$\sigma_{ec} = \frac{\pi^2 E}{12(1-\upsilon^2)}\left(\frac{t}{b}\right)^2 \cdot k_c$$

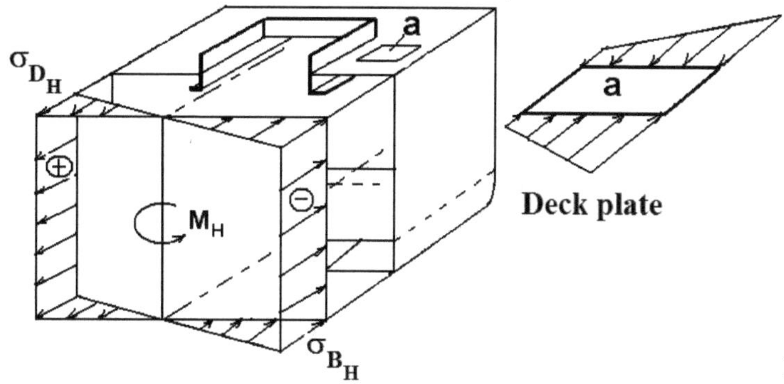

Fig. 12.42. A deck plate panel subjected to compressive stresses induces by hull girder horizontal bending

The primary stress in the deck plating due to a sagging moment is given by, see Fig. (12.43):

$$\sigma_D = \frac{M \cdot y_D}{I}$$

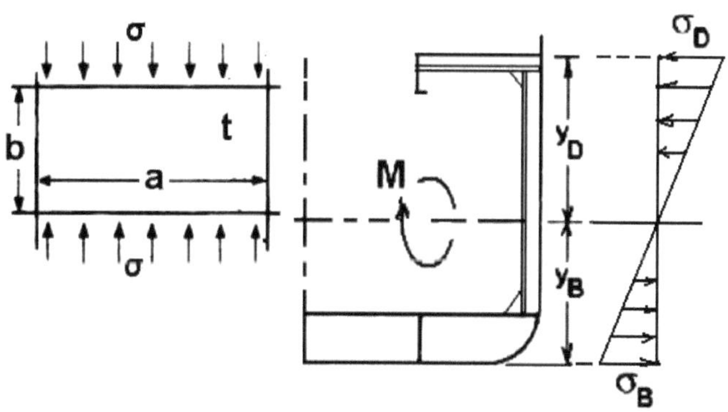

Fig. 12.43 A deck plate in a transversely stiffened deck structure subjected to in-plane compressive stresses induced by a sagging bending moment

The secondary stress in the deck plating is given by, see Fig. (12.44):

$$\sigma_{s1} = \frac{m_2 \cdot y_p}{i}$$

Where: σ_{s1} = secondary stress in the deck plating
y_p = distance of deck plating from neutral axis of longitudinal section

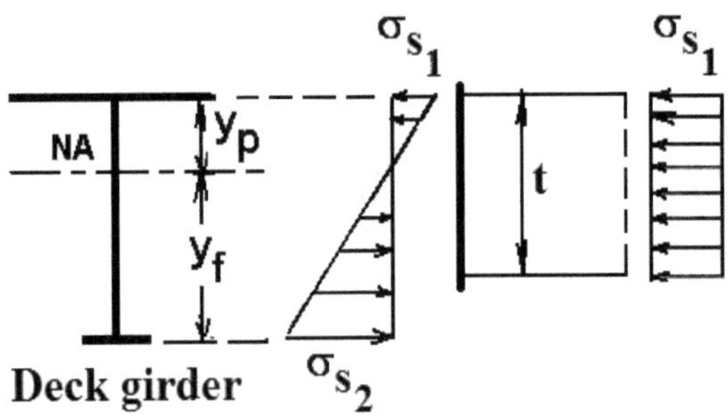

Fig. 12.44 Secondary in-plane stresses in the deck plating

The deck plating could be subjected to non-uniform in-plane compressive stresses resulting from the combined effects of the hull girder vertical sagging bending moment and horizontal bending moment, see Fig. (12.45).

This complex system of stresses could be idealized by a system of in-plane stresses comprising uniform in-plane stresses and pure bending stresses as shown in Fig. (12.46). Because of the difficulty to deal with the complexity of this idealized combined system of stresses, a simplified idealized system of in-plane stresses could be used without deviating significantly from the actual stress system. The assessment of buckling of a plate panel subjected to combined in-plane compressive stresses and pure bending stresses on all edges of the plate could be carried out using the interaction buckling equations, see Fig. (12.47).

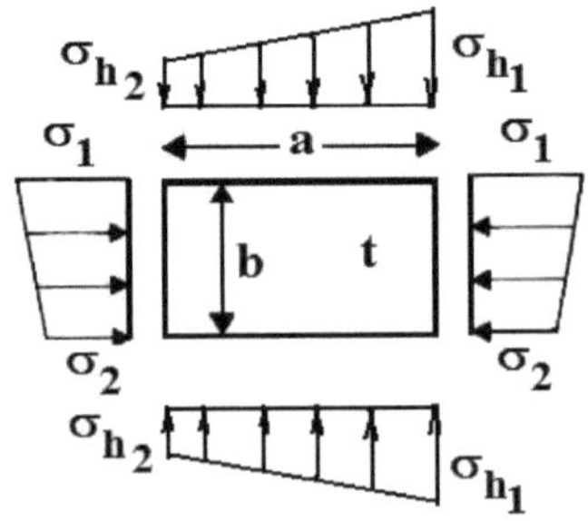

Fig. 12.45 System of in-plane stresses over a deck plate panel subjected to vertical and horizontal bending moments

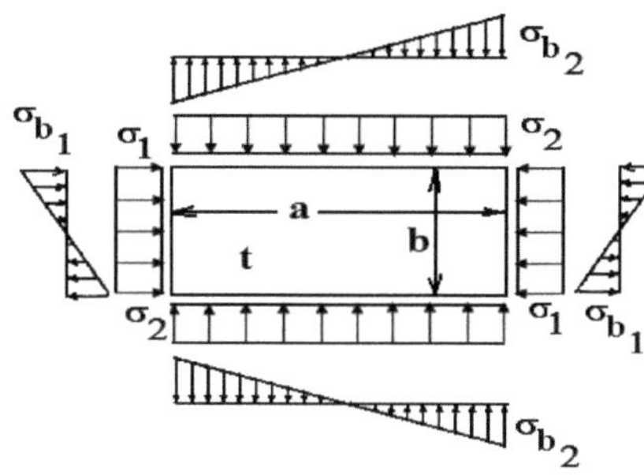

Fig. 12.46 A general idealization of the system of in-plane stresses over a deck plate panel subjected to vertical and horizontal bending moments

$$\frac{\sigma_{c1}}{\sigma_{c1_{ec}}} + \frac{\sigma_{c2}}{\sigma_{c2_{ec}}} + \left(\frac{\sigma_{b1}}{\sigma_{eb_1}}\right)^2 \leq 1.0$$

Where: σ_{c1} = uniform in-plane compressive stress over short edge
σ_{c2} = uniform in-plane compressive stress over long edge
σ_{c1ec} = critical buckling stress for uniform compression
σ_{c2ec} = critical buckling stress for uniform compression
σ_{b1} = pure bending stress over the short edge
σ_{eb1} = critical buckling stress for pure bending

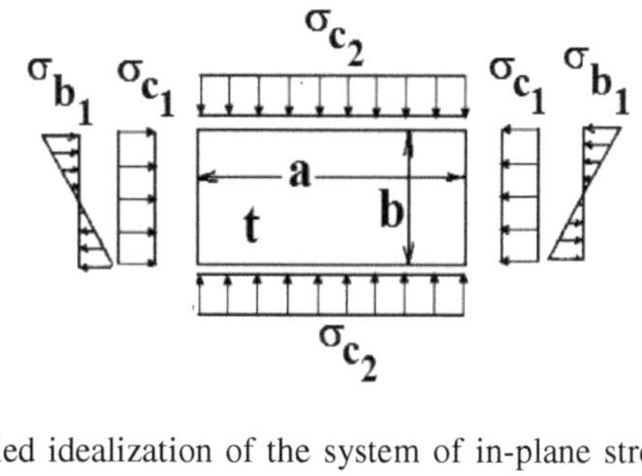

Fig. 12.47 A simplified idealization of the system of in-plane stresses over a deck plate panel subjected to vertical and horizontal bending moments

If the magnitude of the shear stress induced in the deck plate is significant, then the interaction equation is given by:

$$\frac{\sigma_{c1}}{\sigma_{c1_{ec}}} + \frac{\sigma_{c2}}{\sigma_{c2_{ec}}} + \left(\frac{\sigma_{b1}}{\sigma_{eb_1}}\right)^2 + \left(\frac{\tau}{\tau_{es}}\right)^2 \leq 1.0$$

10.3 Buckling of Side Shell Plating

10.3.1 Configurations of Side Shell Plate Panels

Side shell plate panels are stiffened either transversely or longitudinally, see Fig. (12.48).

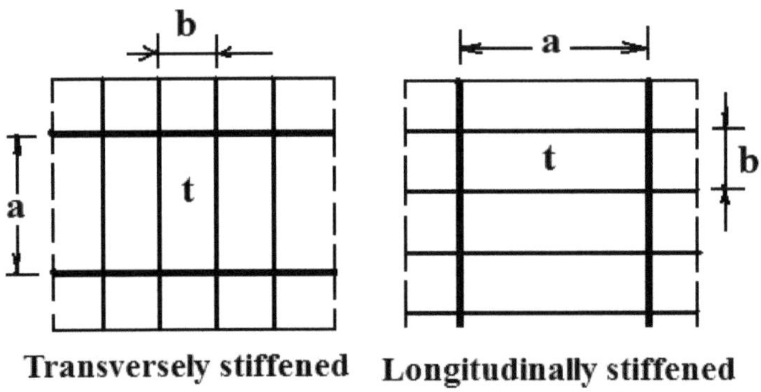

Fig. 12.48 Longitudinally and transversely stiffened side shell plate panels

10.3.2 Induced Stresses in the Side Shell Plating

The main stress components induced in the side shell plating are:

- In-plane stresses induced by the vertical hull girder bending
- In-plane stresses induced by hull girder horizontal bending
- Shear stresses induced by the hull girder vertical and horizontal shear forces
- Shear stresses induced by the hull girder torsional moments

The hull girder in-plane stresses induced in the side shell plating could be either tensile or compressive depending on the direction of the bending moment and the location of the plate panel with respect to the neutral axis of the ship section.

The primary hull girder bending stress induced in a panel of plating "abcd" above the neutral axis of the ship section in the side shell plating of a general cargo ship subjected to a sagging bending moment is given by, see Fig. (12.49):

$$\sigma_1 = \frac{M.y_1}{I}$$

And $$\sigma_2 = \frac{M.y_2}{I}$$

The in-plane compressive stress induced in the side shell plate panel by a horizontal bending moment is given by, see Fig. (12.50):

$$\sigma_S = \frac{M_H . B/2}{I_y}$$

Where: M_H = horizontal bending moment
I_y = moment of inertia of ship section about the y-axis
B = ship breadth

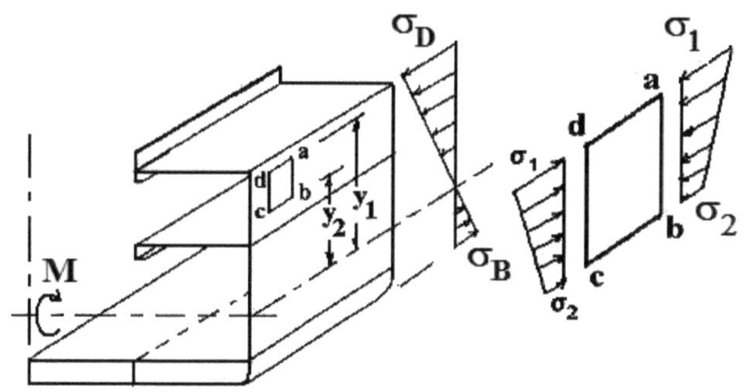

Fig. 12.49.Side shell stresses due to vertical bending moment

The shear stress distribution over a ship section of a general cargo ship and a bulk carrier due to a longitudinal vertical shear force is shown in Fig. (12.51). The maximum shear stress occurs at the shell plating near the neutral axis of the ship section. However, high values of shear stresses occur at several other locations over the ship section.

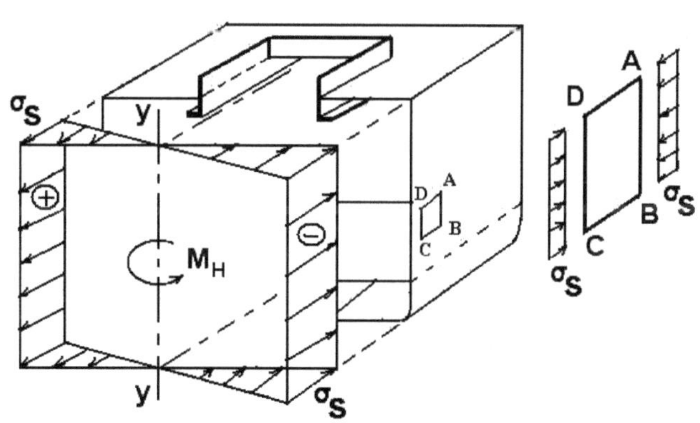

Fig. 12.50 Side shell stresses due to horizontal bending moments

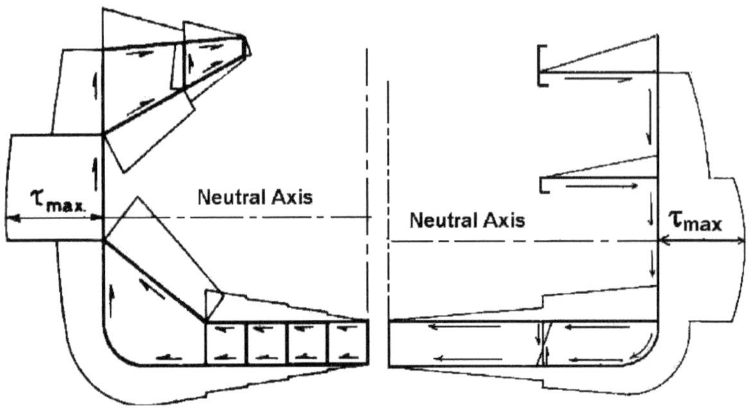

Fig. 12.51 Shear stress distribution over ship sections of a general cargo ship and a bulk carrier

10.3.3 Compounding of Stresses in Side Shell Plate Panels

The compound in-plane compressive stress induced in plate panels of the side shell above the neutral axis of the ship section is composed of the primary compressive stress (σ_1) induced by the vertical hull girder sagging moment, stress (σ_H) induced by the horizontal hull girder bending moment and shear stress (τ) induced by the vertical hull girder shear force. The magnitude and distribution of each component of these stresses depend on the location of the plate panel in the side shell structure relative to the neutral axis of the ship section.

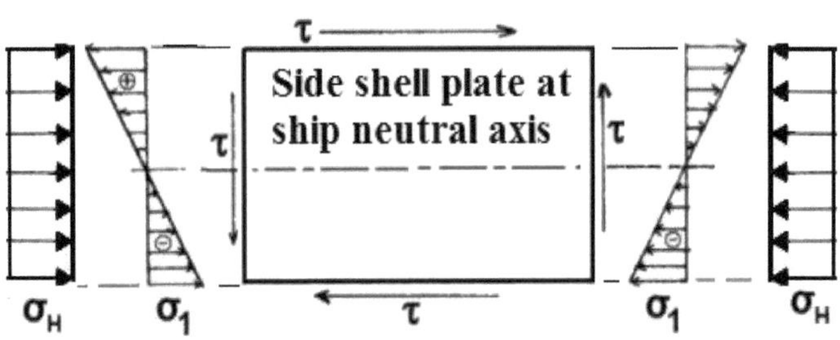

Fig. 12.52. Shear and vertical and horizontal bending stresses in a side shell plate at the ship section neutral axis

For a plate panel at the neutral axis of the ship section, the main in-plane stress components are shown in Fig. (12.52). For a plate panel in the side shell above the neutral axis of the ship section, the in-plane stress components induced by hull girder vertical and horizontal bending moments and vertical shear force are shown in Fig. (12.53).

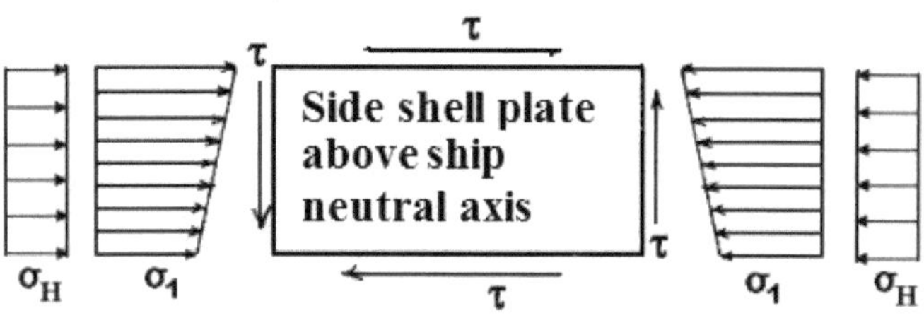

Fig. 12.53 Shear and bending stresses in a side shell plate above ship section neutral axis

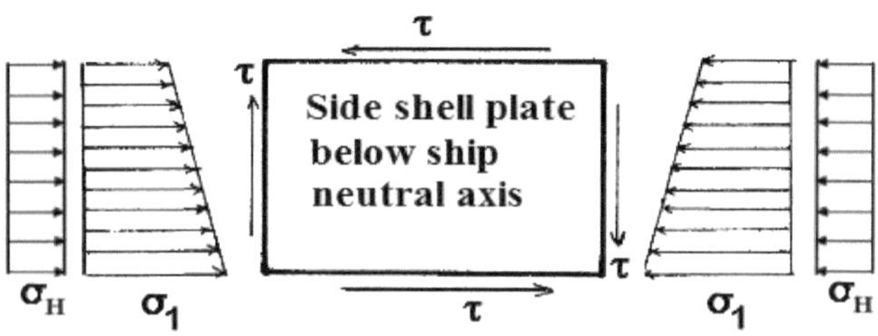

Fig. 12.54 Shear and bending stresses in a side shell plate below ship section neutral axis

Similarly, for a plate panel below the neutral axis of the ship section, the compounded stresses induced by hull girder vertical bending and vertical shear force are shown in Fig. (12.54).

10.3.4 Assessment of Buckling of Side Shell Plating

The assessment of buckling of a plate panel of the side shell at the neutral axis of the ship section could be carried out using the idealized combined system of stresses shown in Fig. (12.55). The stresses shown in Fig. (12.55) are defined as follows:

σ = in-plane compressive stress due to horizontal bending of hull girder
σ_b = in-plane bending stress due to vertical bending of hull girder
τ = average shear stress over the plate edge

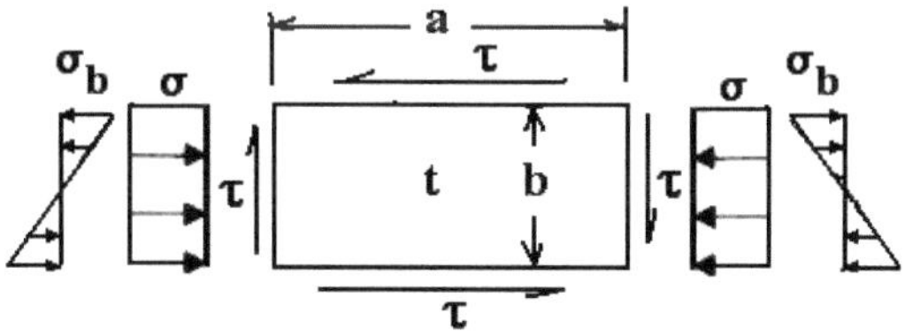

Fig. 12.55 Shear and bending stresses in a side shell plate at the ship section neutral axis

The equation governing the critical buckling stress is given by:

$$\frac{\sigma}{\sigma_{ec}} + \left(\frac{\sigma_b}{\sigma_{eb}}\right)^2 + \left(\frac{\tau}{\tau_{es}}\right)^2 \leq 1.0$$

Where: $\sigma_{ec} = \frac{\pi^2 E.k_c}{12\left(1-\upsilon^2\right)}\left(\frac{t}{b}\right)^2$

$$\sigma_{eb} = \frac{\pi^2 E.k_b}{12\left(1-\upsilon^2\right)}\left(\frac{t}{b}\right)^2$$

$$\sigma_{es} = \frac{\pi^2 E.k_s}{12\left(1-\upsilon^2\right)}\left(\frac{t}{b}\right)^2$$

The value of the constant "k" depends on the type of loading and boundary support conditions.

The assessment of buckling of a plate panel of the side shell above the neutral axis of the ship section when subjected to the combined system of stresses shown in Fig. (12.56) could be idealized by the system of stresses shown in Fig. (12.57). The stress systems shown in Figs. (12.56, 12.57) are defined as follows:

σ_1 = in-plane compressive stress at the bottom edge of the plate
σ_2 = in-plane compressive stress at the top edge of the plate
τ = average shear stress over the plate edge
σ = in-plane uniform compressive stress
σ_b = in-plane pure bending stress
τ = average shear stress over the plate edge

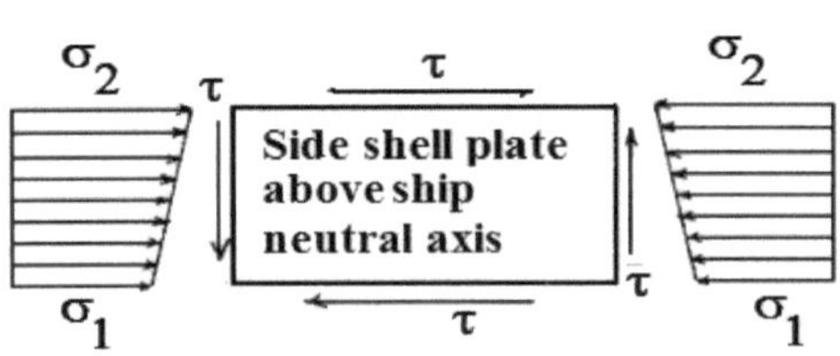

Fig. 12.56 Shear and bending stresses in a side shell plate above ship section neutral axis

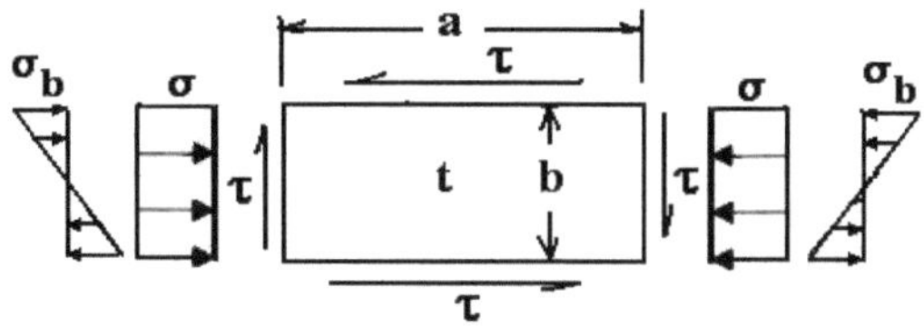

Fig. 12.57 The idealized stress system for a plate panel above neutral axis of ship section

The equations governing the critical buckling stress in this case are as given before for the plate panel at the neutral axis of the ship section. The same procedure could be used for the assessment of buckling of a plate panel below the neutral axis of the ship section.

Chapter 13
Control of Buckling Failure of Ship Structure

1 Introduction

This chapter presents the various measures commonly used to control buckling of ship structural members. The reliability basis of ship structural safety is introduced. The concept of safety margin for the buckling strength of a structural member is explained. The role of Classification societies in controlling failure of ship structural members is explained. The deleterious effects of deterioration of strength of ship structural members with time are clarified. The variation of strength and probability of failure with time is illustrated. The effect of corrosion on the flexural rigidity and buckling of ship strength members are explained. Linear and exponential equations are used for modeling the variation of the rate of corrosion with time.

Improved designs of ship structural details and connections to control buckling failure are presented. These improved designs include the use of symmetrical straight and curved face plates. Commonly used owners approach for improving ship safety is presented. The most common design and construction measures adopted to control welding distortions are given. Measures to control deformations, warping and lateral deviations of face plates, flanges and web plates of steel sections are highlighted. The importance of improving control on tolerances of ship structure members and quality of ship fabrication processes is stressed.

2 Reliability Basis of Ship Structural Safety

A major requirement for any marine structure is to have low initial and operational costs, to be reasonably safe, not to have catastrophic failure nor to have much trouble in service due to frequent minor failures. Safety is concerned not only with the structure itself, but also with external damage that may result as a consequence of failure. For any system, the benefits should exceed the penalties, or costs. The selection of a design for a ship, therefore, should be based on the achievement of maximum utility and minimum expected loss in case of failure. The optimality concept includes high reliability, minimum weight, minimum cost, high utilization, long service life, low risk to human life and to environment. The achievement of these desirable and perhaps conflicting features at the same time is rather difficult. This could be practically realized by the minimization of the expected loss associated with failure, while imposing certain limiting conditions on the utility, or benefits and on structural reliability. Therefore, the required safety level should be determined from economic considerations (minimization of cost of failure) for the assumed service life. The variation of the initial cost and the cost of failure with structural reliability is shown in Fig. (13.1). It should be realized that structural reliability reduces with time by virtue of deterioration of structure due to corrosion and damage accumulation.

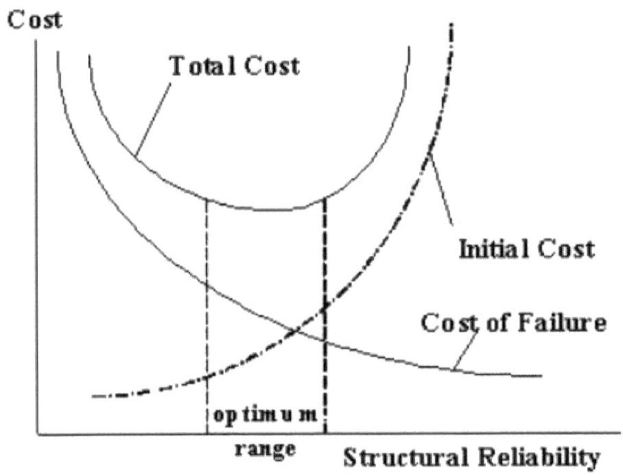

Fig. 13.1 Variation of initial cost and cost of failure with structural reliability

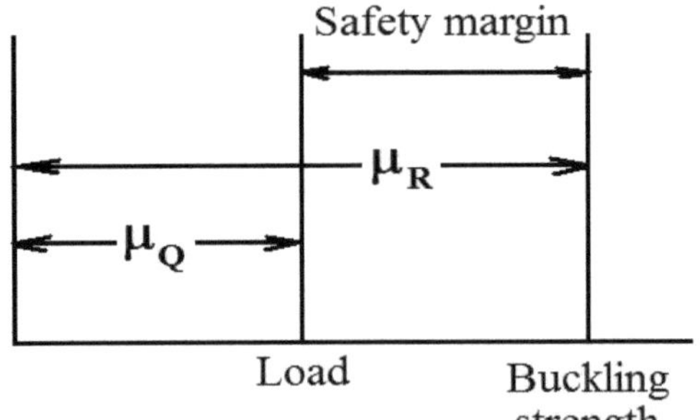

Fig. 13.2 The concept of the safety margin

The basic concept of safety is to ensure that the strength of the structure has adequate safety margin. The concept of safety margin for the buckling strength of a structural member is shown in Fig. (13.2).

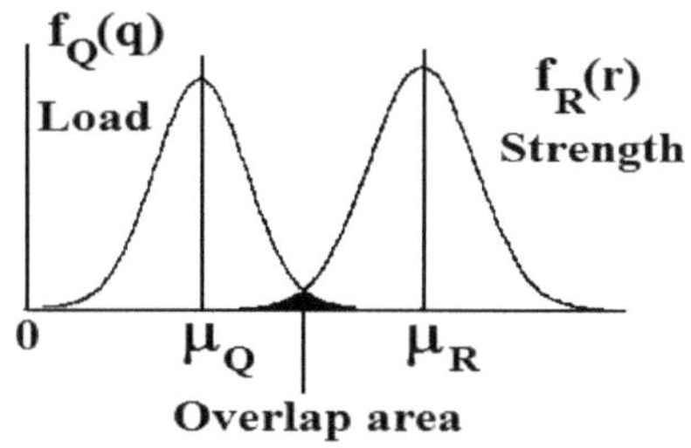

Fig. 13.3 Distribution functions of load and strength

The basic equation for calculating the probability of failure of a structural member is given by, see Fig. (13.3):

$$P_F = \int_{-\infty}^{\infty}\left[\int_{s}^{\infty} f_Q(q)dq\right]\cdot f_R(r)dr$$

Where: $f_Q(q)$ = probability density function (p.d.f) of the load

$f_R(r)$ = p.d.f of the strength

Ship structural failure should be interpreted with respect to some predefined limit state. Depending on the design limit state, the concept of a failure probability is applicable to both the performance and safety of ship structure. Since the probability of failure of ship structure depends mainly on the shape of the upper tail of the loading function and the lower tail of the resistance (strength) function, its magnitude will be very sensitive to the type and shape of the assumed functions, see fig. (13.3). The overlap area of the load and strength distribution functions indicates some measure of the probability of failure. It is important to recognize that regardless of the distributions of the individual variants, it is the magnitude of the mean and distribution of the safety margin M that is important in the calculation of probability of failure of the structure (P_F).

$$P_f = \int_{-\infty}^{m} f_M(m)dm$$

Where: M = safety margin

P_f = probability of failure

$f_M(m) = \textit{probability density function of } M$

3 Deterioration of Structural Capability with Time

Structural capability deteriorates with time as a result of inadequate maintenance, damage accumulation, material wastage due to corrosion and wear and tear. The variation with time of the statistical parameters of strength is shown in Fig.(13.4). The main characteristics of the distribution function of strength are the mean, the shape of the distribution function and the probability of failure. The size of the overlap area increases significantly with the deterioration of strength with time. The variation of strength and probability of failure with time is illustrated in Fig. (13.5).

Therefore, the ability of a structure to maintain its original level of structural reliability over its service life can be significantly improved by corrosion control, using materials having good corrosion resistance, provision of good drainage system, provision of access for cleaning, inspection and painting.

U_{R_0} = Mean value of strength at time 0 R_0 = Initial strength
U_{R_t} = Mean value of strength at time t R_t = Strength at time t
U_{R_T} = Mean value of strength at time T R_T = Strength at time T

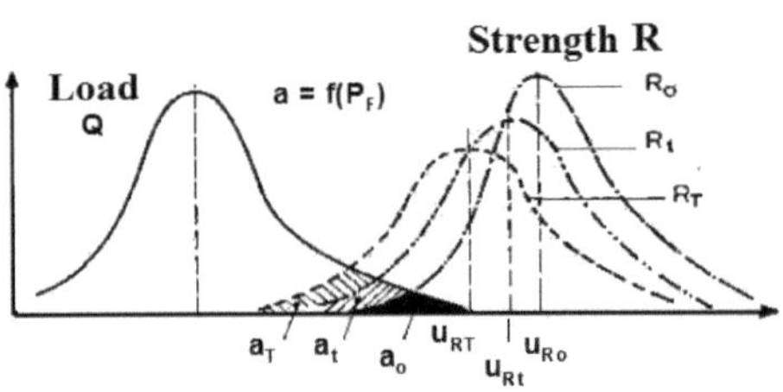

Fig. 13.4 Effect of mean and standard deviation of strength on the probability of failure

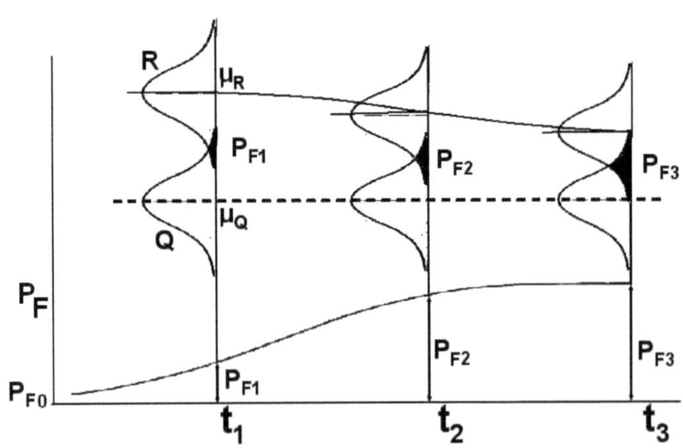

Fig. 13.5 Variation of structural strength and probability of failure with time

4 Responsible Authorities for Ensuring Structural Safety

Most of ship structural failures result from:

- Overloading
- Design errors
- Material errors
- Fabrication errors
- Excessive corrosion (lack of proper maintenance).

Classification societies play a major role in ensuring adequate safety of ship hull girder and local structural details, see Fig. (13.6). For conventional ships, classification societies rules are based on long experience and gradual development. They ensure an acceptable safety level of the structural performance. Structural reliability is based on data collected from ships in service, such as damage statistics. These statistical data are usually condensed and then treated deterministically for developing rules for the structural design of ships. Safety assurance in this deterministic approach is based on the concept of the "factor of safety". Because of the inconsistency and lack of uniformity of these safety factors, marine structures designed according to these rules are generally over designed. For unconventional marine structures, the extrapolation from available data may be extremely difficult. Therefore, a rational procedure is required for ensuring adequate structural safety and reliability.

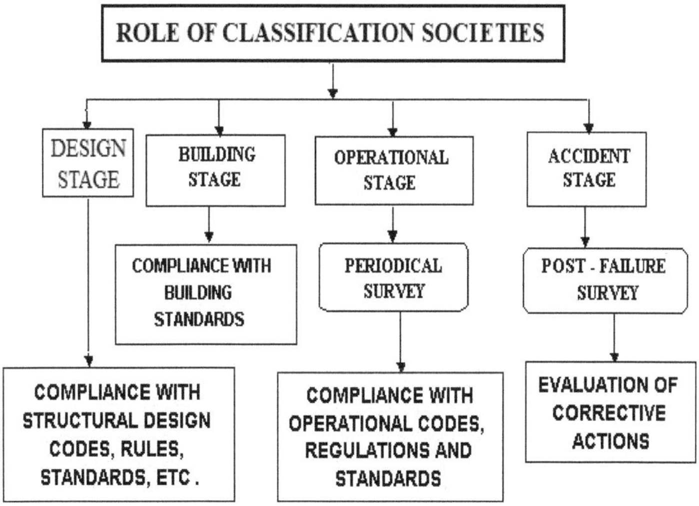

Fig. 13.6 Role of Classification Societies in controlling ship structure failures

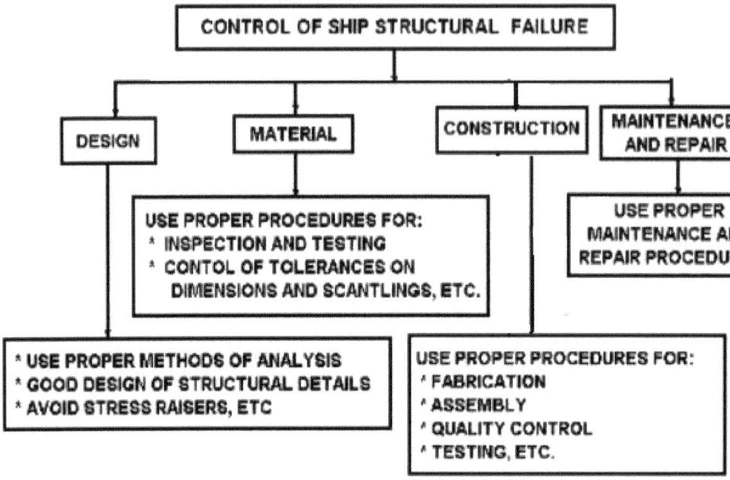

Fig. 13.7 Main elements affecting the control of ship structure failure

The causes of ship structural failures are controlled through the different stages of ship design, material selection, construction, operation, maintenance and repair. The methods commonly used for ensuring adequate structural safety are, see Fig. (13.7):

- Control of structural design by specifying procedures, safety margins and constraints.
- Control of material deterioration by the provision of corrosion margin to ensure adequate strength over the ship service life.
- Control of quality of material and construction.
- Control of quality of maintenance and repair by regular and special surveys.

5 Main Causes of Buckling Failure

The main causes of buckling failure are, see Fig. (13.8):

- Presence of high compressive stresses due to overloading or under design
- Presence of high shear stresses
- Presence of high compressive residual stresses
- Lack of adequate stiffening
- Lack of adequate flexural rigidity
- Excessive general/local material deterioration due to corrosion and/or wear and tear
- Presence of unacceptable fabrication defects, see Fig. (13.9)
- Extensive use of HTS

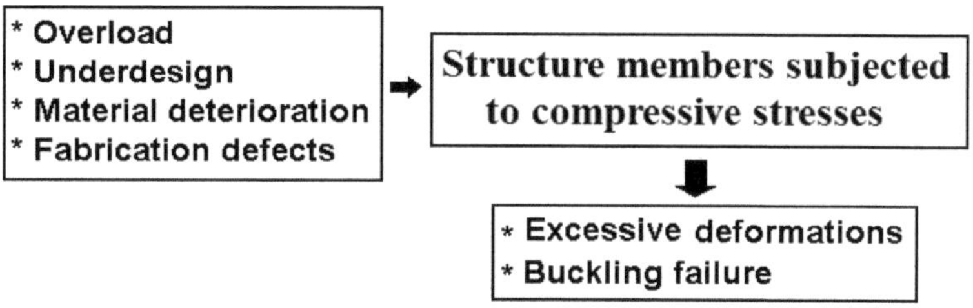

Fig. 13.8 Main causes and consequences of buckling failure

Fabrication defects are numerous and a typical example of welding distortions of stiffened panels is represented by the deformation "w_o" and the angle of slope "β" as shown in Fig. (13.9). These distortions have significant deleterious effects on the buckling strength of plate panels.

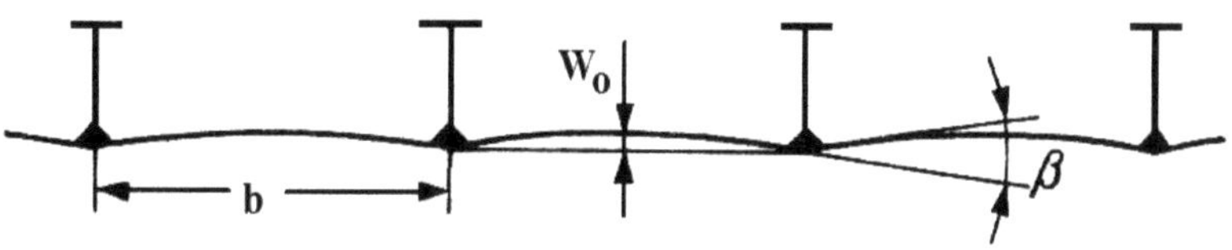

Fig. 13.9 Typical welding distortions of stiffened panels

6 Control of Ship Structure Failure by Improving Design

Most ship structure failures occur in the structural connections of the various strength members of ship structure. Improving design of these structure connections will have significant effects in preventing the occurrence of structure failures. Ship designers should pay much attention to the design of ship structural details, as poor design of these structural details is responsible for most of the local structural failures.

6.1 Improving Design of Plate and Tripping Brackets

Improving design of plate brackets could be achieved by improving the shape of the toes of the bracket, see Fig. (13.10). The improved design of the toes reduces the stress concentration factors and thus reduces the possibility of crack initiation and fatigue failure.

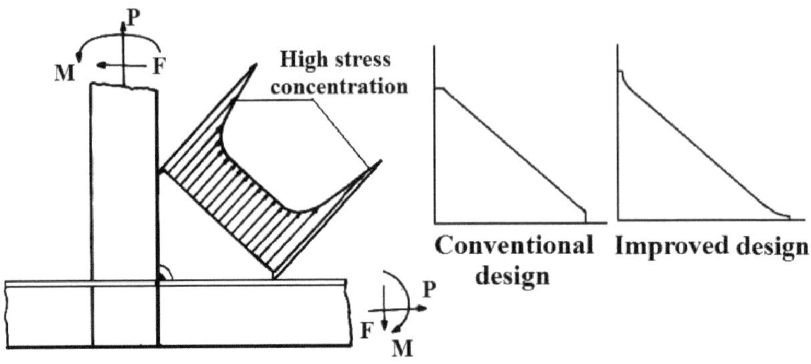

Fig. 13.10 Improving design of plate brackets

Improving design of tripping brackets could be achieved by improving the shape of the toe of the bracket connected to the plating, see Fig. (13.11).

The buckling of the free edge of plate brackets could be greatly improved by adding a stiffener along the free edge, see Fig. (13.12)

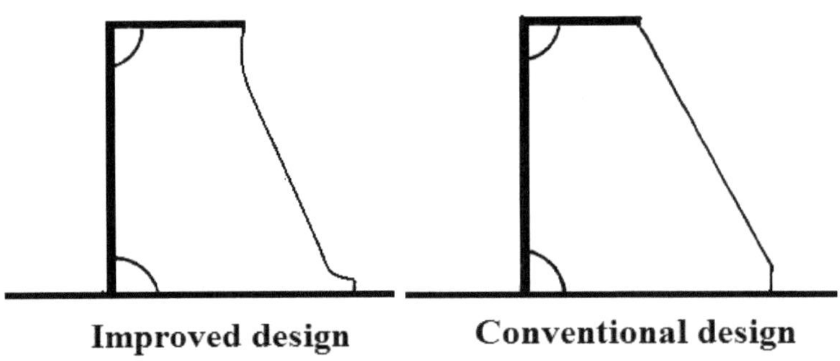

Fig. 13.11 Conventional and improved designs of tripping brackets

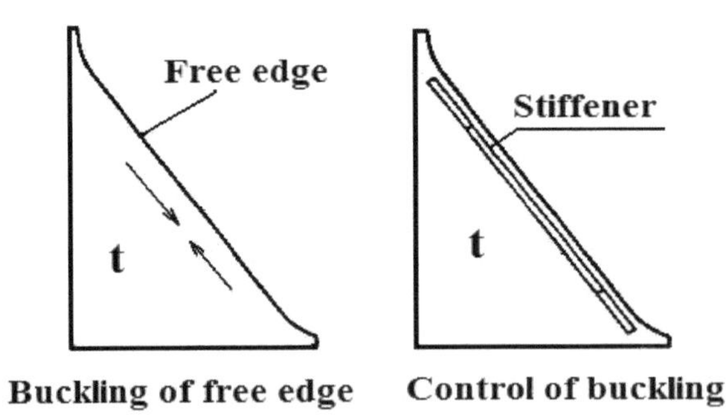

Fig. 13.12 Buckling control of the free edge of brackets

6.2 Using Symmetrical Face Plates of Girders

Beams and girders fitted with asymmetrical face plates are less efficient strength members to those fitted with symmetrical face plates because of the low effectiveness of these asymmetrical face plates. The ratio of width/thickness of the face plate is the main factor affecting the effectiveness of the face plates.

The control of buckling of symmetrical and asymmetrical face plates could be achieved using the following conditions, see Fig. (13.13):

$$\frac{b}{t_f} \leq 0.4\sqrt{\frac{E}{\sigma_y}}$$

The control of buckling of the web plate could be achieved using the following condition:

$$\frac{h}{t_w} \leq \left(3.35 - \sqrt{2(1+\psi)}\right)\sqrt{\frac{E}{\sigma_y}}$$

Where: E = Young's modulus

σ_y = yield stress of the material

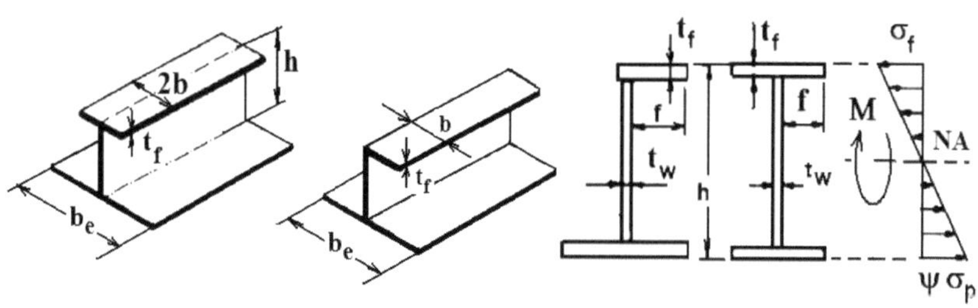

Fig. 13.13 Symmetrical and asymmetrical face plates

6.3 Improving Design of Curved Part of Web Frame Brackets

Strength of curved brackets with symmetrical face plates is superior to asymmetrical ones, see Fig. (13.14). Because of the low effectiveness of the curved face plates due to cross bending, it is recommended to use symmetrical face plate in the design of the curved brackets. The effectiveness of the curved asymmetrical face plate is less than 25%.

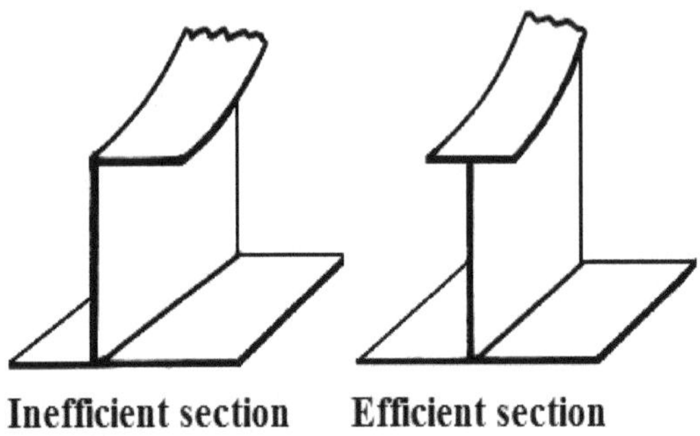

Fig. 13.14 Symmetrical and asymmetrical face plates of curved members

6.4 Improving Design of Plate Panels Loaded by Compressive Forces

The buckling strength of a wide plate could be much improved by fitting a stiffener at mid length in the same direction of the applied compressive loads, see Fig. (13.15).

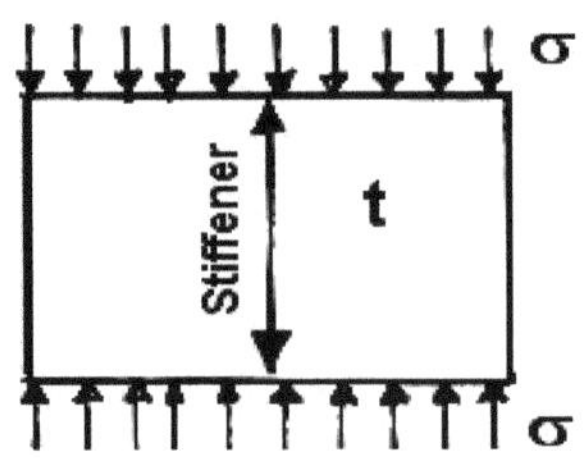

Fig. 13.15 Improving plate buckling by adding a stiffener in the direction of the compressive loads

6.5 Improving Design of Plate Panels Loaded by Shear Forces

A plate panel subjected to in-plane shear stresses over its edges will be distorted and the plate will buckle, see Fig. (13.16). The buckling strength of plate panels subjected to shear loading could be much improved by fitting a stiffener in the diagonal subjected to compressive forces, see Fig. (13.16).

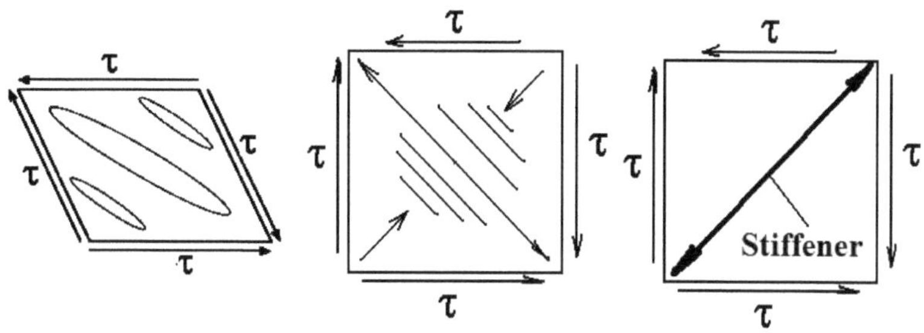

Fig. 13.16 Control of buckling failure of plate panels under shear loading

6.6 Improving Design of Local Structural Connections

The strength of the connection between longitudinals and web plating of transverses is much superior when using direct welding connection instead of fitting a lug, see Fig. (13.17).

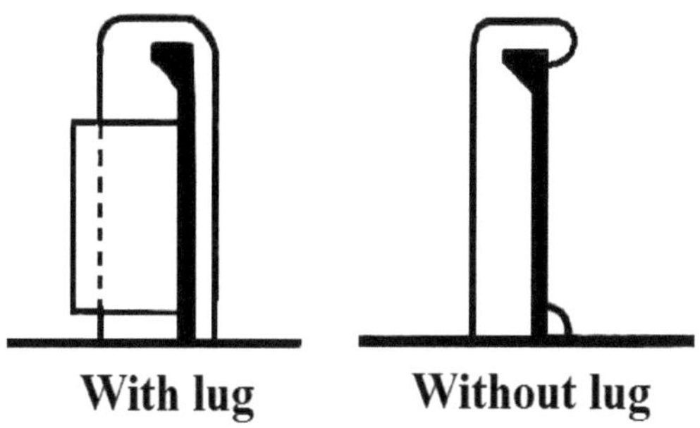

Fig. 13.17 Two different connections between longitudinals and web plating of transverses

6.7 Improving Design of the Connection between the Web Plate Stiffeners and Longitudinals

When the force to be transferred from the longitudinal to the heavy transverse member exceeds the carrying capacity of the lug, a vertical stiffener is fitted to connect the web plate with the face plate of the longitudinal. This connection is better designed by adding a bracket on the other side of the web plate to avoid eccentricity of the loading and the creation of stress concentration and initiation of cracks, see Fig.(13.18).

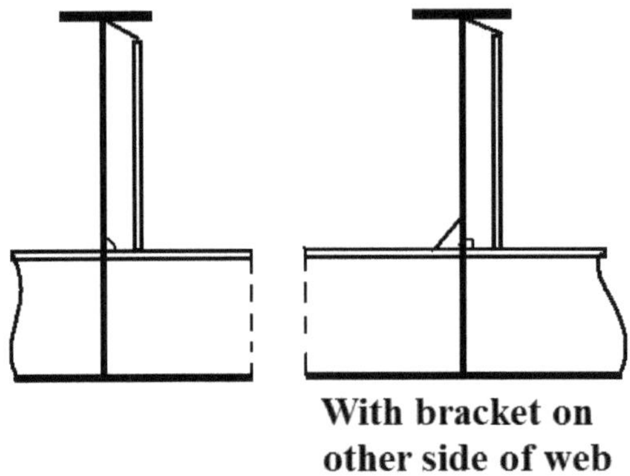

Fig. 13.18 Improved design of the connection of web plate with longitudinals

6.8 Improving Design of Web Plating of Top Wing Tanks

In bulk carriers, the presence of man holes and lighting holes reduces significantly the buckling strength of the web plating of top wing tanks. Fitting stiffeners around these man-holes significantly improve the buckling strength of these web plates, see Fig. (13.19).

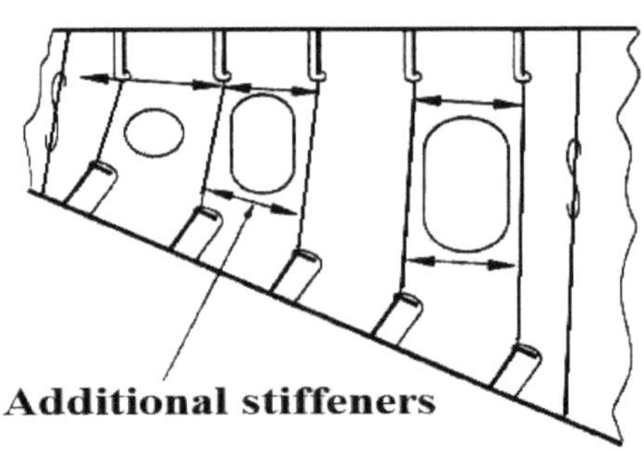

Fig. 13.19 Buckling of the web frames of the top wing tanks and correction measures

6.9 Improving Design of the Ends of Side Girders in Oil Tankers

Improving design of the ends of side girders in oil tankers by increasing the size and thickness of the end part of the web plating and adding diagonal stiffeners improves greatly the buckling strength of these structures, see Fig. (13.20).

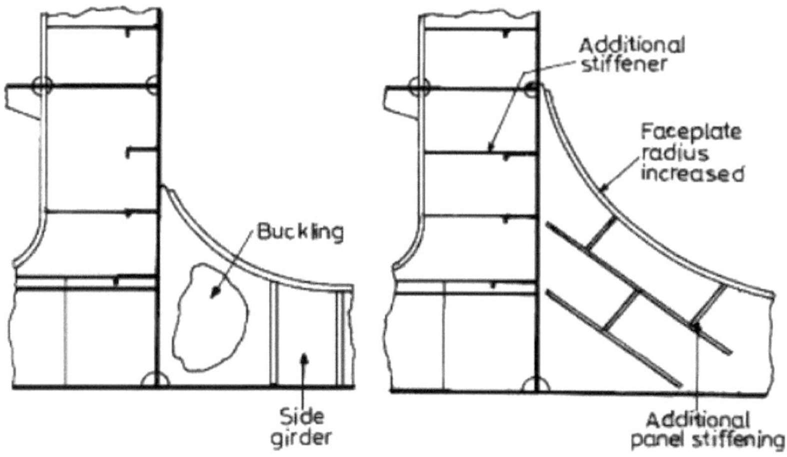

Fig. 13.20 Additional panel stiffening of the ends of side girders in oil tankers

6.10 Improving Design of Web Plating of Deep Girders

Improving design of deep girders subjected to lateral and shear loading could be achieved by fitting stiffeners to the web plating, see Fig. (13.21). These stiffeners increases the buckling strength of the web plate.

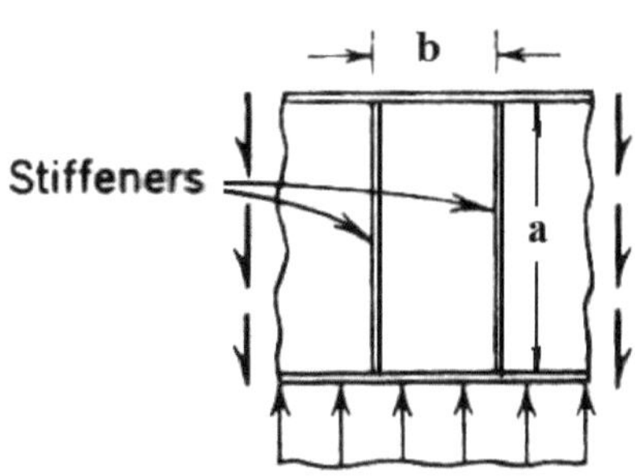

Fig. 13.21 Fitting stiffeners to the web plating of deep girders

7 Owners Approach for Improving Ship Structure Operational Life and Safety

Because of the reduction in thickness of shell plating with time due to corrosion, ship owners usually require an increase of thickness of shell plating by a 10% or 1.0 – 2.0 mm, see Fig. (13.22). The additional increase in thickness not only has significant positive effect on strength and flexural rigidity of ship structural members but it also has significant effect on ship structure safety and the economy of operational life, see Fig. (13.23).

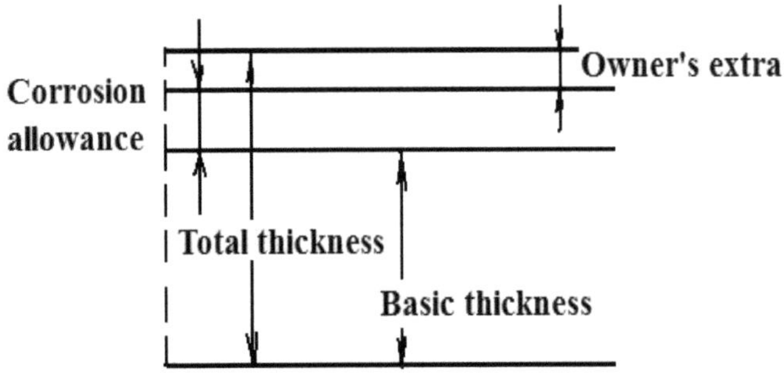

Fig. 13.22 Basic elements of shell plating thickness

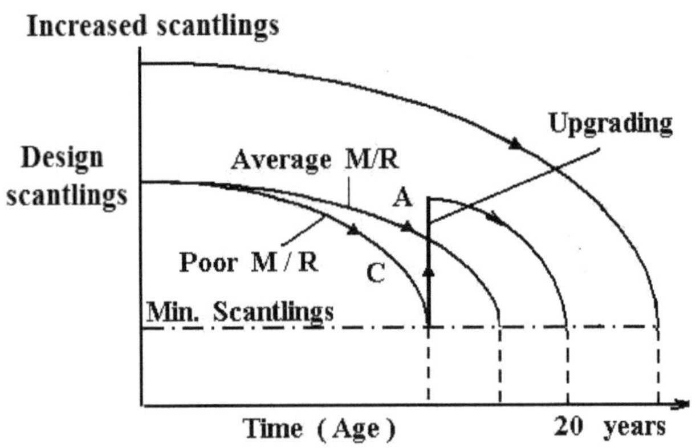

Fig. 13.23 Effect of increased scantlings on ship operational life

7.1 Impact of Corrosion on Strength of Ship Structure Members

The impact of corrosion on strength of ship structure members could be summarized as follows:

- Reduction of load carrying capacity of ship strength members because of the reduction of local and hull girder scantlings
- Increase of hull girder and local stresses
- Reduction of hull girder and local flexural rigidity
- Reduction of buckling strength
- Reduction of fatigue strength

The consequences of corrosion of ship steel plates and sections due to inadequate corrosion protection or defective coatings is the deterioration of flexural strength and the likelihood of buckling failure of these strength members, see Fig. (13.24).

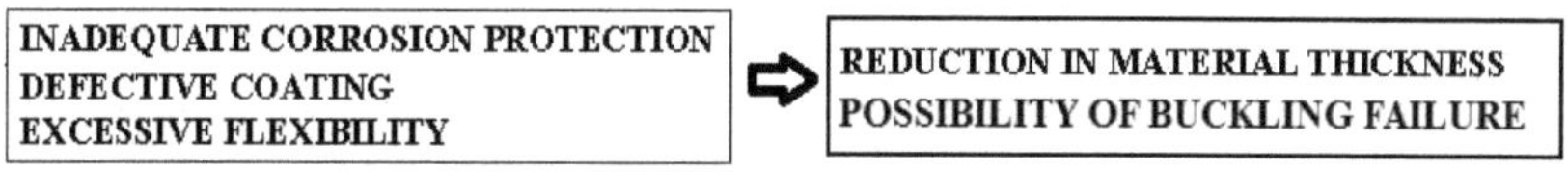

Fig. 13.24 Consequences of corrosion of ship steel plates and sections

The effect of corrosion of deck plating on the section modulus at deck and on the ultimate bending moment of ship section is shown in Fig.(13.25).

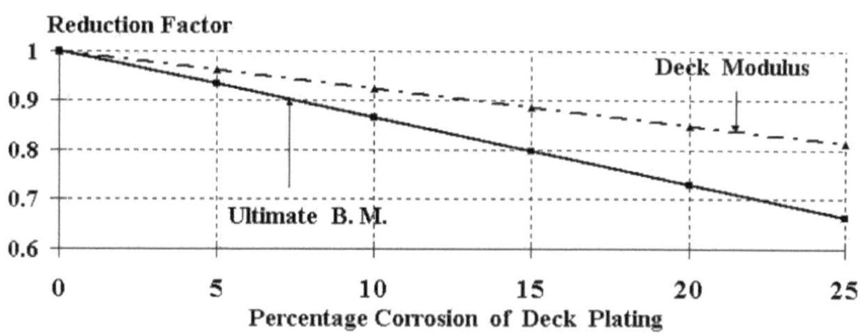

Fig. 13.25 Effect of corrosion of deck structure on deck section modulus and ultimate bending moment

The effect of corrosion on the flexural rigidity of a panel of plating for different plate thicknesses is shown in Fig.(13.26).

The effect of material wastage due to corrosion on the magnitude of the section modulus and inertia of an angle section with attached plating is shown in Fig.(13.27) and on an OBP section is shown in Fig. (13.28).

In order to determine the effect of material deterioration with time due to corrosion on the buckling strength of ship plating, an exponential model is assumed to represent the rate of corrosion of plate thickness with time as follows:

$$w_t = w_o.e^{-at}$$

Where: w_o = plate thickness before corrosion
w_t = plate thickness after corrosion
t = time
a = a constant defining the r4ate of corrosion

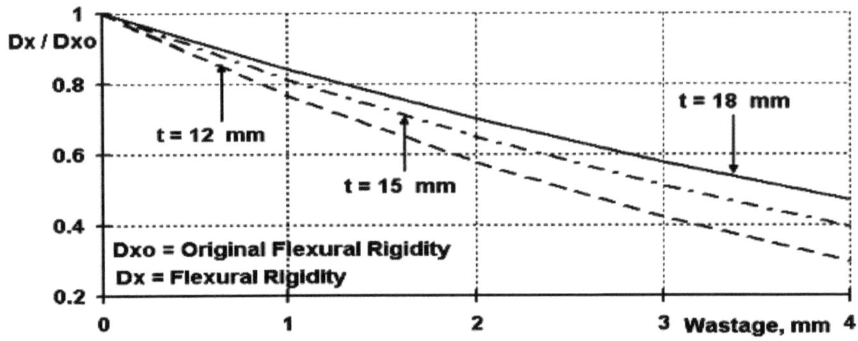

Fig. 13.26 Effect of wastage on plate flexural rigidity

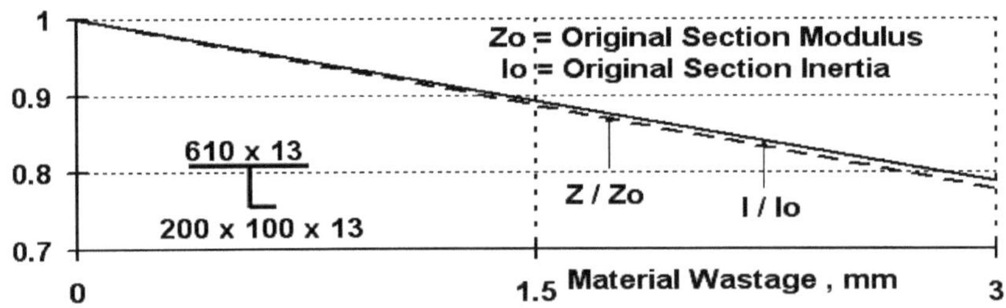

Fig. 13.27 Effect of material wastage on Z/Zo and I/Io of an angle section with attached plating

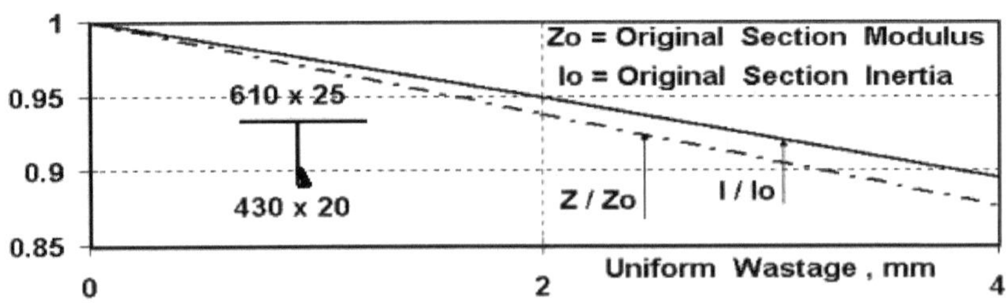

Fig. 13.28 Effect of material wastage on Z/Zo and I/Io of an OBP section with attached plating

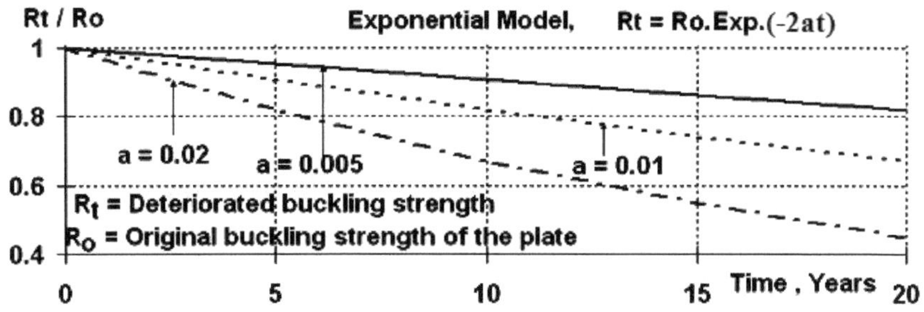

Fig. 13.29 Effect on plate buckling strength of material wastage (exponential model)

The effect on plate buckling strength of material wastage using an exponential model is shown in Fig. (13.29).

The variation of plate thickness with time could be also represented by the following parabolic model:

$$w_t = w_o.\left(1 - bt^2\right)$$

Where: b = a constant defining the rate of corrosion

The effect on plate buckling strength using the assumed parabolic model is shown in Fig. (13.30):

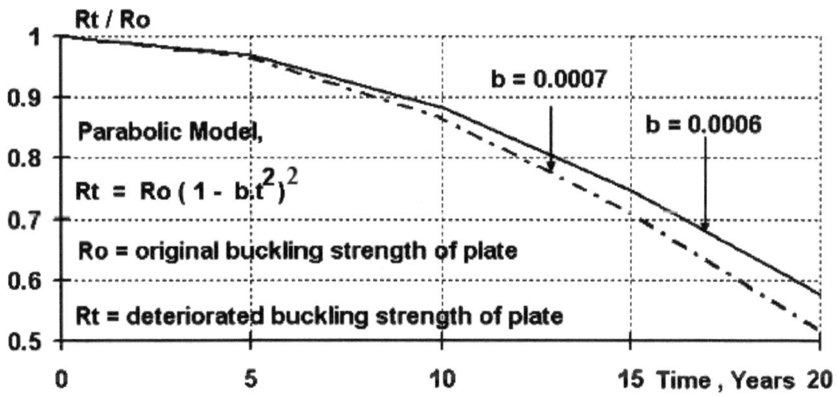

Fig. 13.30 Effect on plate buckling strength of material wastage (Parabolic Model)

8 Control of ship Structure Failure by Improving Quality of Ship Fabrication Processes

Ship structure members are subjected during their fabrication processes (cutting, welding and assembly) to a variety of defects and deviations from required geometry, shape and scantlings. The buckling strength depends on the scantlings and geometry of the structural member as well as the size and shape of the geometric imperfections in these members. When buckling is the governing mode of failure, the buckling strength is evaluated using the Euler buckling equations. These buckling equations are only applicable if the tolerances on the scantlings and geometry of the strength members are within the permissible limits.

8.1 Control of the Out-of Straightness of Stiffeners/Girders and Plate Panels

Controlling the out-of straightness of stiffeners or girders relative to the plate plane should have significant effect in improving the buckling strength of these members, see Fig. (13.31). Similarly, controlling the out-of-plane displacement of plate panels between stiffeners should improve the buckling strength of the plate panel, see Fig. (13.32).

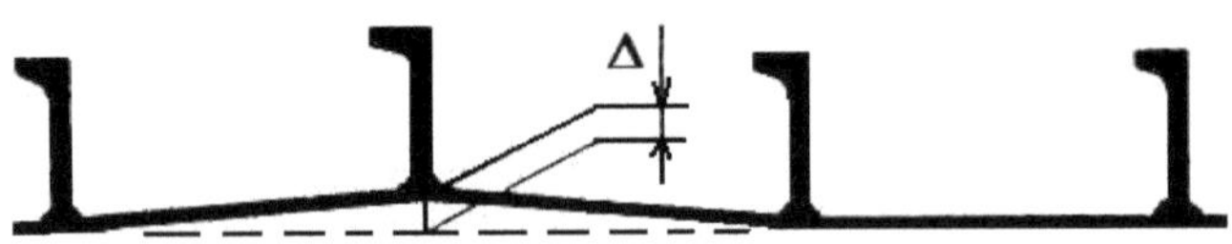

Fig. 13.31 Out-of straightness of stiffeners relative to the plate plane

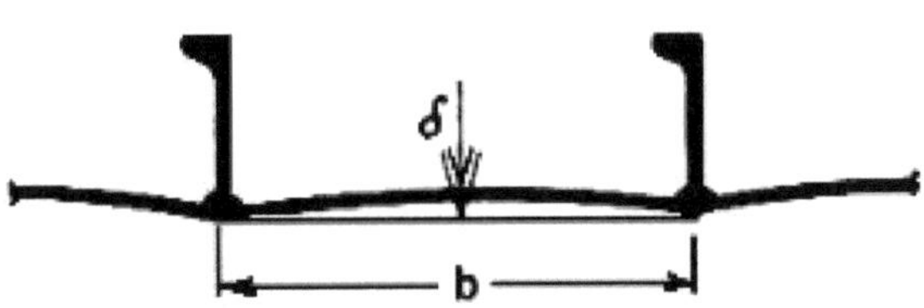

Fig. 13.32 Out-of-plane displacement of plating between stiffeners

8.2 Control of Fabrication Deformations of Ship Structure Members

Control of deformations, warping and lateral deviations of face plates, flanges and web plates of steel sections should have significant effects on the buckling strength of these sections. The main deformations occurring in the fabrication of ship structural members are:

8.2.1 Out-of-Straightness

The out-of-straightness of the stiffener or girder flange relative to the web plane is shown in Fig. (13.33).

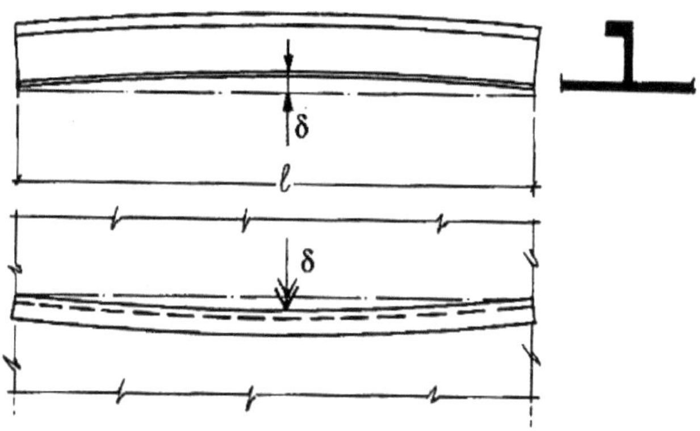

Fig. 13.33 Out of straightness of stiffeners and girders

The out-of-straightness of compression members is shown in Fig. (13.34)

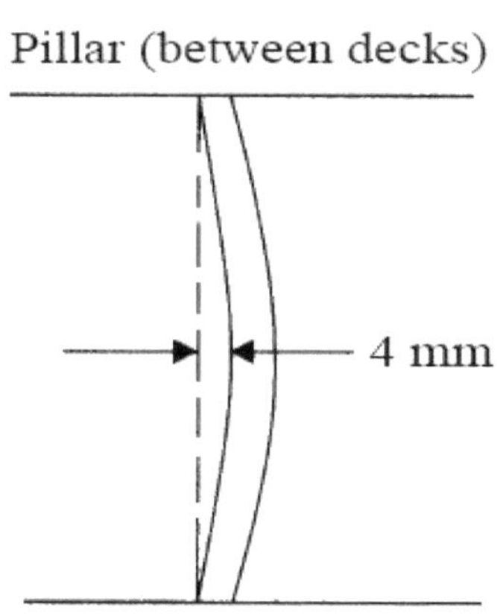

Fig. 13.34 Out-of-straightness of a compression member

8.2.2 Warping of the Whole Section of the Strength Member

The warping of an angle section is shown in Fig. (13.35).

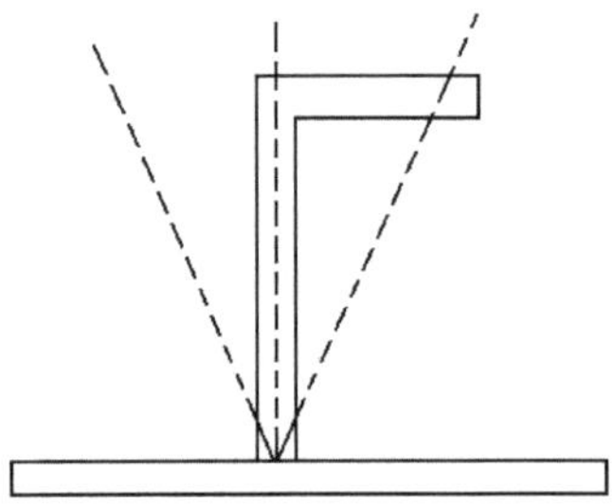

Fig. 13.35 Warping of the whole section

8.2.3 Warping of Face Plate

The warping of face plate of an angle section is shown in Fig. (13.36)

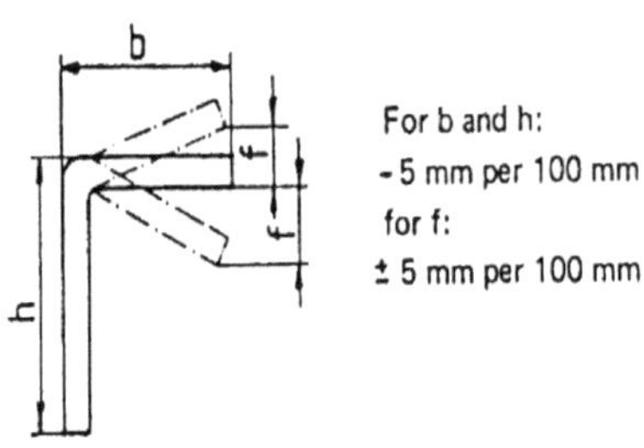

Fig. 13.36 Warping of face plates, flanges of steel sections

8.2.4 Lateral Deviations between Centerline of the Web and Centerline of the Flange

The lateral deviation of the face plate with respect to the centerline of the web plate is shown in Fig. (13.37).

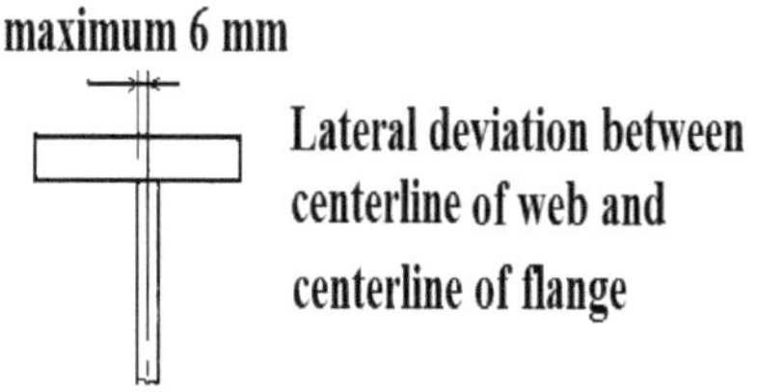

Fig. 13.37 Lateral deviations between centerline of the web and centerline of the flange

8.2.5 Inclination of the Web Plate of Section with Respect to the Attached Plating

The inclination of the web plate of an OBP section and a T-section with respect to the attached plating is shown in Fig. (13.38).

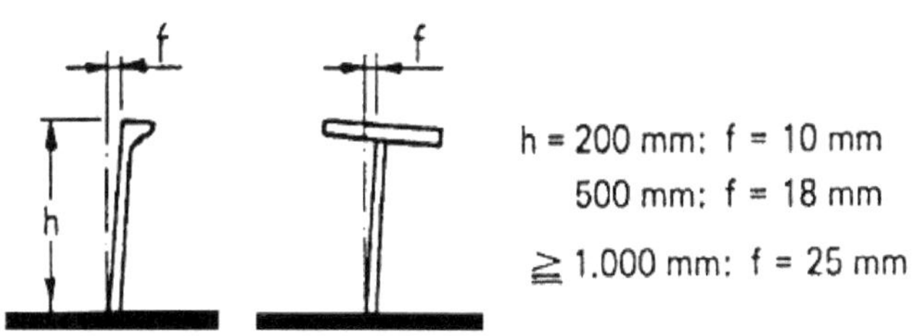

Fig. 13.38 Inclination of the web plate of section with respect to the attached plating

8.2.6 Deformations and Deviations of Face Plates or Flanges

Typical forms of deformations and deviations of face plates or flanges are shown in Fig. (13.39).

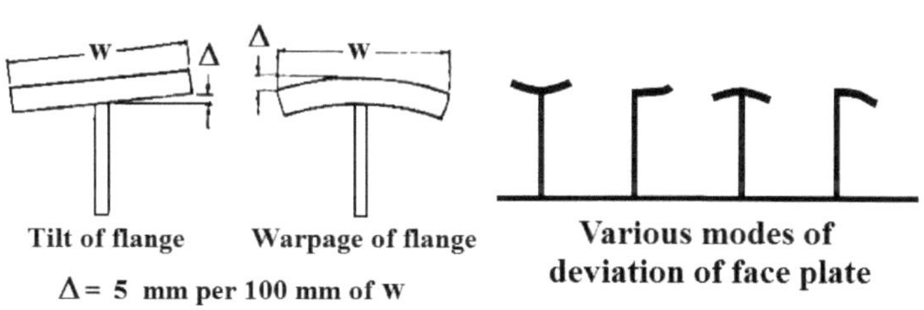

Fig. 13.39 Deformations and deviations of face plates or flanges

8.2.7 Gap Between Beam and Frame

The beam and frame gap resulting from fabrication and assembly errors should be controlled and eliminated so as to ensure adequate strength of these connections, see Fig. (13.40). The increase in this gap weakens the strength of this connection.

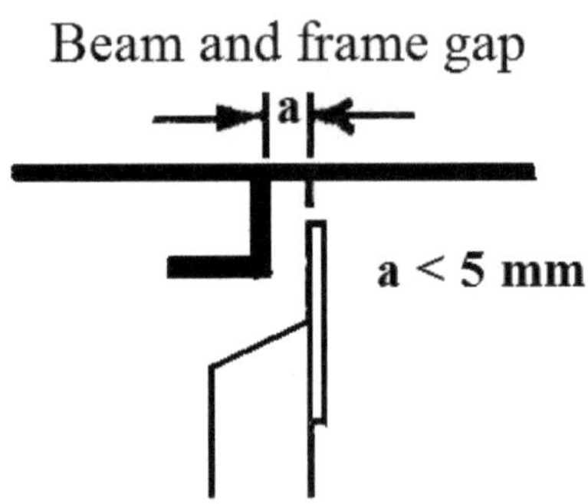

Fig. 13.40 Gap in the connection between Beam and frame

8.3 Control of Welding Distortions

Welding distortions has deleterious effects on the buckling strength of plating and stiffeners. Shipyards should take all necessary measures to prevent/control these distortions. Welding distortions can be considerably reduced by using the following procedures:

8.3.1 Avoiding Over-Welding

Over-welding causes greater shrinkages and distortions and therefore should be totally avoided. Correctly sizing a weld not only minimizes distortion, but also saves weld metal and time.

8.3.2 Placing Welds near the Neutral Axis or the Center of the Part

Welding Distortion is reduced by providing less leverage for the shrinkage forces to pull the plates out of alignment.

8.3.3 Balancing Welds Around the Neutral Axis

Welding on both sides of the plate offsets one shrinkage force with another. Balanced welding on both sides of the plate should minimize distortion.

8.3.4 Control of Alignment of Butt and Fillet Welds

Buckling strength of welded members can be significantly reduced due to the presence of excessive errors in the alignment of butt and fillet welds beyond the permissible tolerances. Assembly work of ship structure members can be considerably improved by controlling the fabrication errors particularly the alignment of butt and fillet welds, see Fig. (13.41). Fabrication errors are controlled by adhering to the permissible fabrication allowances.

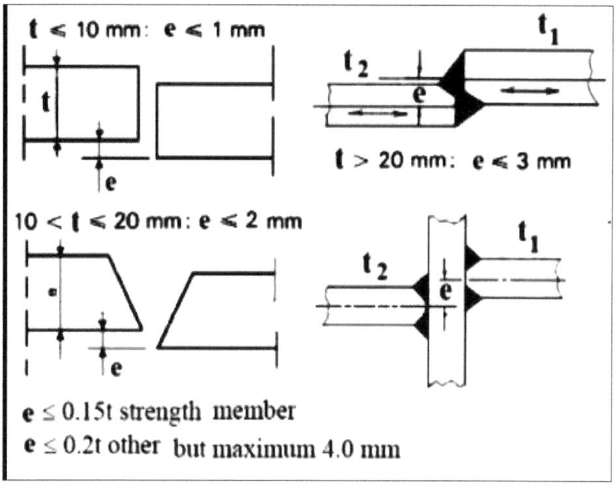

Fig. 13.41 Allowances for alignment of butt and fillet welds

9 Improving Control of Corrosion

Corrosion of steel plates and sections causes deterioration of scantlings and strength of ship structural members and as such affect ship structural safety. The rate of deterioration is reduced significantly when using effective methods of corrosion control, see Fig. (13.42).The flexural rigidity of steel plates depends on the plate thickness. Any reduction to the plate thickness due to corrosion will cause significant reduction in the flexural rigidity and buckling strength of the plate panel.

Reducing the rate of corrosion of steel ships could be achieved by:

- Using high quality surface preparation and protective coatings
- Providing ease of access of the various parts of cargo holds and ballast tanks
- Using proper hull maintenance strategies see Fig. (13.42).

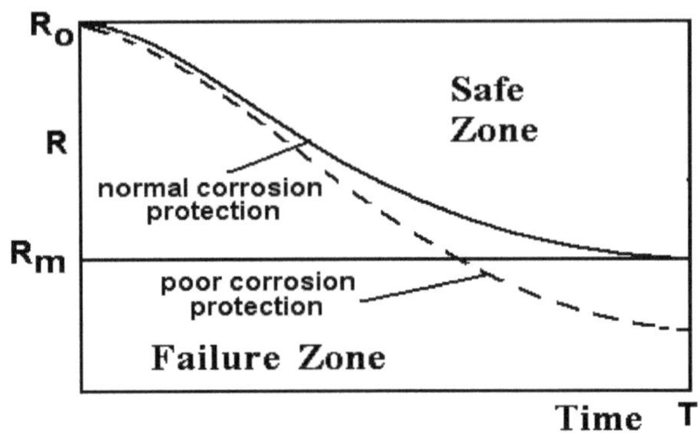

Fig. 13.42 Variation of buckling strength with time with and without corrosion control

Reducing rate of corrosion of ship strength members should have measurable influence on the factor of safety of these strength members, see Fig. (13.43)

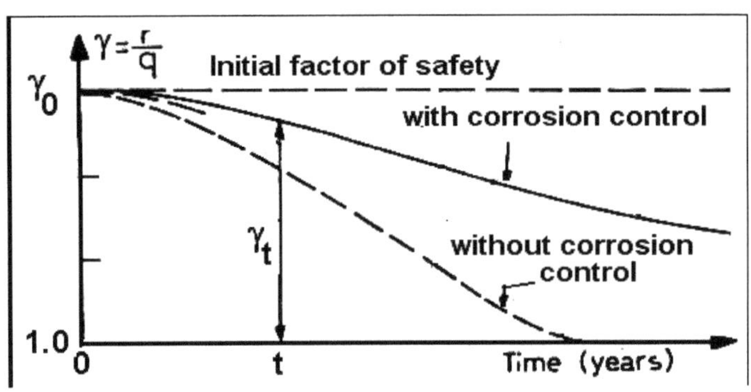

Fig. 13.43 Variation of the factor of safety with time and quality of maintenance

Chapter 14
Problems

1-The steel girder shown in Fig. 14.1 has the following particulars:

The yield stress of steel is 2.4 t/cm2, Flange section area a_f = 30 cm2, The attached plate section area a_p= 60 cm2,
The web plate of the section is 44 x 10 cm.
The factor of safety against yielding of the flange: γ_f = 1.5
Make all necessary assumptions to determine:

a. The factor of safety against buckling of the face plate
b. The factor of safety against buckling of the web plate
c. The bending stress in the attached plating
d. The maximum shear stress in the web

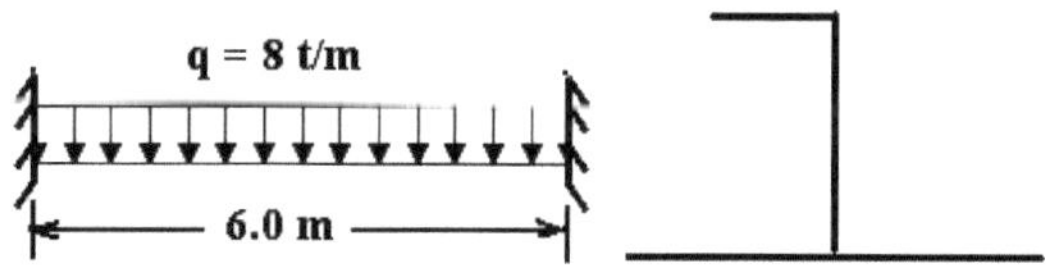

Fig. 14.1

2- The deck structure and midship section of a general cargo ship is shown in Fig. 14.2. The ship has the following particulars: L = 118 m, B = 16.8 m, D = 8.0 m, T = 5.2 m, C_b = 0.7, V = 15 knots, hold length = 18.0 m, Ms = 10000 tm sagging, Mw = 15000 tm. Side frame spacing = 650 mm, Longitudinal spacing = 700 mm. Hatch opening = 12 x 12.6 m. The deck loading = 2.5 t/m2. Cargo loading on tank top = 4.0 t/m2, Deck transverses T 280 x 8 + 120 x 10, Deck center girder T 280 x 8 + 120 x 10

Make all necessary assumptions to calculate:

1-Tertiary loading and stresses of a deck longitudinal
2-The secondary stress in the flange of the bottom center girder at its end at the transverse bulkhead
3-The hull girder stress at the deck
4-the factor of safety against buckling of deck plating
5- The factor of safety against buckling of the tank top plating

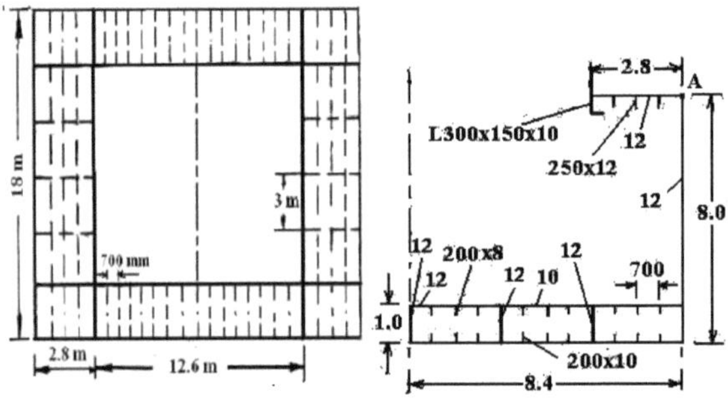

Fig. 14.2

3- For the grillage structure shown in Fig. 14.3, calculate the bending moments at points K and F.

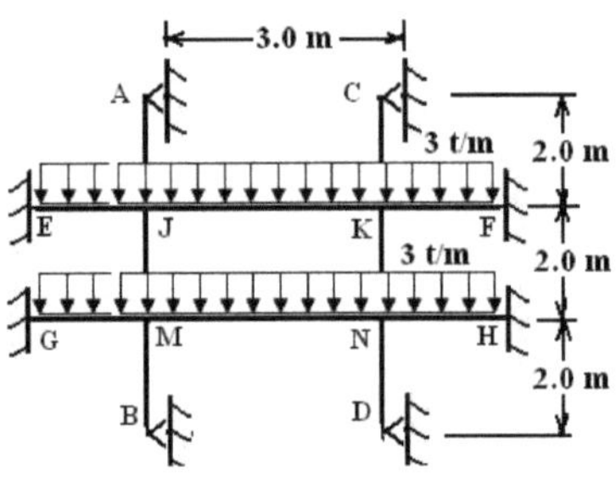

Fig. 14.3

4 - For the girder shown in Fig. 14.4 the external pressure on the bottom plating = 2.0 t/m^2. The breadth of plating subjected to the external pressure = 2000 mm. Calculate:

a. The effective breadth of the face plate
b. The effective section modulus
c. The idealized structure as a beam fixed at both ends
d. The maximum stress in the face plate of the section

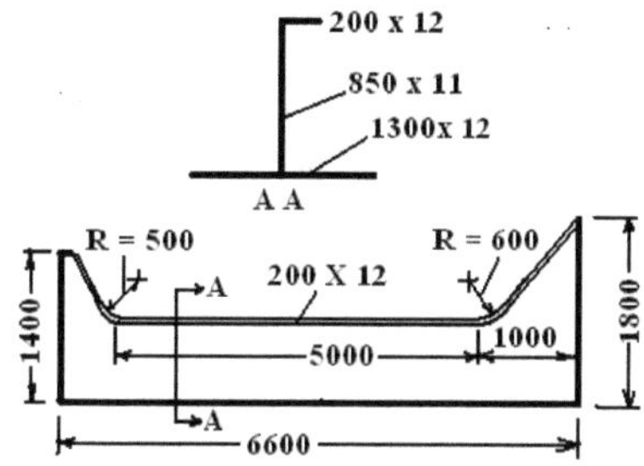

Fig.14.4

5 - The deck structure of a general cargo ship is shown in Fig. 14.5
Longitudinal spacing = 700 mm. Deck loading = 2.5 t/m2, , deck long. L 200 x 120 x 8 mm, deck transverses T 280 x 8 + 120 x 10, hatch girders L 300 x 10 + 120 x 10, deck center girder T 400 x 10 + 150 x 12, deck plate thickness = 10 mm, Make all necessary assumptions to calculate:

1- Loading on a deck longitudinal
2- Loading on the hatch end beam
3- Bending stresses in the flange of a deck longitudinal
4- Bending stresses in the flange of the hatch girder
5- Tertiary stress in a panel of deck plating

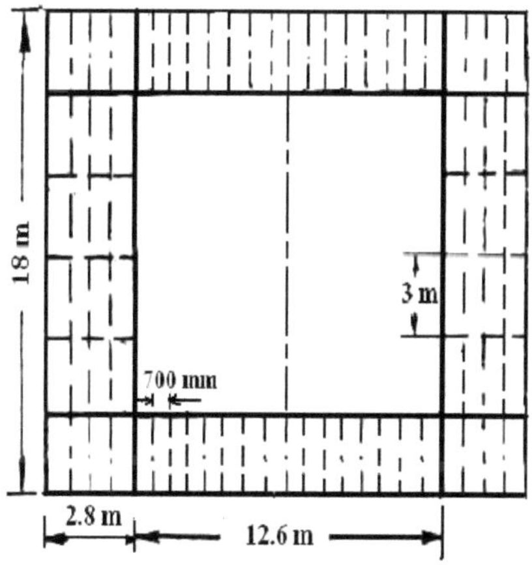

Fig. 14.5

6- A longitudinally stiffened oil tanker has the following particulars, see Fig. 14.6: LBP = 300.0 m, B = 48.0 m, D = 22.5 m, T= 19.0 m, Transverse web Spacing = 3.0 m Deck and bottom longitudinal spacing = 1000 mm Side shell and longitudinal bulkhead spacing = 900 mm Deck longitudinals = FB 600 × 20 mm Bottom longitudinals = T 1900 × 20 +400 × 25 Side shell longitudinals = L 600× 200× 15 Bulkhead longitudinals = L 600× 200× 15 Yield stress $\sigma_y = 2.4$ t/cm^2.
Make all necessary assumptions to determine:

a. The idealized ship section as an I. Beam
b. The flexural characteristics of the ship section
c. The magnitude of the local bending stress of bottom plating
d. The factor of safety of the bottom plating against buckling when the ship is subjected to M_s = 150,000 t.m sagging and M_w = 180,000 t.m,

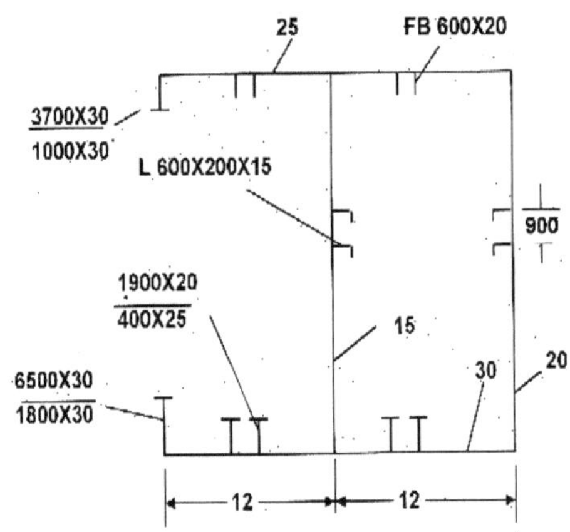

Fig. 14.6

7-The following data are given for the deck structure shown in Fig. 14.7: L = 16 m, B = 9.8 m, hatch opening = 4.2 x 8 m, deck loading = 2.0 t/m^2, longitudinal spacing = 700 mm, web frame spacing = 2.0 m, deck plate thickness = 10 mm,
Calculate:

1- Loading on a deck longitudinal
2- Loading on the hatch girder
3- Local bending moment on a deck longitudinal
4- Tertiary stress in deck plating

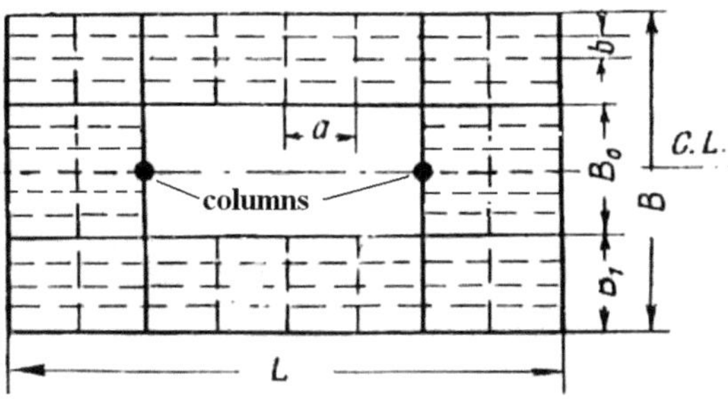

Fig. 14.7

8-The midship section of the general cargo ship shown in Fig. 14.8 has the following particulars: L = 128 m, B = 18 m, D = 9.2 m, T = 7.2 m, C_b = 0.7, V = 15 knots, hold length = 18 m, beam spacing b = 600 mm, hatch opening = 12 x 12 m The deck loading p = 2.5 t/m2. Deck beams L 200 x 120 x 8 mm The Stillwater bending moment Ms = 30000 tm The wave induced bending moment Mw= 50000 tm

Make all necessary assumptions to calculate:

a – The primary stress in the tank top plating
d- The hull girder stress in the hatch girder of the second deck
e- The local stress in the bottom plating

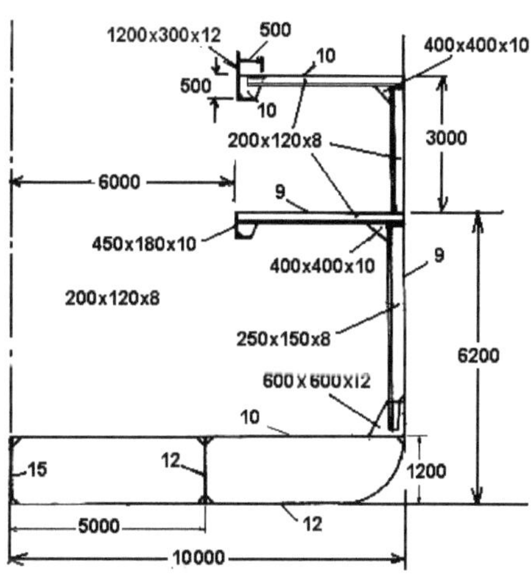

Fig. 14.8

9-The general cargo ship shown in Fig. 14.9 has the following data: Ship length LBP=132 m, B = 22.4 m, D = 13.4 m, T = 9.0 m, hold length = 21m, hatch opening length = 15 m, Floor spacing = 3.0 m, Frame spacing = 700 mm, longitudinal spacing = 700 mm, spacing of upper deck heavy beams = 3.0 m, max. upper deck loading = 2.8 t/m^2, max. second deck loading = 2.5 t/m^2, max. tank top loading =

8.0 t/m^2.upper deck transverses: T 300 ×10+150×12, 2nd deck heavy beams T 300 ×10+150×12, upper web frame: L 280 ×140 ×10 mm, lower web frame: L 320 ×160 ×12 mm. The ship is subjected to a sagging stillwater bending moment, M_S = 18000 tm, a wave BM, M_w = 32.000 t.m. Make all necessary assumptions to determine:

1- The design load for the following structural members:

a. A deck longitudinal,

b. the 2nd deck hatch girder

2- The magnitude of the local bending stresses of:

a. A deck longitudinal

b. The 2nd deck girder

c. The local bending stresses of bottom plating.

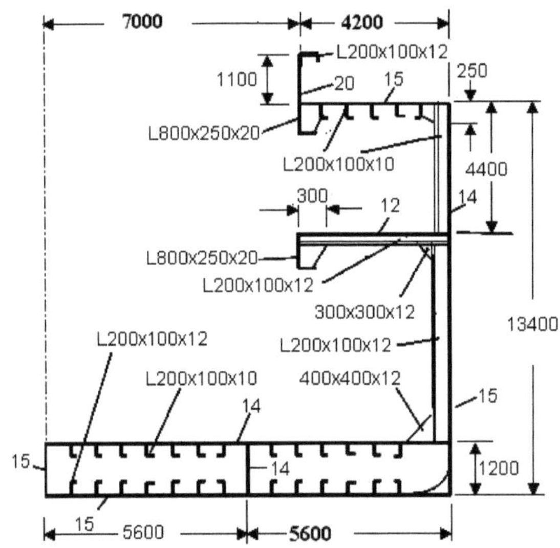

Fig. 14.9

10- The coastal tanker shown in Fig. 14.10 has the following particulars: L = 60 m, B = 13 m, D = 11 m, draft T = 9 m, tank length = 12 m, longitudinal spacing = 650 mm, spacing of transverses = 3 m. The ship section is subjected to a vertical hogging bending moment M = 5000 tm

Make all necessary assumptions to calculate:

a- The tertiary stresses in a bottom longitudinal

b- The local stresses in the bottom plating

c- The buckling strength of a plate panel in the side shell

d- The factor of safety of bottom plating against buckling

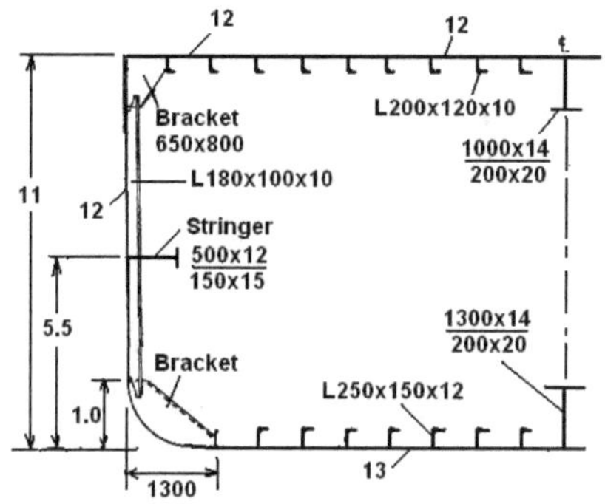

Fig. 14.10

11-The idealized barge section shown in Fig.14.11 has the following particulars: L = 60 m, B = 12 m, D = 6 m, a = 2 m, t_B = 10 mm, t_D = 12 mm, t_S = 10 mm. The ship section is subjected to a hull girder bending stress at the deck = 0.8 t/cm2 and a vertical shear force F =50 ton

Make all necessary assumptions to calculate:

a - The shear stresses at B, C, D

b - The equivalent stresses at B and D

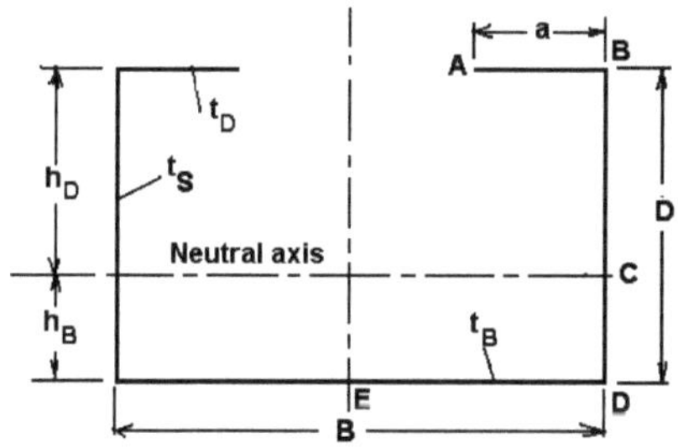

Fig.14.11

12- The deck structure of a general cargo ship is shown in Fig.14.12 the longitudinal spacing = 700 mm. The deck loading = 2.5 t/m2, deck long. L 200 x 120 x 8 mm, deck transverses T 280 x 8 + 120 x 10, hatch girders L 300 x 8 + 120 x 10, deck center girder T 400 x 10 + 150 x 12, deck plate thickness = 10 mm,

Calculate:

1- Loading on a deck longitudinal

2- Loading on the hatch end beam

3- The bending stresses in the flange of a deck long.
4- Local stresses in a panel of deck plating
5- The maximum allowable hull girder stress in the deck plating when the factor of safety against buckling of deck plating is 1.8

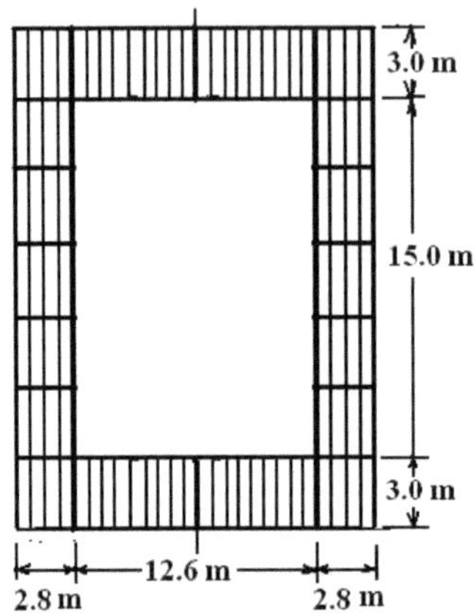

Fig. 14.12

13-The main ship section particulars of the RO/RO ship shown in Fig.14.13 is given by: L = 128m, B = 22 m, D = 12 m, D_2 = 7 m, h = 1.2 m, T = 6 m. The upper deck load p_{UD} = 2.5 ton/m2. The second deck loading p_{SD} = 3 ton/m2. Length of deck girder = 12 m. Spacing of upper and second deck transverses = 3 m. Spacing of side frames = 700 mm. Spacing of longitudinals in the upper and second deck = 650 mm. The upper deck longitudinals are flat bars 250x12 mm. Bottom plating thickness =12 mm. Calculate;

a- The load on the side frame of the side tank, the vent pipe of the side tank is at 2.5 m. above the main deck. The side stringer is located at mid-depth of the tank.
b- The load on the deck girder
c- The load on the second deck longitudinals.
d- The bending stresses of the upper deck longitudinals
e- Local stresses in the bottom plating

14-For the hatch cover shown in Fig.14.14 calculate the bending stresses at the plating and at the flange of the stiffeners when the deck loading over the hatch cover plating p = 2.5 t/m^2.

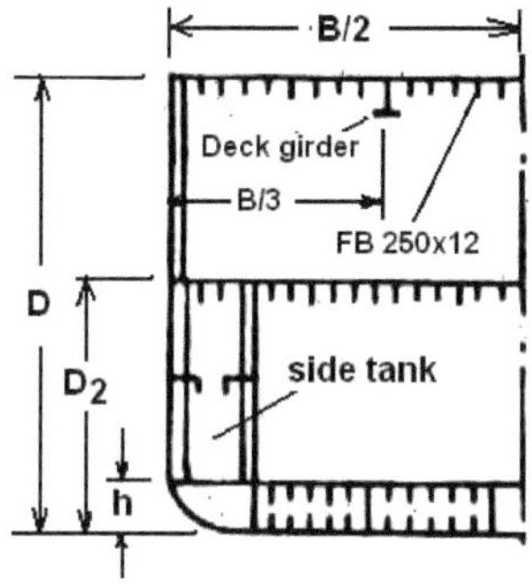

Fig.14.13

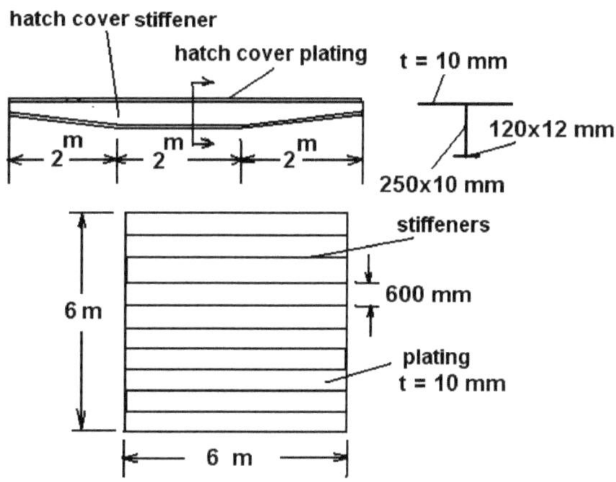

Fig.14.14

15- The coastal tanker shown in Fig.14.15 has the following particulars: L = 60 m, B = 13 m, D = 11 m, draft T = 9 m, tank length = 12 m, longitudinal spacing = 650 mm, spacing of transverses = 3 m. The ship section is subjected to a longitudinal vertical sagging bending moment = 6000 tm.

Make all necessary assumptions to calculate:

a. The local stresses in the bottom plating
b. Tertiary stresses in bottom longitudinals
c. The tertiary stresses in a bottom longitudinal
d. The factor of safety against buckling of deck plating

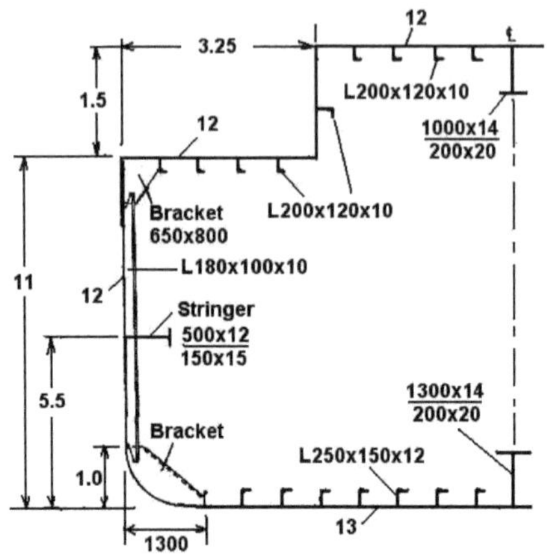

Fig.14.15

16-For the bottom transverse structure shown in Fig.14.16 the transverse spacing = 3 m, the outside pressure p = 4.0 t/m2, make all necessary assumptions to calculate:

a. The effective breadth of the flange over the length of the web section
b. The idealized bottom transverse as a beam element
c. The bending stress at the flange of the section at mid span

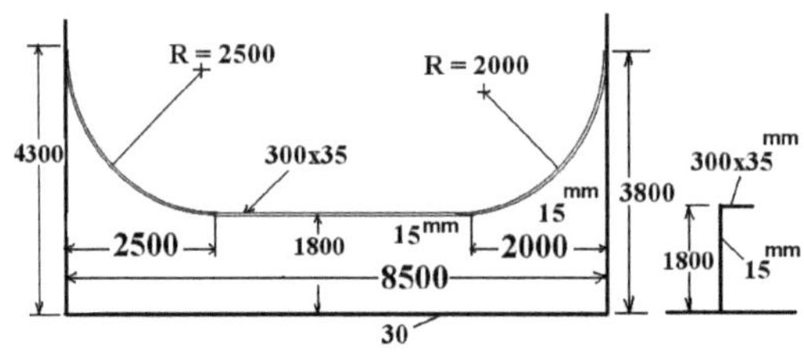

Fig.14.16

17-For the horizontal girder shown in Fig.14.17, make all necessary assumptions to:

a - Idealize the girder as a beam element
b - Calculate the bending stresses at midspan when the external water pressure p = 5.0 t/m2

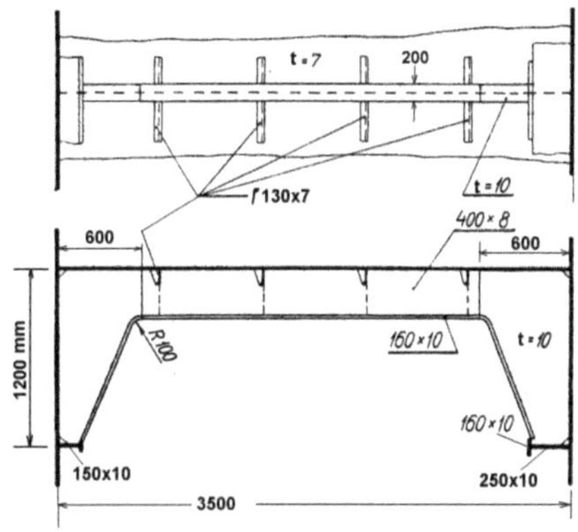

Fig.14.17

18- For the grillage structure shown in Fig.14.18 calculate the bending moments at points K, F and H.

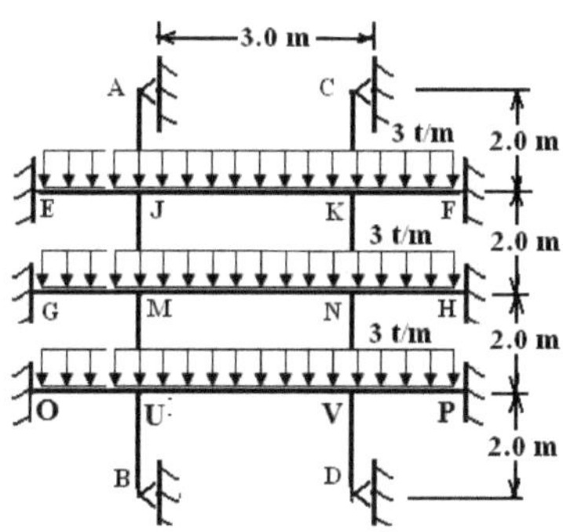

Fig.14.18

19-Determine the maximum stress and maximum deflection of the beam column shown in Fig. 14.19 given the yield stress of the material is 2.35 t/cm2 and Young's modulus is E = 2.08 x 106 kg/cm2

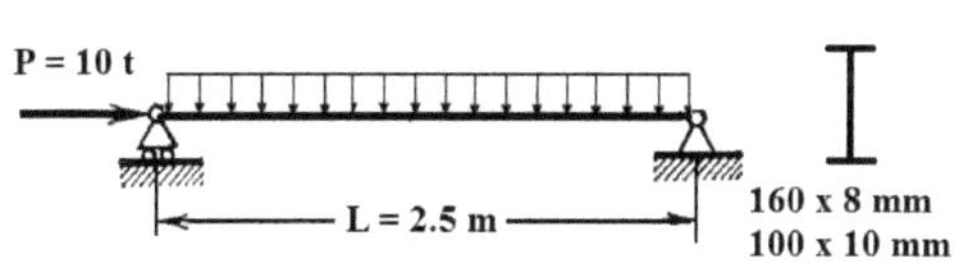

Fig. 14.19

20-Determine the load carrying capacity of the column shown in Fig. 14.20, given yield stress of the material = 2.35 t/cm^2

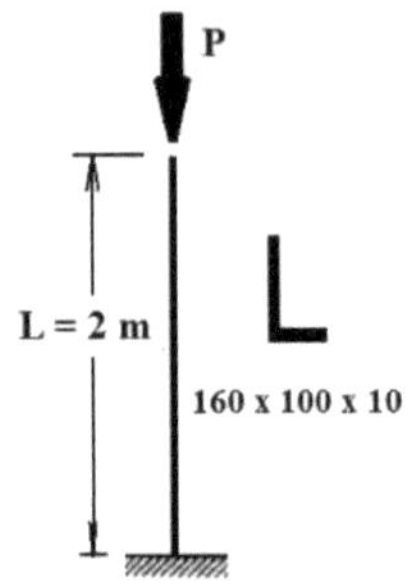

Fig. 14.20

21-Determine the magnitude of the maximum load the column shown in Fig. 14.21 can carry without buckling, the yield stress of the material is 2.4 t/cm^2, neglect torsional effects.

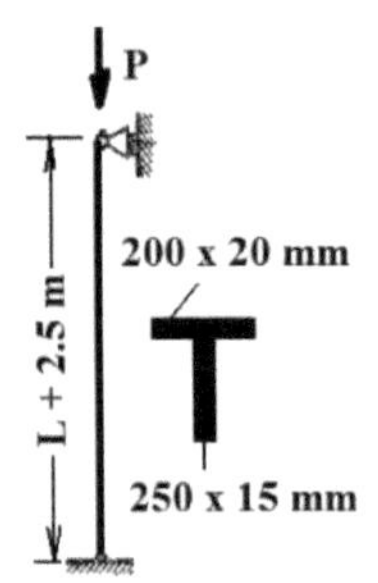

Fig. 14.21

22-Determine the factor of safety against buckling of the column shown in Fig. 14.22, the yield stress of the material is 2.35 t/cm^2.

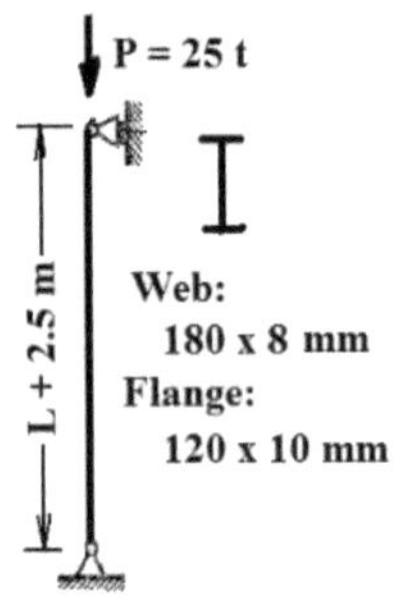

Fig. 14.22

23-Determine the critical buckling load for the column shown in Fig. 14.23, the yield stress of the material is 2.4 t/cm^2.

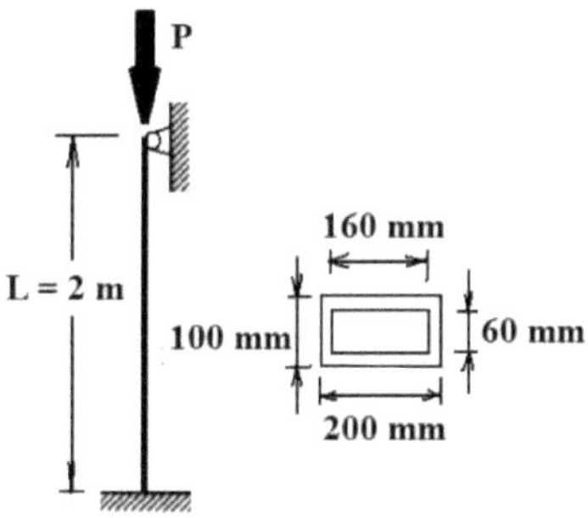

Fig. 14.23

24. For a containership L = 145 m, B = 21.6 m, D = 12.7 m, T = 9.4 m, the following data is given:

M_s (sagging)	= 4.045×10^5 KNm
M_w	= 3.48×10^5 KNm
M_H	= 2×10^5 KNm
y_B	= 5070mm
I_x	= 41.55 m^4
I_y	= 123.76 m^4

Determine the maximum stresses at the top end of the sheer strake at the Port & SB sides and the stresses at the keel.

References

1. Shama, M.: Torsion and Shear Stresses in Ships. Springer, Heidelberg (2010)
2. ABS, Rules for Building and Classing Steel Vessels (2010)
3. Shama, M.A.: Basic concept of the factor of safety in marine structures. SAOS (2009)
4. Veritas, B.: Common structural rules for bulk carriers (2009)
5. DNV Rules (2008)
6. ABS, Rules for building and classing steel vessels (2008)
7. Shama, M.A.: Lecture Notes, Post-Graduate Course 08611: Advanced Ship Structure Analysis, Dept. of Naval Architecture and Marine Engineering, Faculty of Engineering, Alexandria University (2007)
8. Lloyds, G.: Rules for classification and construction. Ship Technology, Sea going ships, Hull structure (2006)
9. Shama, M.A., et al.: Reliability of Double Hull Tanker Plates Subjected to Different Loads With Corrosion Effects. AEJ (2002)
10. Jensen, J.J.: Load and Global Response of Ships. Elsevier Ocean Engineering Book Series (2001)
11. Beck, R., Reed, A.M.: Modern Computational Methods for Ships in a Seaway. SNAME Transactions (2001)
12. Jensen, J.J., Beck, R.F., Du, S., Faltinsen, O.M., Fonseca, B., Rizzuto, E., Stredulinsky, D., Watanabe, I.: Extreme Hull Girder Loading. ISSC Committee Report (2000)
13. Zhou, Z., De Kat, J.O., Buchner, B.: A Non-linear 3-D Approach to Simulate Green Water on Deck. In: Seventh International Conference on Numerical Ship Hydrodynamics, Nantes (1999)
14. Lin, W.M., Shin, Y.S., Chung, J.S., Salvesen, N.: Nonlinear Predictions of Ship Motions and Wave Loads for Structural Analysis. OMA (1997)
15. SSC-394, Strength assessment of pitted plate panels (1997)
16. Shama, M.A.: Impact on Marine Environment of Ship Structural Failures and Casualties. AEJ (January 1997)
17. SSC-397, Commercial ship design for corrosion control (1997)
18. SSC-399, Strength and stability of stiffened plate components (1997)
19. Shama, M.A.: Impact on marine environment of ship structural failures and casualties. AEJ (January 1997)

20. Shama, M.A., et al.: Shear strength of damaged seagoing oil taker under vertical shear loading. AEJ (January 1996)
21. Shama, M.A.: Impact on Ship Strength of Structural Degradation Due to Corrosion. In: Second Arabian Corrosion Conference (1996)
22. Shama, M.A.: Ship Structural Failures: Types Causes and Environmental Impact. Alexandria Engineering Journal, AEJ (July 1995)
23. Shama, M.A.: Impact on Ship Strength of Structural Degradation Due to Corrosion. AEJ (July 1995)
24. Shama, M.A.: Ship Structural Failures: Types Causes and Environmental Impact. AEJ (July 1995)
25. Shama, M.A.: Shear strength of damaged coastal oil taker under vertical shear loading. AEJ (April 1995)
26. SSC–379, Improved ship hull structure details relative to fatigue (1994)
27. Gavin, A.G.: Double hull tankers- Paper no. 1, Lloyds Register Technical Association, Session (1994-1995)
28. IACS, Bulk carriers, Guidelines for survey, assessment and repairs of hull structure (1994)
29. Lloyds Register of Shipping, On Bulk carrier safety, Paper No. 5, Lloyds Register Technical Association, Session (1992-1993)
30. Shama, M.A.: Marine Structural Safety and Economy. In: SNAME, Symp. Marine Structural Inspection, Maintenance, and Monitoring (March 1991)
31. Faltinsen, O.M.: Sea Loads on Ships and Offshore Structures. Cambridge Ocean Technology Series. Cambridge University Press (1990)
32. SSC-332, Guide for ship structural inspection (1990)
33. ABS, Structural analysis of tankers., Far East Technical Seminars (November 1989)
34. Hughes, O.F.: Ship Structural Design: A Rationally-Based, Computer-Aided, Optimization Approach. SNAME (1988)
35. ABS, Compressive strength assurance of flat plate structures in ships., Research and Development Division (1985)
36. Gray, T.G.F., Spence, J.: Rational welding design, 2nd edn. Butterworths (1982)
37. Shama, M.A.: Safety Assurance: Methods of Assessment for Ship Structures. In: IMAEM, Second Conference, Trieste (September 1981)
38. Ship design and construction, The society of naval architects and marine engineers (1980)
39. Shama, M.A.: ON the Economics of Safety Assurance. Technical Report, Dept. of Naval Architecture and Ocean Eng. Glasgow University (1979)
40. ISSC, Int. Ship Structure Congress, Paris (1979)
41. Kirby, P.A., Nethercot, D.: Design of structural stability. John Wiley and Sons (1979)
42. Shama, M.: On the Rationalization of ship Structural Design. Schiff und Hafen (March 1979)
43. Jensen, J.J., Pedersen, P.T.: Wave-Induced Bending Moments in Ships, A Quadratic Theory. RINA 121 (1979)
44. Ochi, M.K.: Wave Statistics for the Design of Ships and Ocean Structures. Trans. SNAME (1978)
45. Shama, M.A.: Flexural Warping Stresses in Asymmetrical Sections. In: International Conference on Practical Design in Shipbuilding PRADS77. SNAJ, Tokyo (October 1977)
46. Shama, M.A.: The Cost of Irrationality in Ship Structural Design. In: Int. Conference on Practical Design in Shipbuilding PRADS. SNAJ, Tokyo (October 1977)

47. Shama, M.A.: Economic Consequences of Irrational Structural Design of Ships, vol. 2(1). Bull. of Collage of Eng., Basra University (March 1977)
48. Shama, M.A.: Analysis of Shear Stresses in Bulk Carriers. Computers and Structures 6 (1976)
49. ISSC, Int. Ship Structure Congress (1976)
50. Shama, M.A.: Stress Analysis and Design of Fabricated Asymmetrical Sections. Schiffstechnik (September 1976)
51. Shama, M.A.: Structural Capability of Bulk Carries under Shear Loading. In: Int. Ship Symposium, Rostock Univ. (September 1975)
52. Shama, M.A.: Shear Stresses in Bulk Carriers Due to Shear Loading. J. S. R., SNAME (September 1975)
53. Shama, M.A.: Optimizing Hull Steel Weight for Overall Economic Transportation. Marine Week, UK (May 2, 1975)
54. Ang, A., Cornell, C.A.: Reliability bases of structural safety and design. Journal of Structural Div. (September 1974)
55. Symposium on non-linear analysis of plated structures, Imperial College, London (November 1974)
56. Shama, M.A.: Calculation of Sectorial Properties, Shear center and Warping Constant of Open Sections. In: Bull. of the Fac. of Engineering, Alexandria University (1974)
57. Shama, M.A.: Effective Breadth of Face Plates for Fabricated Sections. Shipping World and Shipbuilder (August 1972)
58. Shama, M.A.: An Investigation Into Ship Hull Girder Deflection. In: Bull. of the Faculty of Engineering, vol. XII, Alexandria University (1972)
59. Shama, M.A.: On the Optimization of Shear Carrying Material of Large Tankers. J.S.R, SNAME (1971)
60. Mathewson, J.I., Viner, A.C.: Buckling as a mode of failure in ship structures, Part I. LR Staff Association (1971)
61. Roberts, W.J.: Strength of Large Tankers. NECIE (January 1970)
62. Goodman, R.A.: Wave excited main hull vibration in large tankers and bulk carriers. RINA (1970)
63. Bresler, N., Lin, T., Scalzi, J.: Design of Steel Structures. Wiley Eastern Private Ltd. (1970)
64. Shama, M.A.: Effect of Variation of Ship Section Parameters on Shear Flow Distribution, Maximum Shear Stresses and Shear Carrying Capacity Due to Longitudinal Vertical Shear Forces. European Shipbuilding 18 (1969)
65. Ship design and construction, The society of naval architects and marine engineers (1969)
66. Shama, M.A.: Effect of Ship Section Scantlings and Trans. Position of Long. Bulkheads on Shear Stress Distribution and Shear Carrying Capacity of Main Hull Girder. Int. Shipbuilding Progress 16(184) (1969)
67. Feodosyev, V.: Strength of Materials. Mir Publishers (1968)
68. Pedersen, B.: Wave loads on the fore ship of a tanker: European shipbuilding (6) (1968)
69. Oden, J.T.: Mechanics of Elastic Structures. McGraw-Hill Co. (1967)
70. Gerritsma, J., Beukelman, W.: The Distribution of the Hydrodynamic Forces on a Heaving and Pitching Ship Model in Still Water. In: Fifth Symposium on Naval Hydrodynamics, Washington (1964)
71. Terazawa, K., et al.: On the Elastic Bending of Parallel Inverted Angles with Plating. J. Zosen Kiokai 112 (1963)

72. Bleich, F.: Buckling Strength of Metal Structures. McGraw Hill Book Co., New York (1962)
73. Terazawa, K., et al.: On the Elastic Bending of Inverted Angle with Plating. J. Zosen Kiokai, SNAJ 110 (1961)
74. Timoshenko, S., Gere, J.: Theory of Elastic Stability. McGraw-Hill Co., New York (1961)
75. Yuille, I.M.: Application of the Theory of Elastic Bending to the Structural Members of Ships. RINA (April 1960)
76. Murray, J.M.: Longitudinal Strength of Tankers. International Shipbuilding Progress 5(52) (1958)
77. Korvin Kroukovsky, B.V.: Investigation of Ship Motions in regular Waves. Trans. SNAME (1955)
78. Timoshenko, S.: Theory of Bending, Torsion and Buckling of Thin-Walled Members of Open Cross-Section. Journal of the Franklin Institute (1945)
79. Ostenfeld, A.: Exzentrische und zentrische Knickfestigkeit. Zeitschrift des Vereines Deutsoher Ingenieure 42 (1898)

Subject Index

A

allowable stress 64, 67, 68, 109, 192, 184, 185
aspect ratio 77, 79, 133, 274, 278–281, 309
assessment of buckling 273, 288, 305, 306, 308, 312, 316, 319–322, 325, 326, 328, 330, 335–337
asymmetrical face plate 71, 102, 103, 110, 113, 271, 312, 322, 347
asymmetrical sections 45, 60, 71, 95, 98, 101, 103, 106, 110
axial force 93, 251, 252, 254, 258, 263
attached plating 45, 48–59, 64, 72, 79, 99, 105, 107, 158, 176, 181, 189, 209, 210, 211, 214, 218, 223, 270, 312, 314, 324, 353, 359, 363

B

ballast tanks 4, 13, 16, 21, 22, 361
beams 30, 33, 39, 41, 45, 47, 48, 62, 68, 71, 72, 75, 80, 93, 96, 118, 120, 251, 267, 367, 368
beam bracket 25, 39–41
beam columns 237, 251–254
beam deflection 87–89
beam elements 71–77, 79–81, 104, 110, 157
beam stresses 93–96
bilge plating 10
bottom girder 10, 35, 48, 75, 153, 155–159, 161, 165, 168, 194–199, 203, 204, 206–208, 225–227, 305, 319, 320
bottom longitudinals 10, 25, 42, 159, 166, 174–176, 181, 204, 205, 208, 213, 215, 225, 228, 229, 322, 323, 366, 371
bottom shell plating 10, 27, 28, 204
bottom slamming 125, 126
bottom structure 7, 13, 20, 25, 26–29, 38, 39, 75, 76, 126, 132, 133, 143, 147, 153–161, 165–167, 171, 174, 194, 197, 199, 202, 204–215, 225, 229, 305, 326, 327
bottom transverses 25, 42, 232
boundary conditions 38, 62, 74, 80, 88, 267, 274, 282, 287, 289, 297, 306, 309, 310
buckling failure 305, 307, 312, 339, 344, 353
buckling limit state 302
buckling of columns 237, 243
buckling of stiffeners 268, 270, 314
bulk carrier 3–9, 29, 31, 36, 78, 127–139, 131, 132, 152, 154, 333, 350

C

capesize 7, 8
clamped end 82, 88, 96, 100, 107, 155–161, 163–168, 173, 188, 203, 211, 214, 219, 221, 222, 224
classes of perturbations 237, 238
columns 45, 63, 80, 237, 251–254
combined loading 267, 276, 288
compounding of stresses 191, 193–195, 199, 202, 204–206, 212, 215, 217, 219, 222, 225, 226, 228, 305, 334
compound stress 191, 195, 199, 202, 206, 208, 215, 217, 219, 222, 226, 228, 230, 319, 321, 323, 327
containers 14, 16, 17, 129, 376
container ship 3, 14–17, 31, 38, 62, 78, 129
continuous beam 74, 155, 166, 218, 220, 232
control of buckling 339, 347
core radius 239
corrosion 8, 21, 185, 339, 340, 342, 343, 344, 351–355, 361, 362
corrugated bulkhead 36

critical force 237, 240, 241
critical stress 237, 241, 245
cross-stiffened panels 46
crude tankers 19
curved face plate 110, 112–114, 339, 347
curved plates 45, 59

D

deck beams 30, 33, 39, 41, 47, 48, 75, 267, 367
deck girder 33, 34, 41, 47, 75, 162–164, 217, 218, 223, 224, 267, 319, 368, 370
deck loadings 187
deck longitudinals 9, 41, 148, 162, 164, 186, 187, 188, 217, 219–224, 267, 370
deck plating 9, 29, 32, 33, 39, 143, 145, 147–149, 162, 163, 186, 188, 217, 220, 222–225, 302, 303, 328–330, 353, 364–366, 368, 369
deck structure 25, 31–34, 39, 40, 75, 126, 143, 148, 162, 163, 186, 191, 217–225, 267, 269, 302, 305, 306, 329, 363–366
deep girders 351
degree of constraint 90, 91, 102, 187
deterioration of structure 340
distribution functions 303, 341
double bottom 3, 4, 7–10, 13, 16, 19, 21–23, 26–28, 37, 38, 154, 155, 157, 158, 160, 161, 183, 191, 194, 195, 198, 199, 201, 203, 204, 213
double bottom floor 10
double bottom girders 10
double bottom structure 26–28, 38, 154, 155, 157, 158, 160, 161, 183, 194, 195, 198, 199, 204, 213
double bottom tank 4, 9, 26
double hull tanker 3, 21
double side 3, 6, 7, 16, 21–23, 31
duct keel 10
dynamic bending moment 118, 127
dynamic loadings 4, 124, 126, 127
dynamic shear force 127
dynamic stress 8, 141, 193

E

eccentric load 237, 248
effective breadth 49, 56, 71, 74, 101, 106, 107–114, 158, 176, 177, 188, 368, 372
effective length of column 242
effective span 33, 34, 71, 80–82
effective thickness 147, 148, 152
effective width 267, 299, 300, 301, 302
elastic buckling 243, 288, 309
end brackets 9
end supports 4, 83, 85, 89–91, 109, 237, 249,278,
equivalent section 56,57
eulerbuckling 242–244, 247, 248, 264, 273, 274, 282–288, 295, 296, 299, 303, 308, 309, 318, 321, 356
extreme values 139

F

fabrication defects 344, 345
fabricated sections 45, 49, 50, 55, 60
fabrication stresses 192
face plate 33, 49, 55, 56, 60, 64, 65, 67, 68, 69, 71, 97–107, 109–113, 162–166, 176, 177, 185, 226, 267, 270, 271, 305, 306, 310, 312, 321–323, 325, 339, 347, 349, 356, 358, 359, 363, 364
factor of safety 185, 343, 363– 366, 368, 370, 371
FEM 38, 62, 78, 104, 110
fixed end 90, 158, 185, 210, 213, 233, 254, 316, 327
flare slamming 125, 126
flexural buckling 238, 244, 268–270
flexural properties 47, 55, 59, 60, 71, 102, 107, 108, 141, 145–148, 158, 303
flexural rigidity 232, 269, 270, 314, 339, 344, 351, 352, 353, 361
flexural stresses 71, 98, 101, 103, 104, 107, 193
flexural torsional buckling 238
floors 7, 25, 26, 27, 29, 38, 47, 157, 174, 177
floor spacing 172, 370
frame analysis 155
frame bracket 25, 38, 347
frameworks 45, 47, 48

G

general cargo ships 10, 12, 14
geometrical properties 45, 48, 50, 51, 52, 54, 73, 109

girders 7, 10, 26, 27, 29, 33, 34, 35, 36, 37, 41, 45, 49, 64, 65, 68, 71, 75, 76, 79, 106, 107, 110, 146, 153, 155, 157, 158, 162, 163, 165, 182, 185, 191, 194, 195, 196, 197, 204, 217, 218, 223, 225, 227, 267, 305, 306, 309, 312, 316, 319, 321, 322, 325, 347, 350, 351, 356, 365, 369
grillage analysis 155, 158, 220
grillage structure 72, 75, 76, 157, 158, 160, 162, 165, 166, 169, 197, 207, 210, 228, 364, 373

H

handymax 7
handysize 7
hatch girder 92, 364, 365–369
hatch opening 12, 13, 16, 33, 34, 363, 366, 367
hatch side coaming 10
hatch side girder 10
heaving 124, 126, 127
high tensile steel 4, 71, 96, 97
hogging 120, 29, 33, 41, 53, 91, 97, 199, 200, 204, 207, 209, 212, 218, 220, 223, 226, 263, 305, 322, 368
hopper side tanks 3
hopper tank 4, 9, 10
horizontal bending moment 130, 131, 141, 150, 151, 191, 223, 328, 330, 333, 334
hull fittings 48
hull girder 12, 13, 15, 16, 31, 32, 34, 35, 45, 46, 78, 117, 118, 120, 123, 127, 128, 131, 132, 134, 136, 137, 141–147, 149–151, 155, 171, 174, 191–194, 196, 199, 200, 204, 207, 209, 212, 218, 223, 226, 263, 302, 305, 319, 322, 323, 326–328, 330, 332, 334, 335, 343, 352, 364, 367, 369, 370
hull girder bending 32, 118, 141, 144, 147, 150, 155, 196, 207, 209, 212, 223, 323, 332, 334, 369
hull girder loading 117, 137, 191, 204, 223
hull girder shear force 131, 132, 334
hull girder shear stress 151
hull girder stress 141–143, 145, 149, 150, 151, 171, 192, 194, 196, 199, 200, 209, 226, 319, 322, 323, 326, 327, 364, 367, 370

I

IACS 8
ideal column 238, 240
impulsive dynamic shear force 127
inelastic buckling 243, 228, 288, 309
initial eccentricity 241
inner bottom longitudinals 10
inner bottom plating 7, 10, 27, 28
In-plane loading 45, 267–269, 273–275, 278, 283, 305, 314
In-plane stresses 103, 305, 330, 332

L

lateral torsional buckling 238
local buckling 238, 267, 270, 312
local loadings 118, 132, 171, 172, 187, 192, 218, 224, 324
local structural connections 349
longitudinals 9, 10, 25, 31, 33, 35, 41, 42, 48, 62, 126, 146–148, 153, 157, 159, 161, 162, 164–166, 174, 175, 176, 178, 181, 182, 185–189, 191, 204, 205, 208, 213, 215, 217, 219–221, 223, 224, 225, 228, 229, 267, 306, 307, 309, 312, 314, 316, 322, 349, 366, 370
longitudinal framing 7, 8, 28
longitudinal spacing 33, 175, 176, 185, 187, 224, 263, 295, 265, 363, 365, 366, 368
long term prediction 117, 139
lower stools 9

M

mean value 140, 239, 287, 304
modes of buckling 267, 269, 278, 307, 312, 314
modes of deformation 267–269
modulus of elasticity 100, 167, 241, 309
multi-decker 10

N

neutral axis 29, 56, 65, 74, 107, 111, 143–149, 152, 156, 157, 158, 159, 161, 163, 164, 167, 168, 173, 178, 181, 188, 189, 190, 196, 199–203, 207, 209–212, 214, 216, 220–224, 227, 229, 230, 231, 259, 264, 297, 324, 330, 332, 333, 334, 335, 336, 337, 360
non-uniform compression 275, 320, 324

O

offset bulb 28, 30, 54,
offset face plate 99, 100, 106, 111
open floor 25
out-of straightness 356, 357
over-welding 360

P

panamax 7
pillars 33, 34, 244
plate buckling 238, 267, 282, 289, 290, 291, 326, 353, 355
plate element 71, 72, 77, 78, 79, 97, 104
pitching 124, 126, 127
post buckling strength 268, 298
primary stress 141, 142, 191–193, 196, 197, 199, 200, 202, 204, 207, 209, 212, 215, 216, 218–220, 223, 226–230, 319, 329, 367
probability density function 341
probability distribution 138
probabilistic prediction 137
pure bending 89, 237, 267, 275, 277, 285, 290–292, 307, 308, 320–322, 325, 330, 331, 336
pure shear 275, 279, 280, 286, 288, 289, 308

R

ramps 13, 17, 18
RAO 137
reliability 72, 74, 339, 340, 342, 343
residual stresses 191, 192, 241, 306, 344
response spectrum 137
RoRo ship 17

S

safety margin 302–304, 339, 341, 344
sagging 120
scantlings 7, 8, 25, 29, 30, 33, 34, 35, 38, 49, 50, 52, 54, 55, 60–62, 64, 71, 82, 110, 153, 188, 238, 246, 306, 352, 356
secondary structural members 62
secondary stress 153–166, 168, 173, 174, 191, 192, 195, 197, 198, 200–204, 206, 207, 209–211, 213, 217, 218, 220, 221, 223, 226, 227, 229, 231, 232, 319, 322, 326, 327, 329, 330, 364
sectional area 59, 63, 66, 143, 145–149, 239, 241, 245, 250, 263, 264
shear area 109, 121, 133, 145
shear lag effect 106, 145
shear stress 94, 98, 121, 131, 145, 151–153, 193, 275–277, 286, 288, 308, 314, 325, 331–336, 344, 348, 363, 369
ship section 13, 15, 18, 19, 22, 23, 29, 136, 141–149, 151, 152, 189, 196, 200, 202, 207, 209, 212, 216, 220, 223, 229, 231, 263, 303, 323, 332, 333, 334, 335, 336, 337, 353, 363, 364, 367, 371, 372
short term prediction 137
side frame 38, 39, 40, 47, 48, 92, 363, 370
side shell frames 9, 10
side shell longitudinals 9, 365
side shell plating 9, 29, 30, 144, 146, 149, 150, 151, 206, 305, 332, 335
side shell structure 25, 29, 30, 31, 76, 334
simple support 90, 92, 274
single decker 10
single hull tanker 19
slamming 117, 118, 124–127, 134
slenderness ratio 242, 244, 246
sloping plating 10
solid floors 29
space frame 72, 77
span point 71, 80, 81, 83
standard angle sections 52
stability 4, 7, 13, 17, 63, 71, 237, 240, 242, 244, 303
stillwater bending moment 118, 128, 367, 368
stillwater shear force 118, 132

stillwater stress 193
strength deck 9, 29, 31, 33
strength deck plating 9, 29
strength members 7, 9, 25, 45, 60, 71, 141, 143, 146, 147, 152, 153, 154, 155, 162, 165, 171, 174, 186, 191–194, 204, 205, 217, 225, 305, 306, 339, 345, 347, 352, 353, 356, 362
structure assemblies 25, 30, 31, 38, 40, 48, 49, 71, 72, 77, 153, 154, 157, 162, 165, 191
structure modeling 72
symmetrical sections 45, 50, 60, 71, 93–95, 98, 101, 103, 106, 107, 110

T

tank top 10, 38, 144, 153, 157, 161, 171, 174, 178, 185, 191, 194, 196, 202, 203, 204, 215, 216, 324, 363, 364, 367
topside tank 8, 9, 10
topside tank longitudinal 10
top wing tank 4, 350
transverse bulkhead 4, 9, 19, 20, 25, 34–37, 62, 76, 77, 133, 156, 162, 165, 166, 174, 175, 177, 195, 197–199, 201, 203, 205, 232, 309, 364
transverse framing 29, 30, 33
torsional buckling 238, 244, 270
torsional rigidity 16, 100, 182, 269, 270, 310, 314, 315
tripping brackets 25, 41, 50, 345, 346
twin hatches 13

U

ultimate load 90, 240, 299
upper stool 9

W

wave bending moment 120, 128, 129, 130, 134, 142
wave heading 137
web plate 33, 42, 49, 50, 55, 56, 64, 94, 99, 101, 103, 104, 155, 156, 158, 218, 267, 270, 271, 305, 306, 310, 314, 319, 320, 323, 324, 325, 340, 347, 349, 350, 351, 356, 359, 363
welding distortions 339, 345, 360

CV of the Author

Prof. Shama graduated from the Faculty of Engineering, Alexandria University with B.Sc. degree (Honors) in Naval Architecture in I960. He obtained his Ph.D. degree in Naval Architecture from Glasgow University, Scotland, UK in 1965. He has a Patent of Design of Frame Bending Machine, Scotland, UK (1965).

Prof. Shama was Head of the Department of Naval Architecture (1988-1992), Vice Dean for Education and Students affairs (1992-1995), Dean of the Faculty of Engineering, Alexandria University, (1995-1998). Prof. M. Shama is currently Prof. Emeritus of Naval Architecture and Shipbuilding, Department of Naval Architecture and Marine Engineering.

Prof. Shama was granted two years Sabbatical leave at the Research Department of Lloyds Register of Shipping, London, UK (1969-1971). Prof. Shama participated in the Honorary Editorial Advisory Board (HEAB) of the Encyclopedia of Life Support Systems, EOLSS, January (1995-1996).

Prof. Shamais a member of the Supreme Egyptian State Committee for the Promotion of Professors since 1981. He was the Chairman of this committee (1998-2002). Prof. Shama was an External Examiner for Ph.D. Degrees in Reliability Engineering, Maryland University (1998-2000).

Prof. Shama published 85 Papers in the International, Regional and Local Scientific Journals in the fields of Ship Design, Ship Production, Ship Structural Analysis and Ship safety. He is the Author of the published book "Torsion and Shear Stresses in Ships", Springer Verlag, 2010.

Prof. Shama was awarded the Egyptian State Prize for Engineering as well as the Science & Arts Medal of First Grade in 1974. Alexandria University nominated Prof. Shama in 2005 for the State Prize of Recognition. Prof. M. Shama was awarded the Alexandria University 2010 TahaHussainPrize in recognition for his intellectual, managerial and development pioneering achievements.

MIX
Papier aus verantwortungsvollen Quellen
Paper from responsible sources
FSC® C105338

If you have any concerns about our products,
you can contact us on
ProductSafety@springernature.com

In case Publisher is established outside the EU,
the EU authorized representative is:
Springer Nature Customer Service Center GmbH
Europaplatz 3, 69115 Heidelberg, Germany

Printed by Libri Plureos GmbH
in Hamburg, Germany